JOURNEY THROUGH THE ICE AGE

JOURNEY
THROUGH THE
ICE AGE

PAUL G. BAHN
AND JEAN VERTUT

Weidenfeld & Nicolson
London

For Alex and Elaine

First published in Great Britain 1997 by Weidenfeld & Nicolson

Text copyright © Paul G. Bahn, 1988, 1997
The moral right of Paul G. Bahn to be identified as the author of this work has been asserted
in accordance with the Copyright, Designs and Patents Act of 1988
Design and layout copyright © Weidenfeld & Nicolson, 1997

The acknowledgements on p. 240 constitute an extension to this copyright page

A CIP catalogue record for this book is available from the British Library
ISBN 0 297 83588 2

Designed by: Bradbury & Williams
Designer: Bob Burroughs
Typeset in: Perpetua

Weidenfeld & Nicolson
The Orion Publishing Group
5 Upper St Martin's Lane
London WC2H 9EA

TITLE PAGE: Sculpted ibex frieze in the Abri Bourdois
at Angles-sur-l'Anglin (Vienne). Magdalenian. The
animals are carved lifesize. OPPOSITE: Horse engraving
on reindeer antler from Neschers (Puy-de-Dôme)
found by the abbé Croizet c. 1842. Probably
Magdalenian. The antler is c. 27 cm in length, the
horse 9 cm.

CONTENTS

PREFACE AND ACKNOWLEDGEMENTS

Jean Vertut and I began planning this book in 1984. His sudden and untimely death in May 1985 at the age of only fifty-six robbed the world of a kind and remarkable man, and cave art of its foremost photographer. It also meant that the book could no longer be what we had envisaged, since Jean could neither choose the pictures himself, nor describe his methods and techniques as he had wished to do, nor photograph the new sites and objects that we had planned to include.

Nevertheless, with the help and encouragement of his widow, Yvonne, it has proved possible to produce an approximation of the original plan. Naturally, it has been necessary to draw primarily on the range of sites represented in Jean's existing collection, with only a few pictures borrowed from other sources where required to illustrate precise points. Inevitably, therefore, this book concentrates heavily on the Ice Age art of France and Spain.

It was never our intention, however, to produce an encyclopaedic work or a catalogue, or to attempt to replace the great tome by the late André Leroi-Gourhan (which remains one of the glories of Jean's career as well as of Leroi-Gourhan's); we wished to give a survey of past and current theories about the art rather than promote a single favourite interpretation. For reasons explained in the text, it will be some time before another comprehensive volume of that type can appear – if indeed such a thing is desirable. Sadly, our own book has now taken on an extra, unforeseen role, that of serving as a tribute to Jean by displaying the best of his final work. It goes without saying that, since the text was written in 1987, it is entirely my own responsibility; Jean may not necessarily have agreed with all the views expressed here, and the text is the poorer for the absence of his influence and advice.

Since the book is meant for the interested layman as well as for the academic, we decided not to get bogged down in arguments about the definition of 'art', or in speculations as to how and why it arose and disappeared. Much of that debate is redundant, since the earliest art to appear archaeologically is at an already well-developed stage, and we have lost its antecedents. In any case, these subjects have been treated elsewhere, in great depth.[1]

The term 'art' is rapidly losing favour among those researching the Palaeolithic period, since it presupposes an aesthetic function and lumps together a wide range of objects and types of image-making spanning twenty-five millennia; other vaguer terms – 'pictures', 'iconography', 'images', 'pictograms/ideograms', 'symbolic graphisms' or 'decoration' – are now preferred. In this book, therefore, the word 'art' is retained simply as a convenient blanket-term for the decoration of objects, rocks and cave walls, whatever its function or motivation may have been.

Similarly, we chose not to devote the book to long, detailed and tedious discussions of the style and dating of particular caves, and thus to avoid all the entanglements and contradictions those subjective topics lead one into; other books are available which dwell at length on these issues;[2] nor did we wish to devote much space to the equally impressionistic and subjective topic of what it feels like to visit these sites – this too has been done quite vividly in a popular work.[3] Moreover, it is superfluous to describe the beauty of particular depictions – readers can make their own judgements from the illustrations.

Very few books cover both the cave art and the portable art of the last Ice Age, although the two cannot effectively be studied without reference to each other. We were determined to include both, but our book suffers from the usual archaeological vice of displaying only the finest examples; as such, they are not truly representative of the whole, but colour plates are a limited and valuable resource, and it seems a pity to waste them on illustrations which appear mediocre in execution, are in poor condition, or would be unintelligible to the untrained eye.

Lack of space has precluded setting the art within the cultural, economic and social context of Ice Age life, although such aspects inevitably crop up here and there. To integrate these themes properly would require a separate book; meanwhile, a number of texts about Ice Age life and culture are readily available to the layman, while the specialists will have their own views. Palaeolithic art contains so many subjects, and so many erroneous facts and interpretations are already in print about it, that it seemed best to cover it as fully and as factually as possible, in order to correct former errors and to assess the contribution made by recent work and discoveries. References to parallels in other cultures have been kept to an absolute minimum.

Since the text is aimed at the non-specialist as well as the academic, radiocarbon dates have been given as a straightforward figure – they should be taken as only a rough guide to the true age of the object or layer in question; specialist readers can find full details of the plus/minus range, the substance dated, and the laboratory concerned, by following up the references provided.

More and more people are becoming interested in the decorated caves, but fewer and fewer sites can be visited because of the sheer difficulties of access, or the potential damage to the images from accidents, vandals or bacteria – it is generally reckoned that a decorated cave open to the public is probably doomed. There is therefore a clear need for other ways of presenting the art to the public. Lascaux II, a copy

almost indistinguishable from the original, is the ideal answer to this problem, albeit an extremely expensive one; for other sites, videos and television documentaries are a partial solution, and so are books of this kind. Much cave art, however, can be fully appreciated only in three dimensions, so that one can see how the artist was inspired by or utilized the shape of the rock. Among Jean's last photographs is a series of stereoscopic studies which he hoped to place in this volume, together with special glasses to produce a 3-D effect. Such publications will inevitably appear within the next decade, thanks in large measure to the expertise and vision of this great pioneer.

My own acquaintance with Palaeolithic cave art began in June 1963 when, at the age of nine, I heard the exciting story of the discovery of the art in the Volp caves (by the Bégouën brothers in 1912-14) in a BBC radio broadcast for schools (People, Places and Things); at about the same time I acquired – as a free gift from some brand of washing powder – a waste-paper bin, which I use still, decorated with Lascaux-type animals! A more decisive influence was to come at Cambridge University where, during my first undergraduate supervisions with Eric Higgs, I was held spellbound by his tales of visits to the caves of France and Spain. Unlike some of his disciples, and in contrast with his own image as a prehistorian contemptuous of the dilettante art-historian approach to the past, Eric loved Palaeolithic art. My interest was heightened by reading books on the subject, and in lectures at Cambridge given by John Coles and Charles McBurney.

Finally, it was Eric Higgs who suggested that I do the research for a doctoral dissertation on the prehistory of the French Pyrenees – no doubt because he knew I would be fascinated by the region's richness in Palaeolithic art. During expeditions to southern France and Spain with him in the mid-1970s, I was agreeably surprised to find him making detours so that I could visit Gargas, Niaux, Le Mas d'Azil, Candamo, and other sites. After his death in 1976, I pursued my research in France with the active support and encouragement of the late Professor Glyn Daniel, who helped me to obtain my first permission to visit Lascaux, took a keen interest in my work and, I am proud to say, became a close friend. It is therefore to these two men, entirely different yet both admirers of Ice Age art, that I would like to dedicate this volume. I wish they had lived to see it.

During the year 1985/6 I was fortunate enough to be awarded one of the first J. Paul Getty Postdoctoral Fellowships in the History of Art and the Humanities, and used it to improve my knowledge of the cave-art sites and literature in Europe (using Toulouse as base, through the kindness of Claude Barrière), as well as to travel abroad to see as much prehistoric art as I could, especially in Australia. I am profoundly grateful to the Getty Trust for such a wonderful opportunity, from which this book has benefited enormously.

In the course of my 'Getty travels' as well as during visits to France and Spain over the years, I have been helped by a great number of friends and colleagues in my research on Ice Age art. In Paris, the late Léon Pales, the late André Leroi-Gourhan, Arlette Leroi-Gourhan, Claude Couraud, Henri Delporte, Denis Vialou, Michel Garcia, Georges Sauvet, Dominique Buisson, Geneviève Pinçon, Lucette Mons, Suzanne de St Mathurin, Michel Orliac, Sophie de Beaune, Gilles Tosello, Jacques Allain, Nicole Limondin, Marthe Chollot-Varagnac, Marina Rodna. In the Pyrenees, Robert Bégouën, Jean Clottes, Dominique Sacchi, Claude Barrière, Louis-René Nougier, Aleth Plenier, Carol Rivenq, Anne-Catherine Welté, Jacques Omnès, Jean Vézian, Jacques Blot, Robert Arambourou, André Clot, René Gailli, Luc Wahl, Romain Robert, Jean Abélanet, Robert Simonnet, Georges Laplace, Roger Séronie-Vivien, Claude Andrieux. In the Périgord, Brigitte and Gilles Delluc, Paulette Daubisse, Jean Gaussen, Alain Roussot, Jean-Philippe Rigaud, Jean-Marc Bouvier, Jean-Michel Mormone. In Spain, Jesús Altuna, Ignacio Barandiarán, Vicente Baldellou, Reynaldo González García, Alfonso Moure Romanillo, Francisco Jordá Cerdá, Josep Maria Fullola Pericot, Jose Manuel Gómez-Tabanera, Joaquín González Echegaray, Eduardo Ripoll Perelló, José Miguel de Barandiarán. Elsewhere in Europe, Lya Dams, Marylise Lejeune, Yves Martin, André Thévenin, Martine Faure, Jean Combier, Michel Lorblanchet, Gerhard and Hannelore Bosinski, Joachim Hahn, Karl Narr, Hans Biedermann, Bohuslav Klíma, Jan Jelínek, Zoïa Abramova, Mike and Anne Eastham, Pat Winker, Warwick Bray, Alex Hooper, Jonathan Speed; and outside Europe, David Lewis-Williams and Anna Belfer-Cohen. In America, Alex Marshack, Meg Conkey, Ray and Jean Auel, Tom and Alice Kehoe, Glen Cole, John Pfeiffer, Noel Smith, Whitney Davis. In Australia, Robert and Elfriede Bednarik, Geoff Aslin, John Clegg, Pat Vinnicombe, Sylvia Hallam, Andrée Rosenfeld, George Chaloupka, Jo Flood, Rhys Jones, Percy/Steve/Matt Trezise, Paul Taçon, Jo McDonald, Lesley Maynard, Charles Dortch; in China, Zhu Qing Sheng, You Yuzhu; and in Japan, Yuriko Fukasawa, Yoshiaki Kanayama, Gina Barnes and Hideji Harunari.

The original idea for a volume co-authored with Jean Vertut came from Ib Bellew; the project was initiated with the help of Brigitte and Gilles Delluc, and brought to fruition thanks to the unfailing help of Yvonne Vertut.

PAUL G. BAHN
1988

PREFACE TO THE SECOND EDITION

Almost ten years have now passed since *Images of the Ice Age,* the first edition of this book, was written, and during that time many books and several hundred papers have been published on the subject. While the most important of these will be covered in this fully revised and updated edition, the greatest changes of the past decade have come not in the sphere of theory and interpretation, but in the revolutionary developments in pigment analysis and direct dating, as well as through sensational new discoveries such as Cosquer, Chauvet, several major clusters of open-air figures and a few portable objects. All of these combined have ensured that we are currently in the most exciting and progressive phase of Ice Age art studies since the authentication of cave art at the turn of the century.

Looking back at the earlier preface, there is little to add except that the new technology is playing an increasing role, with a 'Virtual Reality' Lascaux and a brief 3-D film of Cosquer already in existence, and quantities of images being digitized and stored on CD-Roms. This phenomenon will certainly accelerate in the next decade, and – together with new replicas such as the state-of-the-art reproduction of the Altamira ceiling in Japan[4] – will provide a new and ever-improving solution to the problem of lack of public access to the original caves.

In addition to all those people thanked in the first preface, some of whom have since passed away, I would like to add the following names of friends and colleagues who have helped me in various ways with my research on Ice Age art around the world:

Mila Simões de Abreu, Juan-María Apellániz, Claude/Jean/Monique Archambeau, Pablo Arias Cabal, Norbert Aujoulat, Dominique Baffier, Hans-Georg Bandi, Rodrigo de Balbín Behrmann, Magín Berenguer, Federico Bernaldo de Quirós, Olga Boiko, Augusto Cardich, Jean-Marie Chauvet, Shirley Chesney, Maria Soledad Corchón, Michèle Crémades, Ron Dorn, Jean-Pierre Duhard, Francesco d'Errico, Nick Evans, Alicia Fernández Distel, Dánae Fiore, Javier Fortea, Céline Fournier, Natalie Franklin, Carole Fritz, Lidia Clara García, Jean-Michel Geneste, Cesar González Sáinz, Carlos Gradin, Haskel Greenfield, Niède Guidon, Mariana Gvozdover, Slimane Hachi, John Halverson, Mauro Hernández Pérez, Dirk Huyge, Ludmila Iakovleva, Ludwig Jaffe, Vitor Oliveira Jorge, Giriraj Kumar, Piero Leonardi, Martí Mas Cornellà, Elena Miklashevich, Desmond Morris, Laurence Ogel, Marcel Otte, Nicole Pailhaugue, Anne-Marie Pessis, Jean Plassard, Nikolai Praslov, Barbara Purdy, Erika Rauschenbach, Sergio Ripoll, Anna Roosevelt, Marvin Rowe, Pam Russell, José Luis Sanchidrián Torti, Viacheslav Shchelinsky, Vladimir Shirokov, Anne and Francis Thackeray, Alice Tratebas, Valentín Villaverde, Christian and Janine Wagneur, Steve Waller, Alan Watchman, Daniela Zampetti, João Zilhão.

For supplying special pictures for this book, I am particularly grateful to Bernadette Arnaud, Dominique Baffier, Robert Bednarik, Robert Bégouën, Gerhard and Hannelore Bosinski, Augusto Cardich, Jean Clottes, Glen Cole, Claude Couraud, Brigitte and Gilles Delluc, Ron Dorn, Javier Fortea, Natalie Franklin, Geneviève Pinçon, Carol Rivenq, Sergio Ripoll, Paul Taçon, Ulm Museum, Valentín Villaverde.

INTRODUCTION: CAVE LIFE

Despite persisting popular belief, fuelled by the best efforts of movie-makers and cartoonists, it is not true that people during the late Ice Age habitually lived in caves. Over the last 100 years, archaeological investigation has made it clear that they primarily occupied cave mouths, sunny rock shelters, and huts or tents in the open air. This is hardly surprising, since caves are not the most pleasant choice of habitat – they are usually dark, wet, slippery, and full of rocks and jagged concretions.

Nevertheless there are cases, many of them in the French Pyrenees, where hearths and other signs of occupation have been found far inside very deep caverns, and even hundreds of metres from the entrance (assuming that the present entrance was also the main one in pre-history). What is the explanation for this? In some cases these are temporary encampments linked to the creation of decoration on the cave walls: for example, in the cave of Tito Bustillo in northern Spain, occupation debris and colouring materials have been found in excavations at the foot of the principal painted frieze (see p.115).

In other cases they may denote the seeking of refuge in particularly harsh conditions: at present the caves of southern France and northern Spain maintain a fairly constant temperature (usually *c.* 14°C) which makes them pleasantly cool in summer and mild in winter. However, recent work by geophysicists has established that during the last Ice Age it was not merely mild but quite warm inside the caves – a fact which helps to explain why the Palaeolithic footprints found in their depths are of bare feet and have no sign of frostbite[1] (but see p.121 for the possibility of frostbite on hands).

The footprints are quite often those of children: for example, in the cave of Aldène (Hérault), or in Fontanet (Ariège) where a child – who also left knee- and handprints – seems to have pursued a puppy or a fox into the cave's depths. Children were clearly not afraid to explore the far depths, narrow passages and tiny chambers of caverns, whether alone (as at Aldène) or with adults; it is possible that the finger-holes and heel-marks found around the clay bison of the Tuc d'Audoubert (Ariège) were made by children playing while the adult(s) made the figures.

It is a tragedy that we have irretrievably lost huge amounts of fascinating information from cave floors: in some cases, the caverns were frequented throughout history, and thus any prehistoric traces on the floors were destroyed centuries ago; in others, the prehistoric

Fig. 1 Prehistoric footprints in Niaux (Ariège), carefully and deliberately made by two young children. Precise date unknown.

galleries were discovered during our own century, but the discoverers inadvertently obliterated the precious evidence because (no doubt with their attention focused on the decorated cave walls) they simply did not notice the footprints, objects and even engravings at their feet. With one important exception – the Tuc d'Audoubert – it is only in recently discovered caves such as Fontanet, Erberua, Chauvet and La Garma that the floors have been carefully preserved intact.

The Tuc d'Audoubert's inner depths were first explored in 1912; as the cave is privately owned by the Bégouën family, visits have been kept to a strict minimum ever since, and nothing has been disturbed. Thus the cave not only has artistic treasures (above all, the unique clay bison figures, as well as engravings), it is also a treasure-house for all manner of evidence about the activities of the Palaeolithic visitors: flint tools, teeth or pieces of bone were carefully placed in rock crevices, or stuck into the floor; stalagmites were deliberately broken; cave-bear jaws were picked up, their canines were removed (presumably for use in necklaces) and then the jaws were thrown down again; the traces in clay around the bison have already been mentioned. In short, since no one visited these galleries between the last Ice Age and 1912, one can actually follow the traces left by Magdalenian people and reconstruct many of their actions.

The same applies at Fontanet, a gallery blocked during the Magdalenian period and only rediscovered in 1972. Here, the back part of the cave has no traces of occupation but does have numerous prints, as mentioned above. The front part near the blocked entrance has no prints, but it has engravings and paintings on the walls, and, on the floor, a series of hearths around which are the animal bones that the occupants probably tossed over their shoulders. Here again, the vestiges and prints are so fresh that one would think they were made a few minutes ago rather than 13,000 or 14,000 years ago.[2] Chauvet Cave has numerous and varied traces of human and animal visitors: in particular, numerous bones of cave-bear are present naturally, but one bear skull has been purposely placed on a natural, isolated rock – this has already caused a great deal of ink to

flow, with predictable speculations about a cave-bear cult, though of course it is just as likely that the skull was placed there by a bored child amusing itself while the adults produced the art.[3]

The Spanish cave of La Garma, whose art was discovered in 1995, has an intact Upper Palaeolithic occupation floor of at least 500 square metres, with thousands of animal bones, sea-shells, and numerous tools of flint, bone and antler on the surface, including two magnificent perforated batons. There are also areas of charcoal, and intentional deposits of bones and stalagmites.[4]

Apart from boulders which may have been used as seats, no cave 'furniture' has survived. Presumably it was all made of wood and has therefore disintegrated through time – as yet, no one has been fortunate enough to find a waterlogged Palaeolithic site with preserved wood: only a few Palaeolithic wooden objects, including a few spears, have survived in Europe

Fig. 2 Prehistoric footprints of three children in the Réseau Clastres (Ariège).

Fig. 3 The left end of the Lion Panel in the Chauvet Cave, showing feline heads at left, and a number of rhinoceros depictions including, at the top, what seems to be a line of them drawn in perspective.

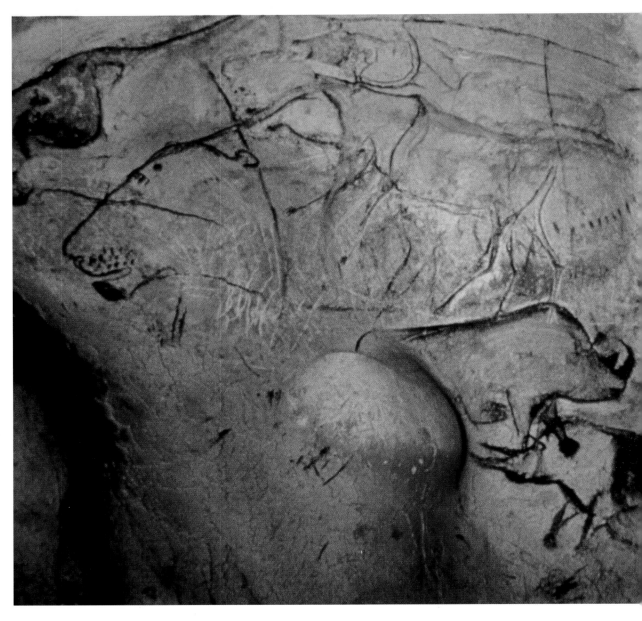

and elsewhere.[5] But we know that Palaeolithic people were perfectly capable of working wood – as will be seen in a later chapter, some of the decorated cave walls definitely required ladders or scaffolding, and the actual sockets for scaffolding-beams survive in Lascaux.

In addition, other forms of evidence show us some improvements and amenities which were brought to the caves: at Enlène (Ariège), for example, thousands of small sandstone and limestone slabs were brought in from local sources and laid down as a kind of pavement; and, as we shall see, many hundreds of them were engraved. Pollen analysis of sediments in

caves such as Lascaux and Fontanet has revealed that great clumps of grasses and summer flowers were brought in, presumably for bedding and seating.[6] So even though few caves were actually inhabited by late Ice Age people for any length of time, it seems they knew well how to adapt this environment to their advantage and comfort.

Life in the Upper Palaeolithic could be harsh or pleasant, depending on climatic and environmental fluctuations. At times there was plenty of food even in the areas close to glaciers, since periglacial environments are rich in flora and fauna, and there is no evidence of rickets in

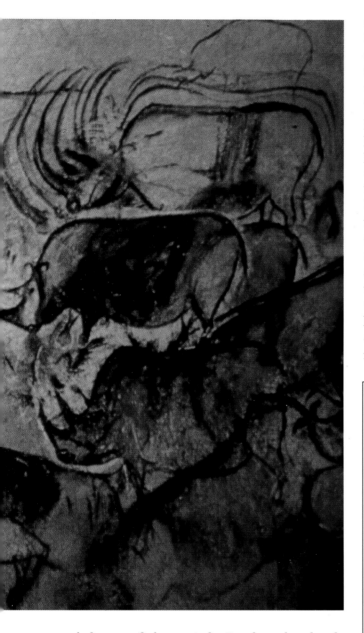

some parts of the world. On the whole, Upper Palaeolithic people were fit and healthy; little evidence has been found for arm or leg fractures, and indeed the only indications of any violence in this period in Europe (apart from the alleged 'pierced humans' depicted in a couple of caves; see p.173–4) are a piece of flint — probably an arrowhead — embedded in the pelvis of an adult woman from San Teodoro Cave, Sicily,[8] and an arrowhead embedded in the vertebra of a child buried in the Grotte des Enfants at Balzi Rossi, Italy ; seven skulls from the upper cave of Zhoukoudian in China have enigmatic depressed fractures.[9] But these are extremely rare exceptions, and it therefore appears that, on the whole, life was peaceful in the Upper Palaeolithic, as far as we can tell. Certainly, no depictions are known of inter-human violence, and the few images interpreted as men holding weapons are confronting animals, not people.[10]

Since this book is primarily about the art of Franco-Cantabria, a brief guide to the cultures and their dates is required. The Middle Palaeolithic (period of Neanderthal people), also called Mousterian after the site of Le Moustier (Dordogne), was succeeded by the Upper Palaeolithic (period of Cro-Magnon people) which has been divided into a series of phases named after different sites in France. The following is a very rough guide to their dates:

Châtelperronian:	starts *c.* 35,000 BC
Aurignacian:	*c.* 30,000 BC
	(though earlier in some areas)
Gravettian:	*c.* 25,000 BC
Solutrean:	*c.* 20,000 BC
Early Magdalenian:	*c.* 15,000 BC
Middle Magdalenian:	*c.* 13,500 BC
Late Magdalenian:	*c.* 10,000 BC
Azilian:	*c.* 8,000 BC

The Magdalenian has also been divided into a sequence of phases (I–VI) which have to be mentioned occasionally in the text, although they are rapidly being abandoned by prehistorians.

skeletons of the period. On the other hand, some skeletons display evidence of phases of slow growth caused by starvation or illness. Only a few cases of illness have been detected, such as hydrocephaly in a child (Rochereil, Dordogne), or a fungal infection in the old man of Cro-Magnon. Whereas the average Neanderthaler died aged about thirty, and very few of them passed their mid-forties, Upper Palaeolithic people had an average age at death of thirty-two, and some were well over fifty.[7]

This may not seem a great age, but death between thirty and forty was normal until only a few centuries ago, and indeed remains so in

THE DISCOVERY OF ICE AGE ART

The saga of how we discovered and came to terms with the fact that our remote ancestors could be brilliant artists is a fascinating one, filled with missed chances, pioneering insights and, above all, the interplay of open and closed minds which has always dogged archaeology, a subject in which schools of thought and personal antagonisms, even today, can play as great a role as the actual evidence.

Portable objects

The first piece of art from the Palaeolithic (Old Stone Age) known to have been unearthed was a piece of reindeer antler decorated with engraved chevrons, which was found by the pioneer prehistorian Paul Tournal during his excavations in the Grande Grotte de Bize (Aude), around 1827/8; unfortunately this specimen was never published, and is now lost.[1] The first finds to be illustrated were recovered in about 1833, in the Magdalenian cave of Veyrier (Haute-Savoie), near France's border with Switzerland, where a certain Dr François Mayor of Geneva encountered a pseudo-harpoon of antler, engraved so as to resemble a budding plant, and a perforated antler baton decorated with a simple engraving which may perhaps be a bird (Fig.1.1);[2] another more famous baton from the same site, with an engraving of a plant on one side and an ibex on the other, may have been found at the same time, but its engravings were only noticed in 1868. A horse head, engraved on a reindeer antler, is said to have been found by the Abbé Croizet in about 1840 at Neschers (Puy-de-Dôme, see p. 5);[3] and a reindeer foot-bone, bearing a fine engraving of two hinds (Fig.1.2), was discovered by Joly Leterme (an architect) and André Brouillet (a notary) in the cave of Chaffaud (Vienne) in 1852.[4] A drawing made of the Chaffaud bone, the first known reproduc-

tion of Ice Age art, was made by the writer Prosper Mérimée, France's Inspector General of Historical Monuments, and sent with a letter in April 1853 to the eminent Danish archaeologist Jens Worsaae. The bone was donated to the Musée de Cluny, Paris, where it was catalogued as part of a series of Celtic objects, since at this time the study of prehistory was in its infancy, and there was as yet no concept of an Old Stone Age, let alone of Palaeolithic art. It was not until some years later that, through comparison with examples excavated in Palaeolithic occupation layers elsewhere, the Chaffaud bone was properly identified.

The existence of Palaeolithic art was first established and accepted through the discovery, in the early 1860s, of engraved and carved bones and stones in a number of caves and rock shelters in south-west France, particularly by Edouard Lartet, a brilliant French scholar funded by Henry Christy, a London banker, industrialist and ethnologist. After finding his first piece of portable art in 1860 (a bear's head engraved on antler, from the cave of Massat in the French Pyrenees), Lartet published a drawing of it in 1861 together with a sketch of the Chaffaud bone; he was thus the first person publicly to identify the latter as Palaeolithic.[5] The depictions he encountered in his excavations came as a great surprise: their quality was astounding, since it had been assumed that prehistoric people were primitive savages with no leisure-time and no aesthetic sense. Yet there could be no doubt that the objects were authentically ancient — they were, after all, associated with Palaeolithic stone and bone tools, and the bones of Ice Age animals; some of them, such as the magnificent mammoth engraved on a fragment of mammoth ivory from La Madeleine (Fig.1.3),[6] accurately depicted extinct species, while others, such as the reindeer from Bruniquel (Fig.3.19), showed animals which

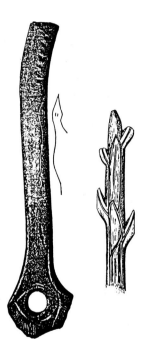

Fig. 1.1 The first decorated Palaeolithic artifacts ever found, from Veyrier (Haute-Savoie). The antler pseudo-harpoon, of which half is shown here, is 11.2 cm long. (After Pittard)

had long since deserted this part of the world.

The phenomenon was so new and unexpected that museums took some time to make up their minds about it. For example, when the Vicomte de Lastic Saint-Jal contacted the French government and offered to the Louvre the art objects which he had dug up at Bruniquel in 1863, the museum hesitated – in part, because of his high price, although he told the local press that he had done everything reasonably possible to assign to his country the results of his first discoveries. The British Museum, on the other hand, sent Richard Owen, one of its foremost experts, to France, and on his recommendation the museum

snapped up the finds, subsequently buying others from this site and elsewhere.[7] By 1867, however, Palaeolithic art was sufficiently well established for fifty-one pieces to be included in the 'History of Work' section of the Universal Exhibition in Paris.

These first discoveries triggered a kind of 'gold rush' which saw numerous people excavating – or, more usually, plundering – likely caves and shelters in search of ancient art treasures. Little attention was paid, except in a few cases, to the exact stratigraphic position of these finds; they were dug up like potatoes. At the same time, however, some diggers were ignoring the art on the cave walls. In some cases

Fig. 1.2 Detail of the reindeer foot-bone from Chaffaud (Vienne), showing engraved hinds. Probably Magdalenian. Total length 13.5 cm, height 3.7 cm.

Fig. 1.3 Mammoth engraved on a large piece of mammoth ivory from La Madeleine (Dordogne), c. 25 cm long, found in 1864. Probably Magdalenian.

Fig. 1.4 The first Ice Age figurine to be found, c. 1864: the 'Vénus impudique' from Laugerie Basse (Dordogne), an ivory statuette of a young girl, 7.7 cm tall. Probably Magdalenian.

they must have seen it, but they did not observe it: in archaeology, one tends to find what one expects to find, see what one expects to see. In other words, the cave art was of no importance in their eyes because it could not be ancient: it was inconceivable that it might have survived so long, and nothing of the sort had yet been found. Nevertheless, a few pioneers had noticed cave art and wondered …

Cave art

Many caves in Europe, as elsewhere, have been frequented since the Ice Age and throughout history by new occupants, by shepherds, by the curious or the adventurous. In some areas, such as the Basque Country, superstitions and religious traditions associated with caves are probably extremely ancient, perhaps even extending back to the period of Palaeolithic art. Certainly there is some evidence that rituals associated with the art may have survived in certain areas until quite recently: in 1458, Pope Calixtus III (a Borgia from Valencia) forbade Spaniards to perform rites in 'the cave with the horse pictures'. We do not know which cave he meant, but from the description it is likely that its pictures dated to the Ice Age.[8]

Many decorated caves also contain graffiti of different periods — Gargas, for instance, has engraved marks on its walls which are probably protohistoric, Gallo-Roman and medieval in date. In some cases, it is certain that people saw the art: in the Pyrenean cave of Niaux, a certain Ruben de la Vialle wrote his name in 1660 on the only panel of the 'Salon Noir' left undecorated by the Magdalenians; later graffiti there are placed within or between animals, carefully avoiding the painted lines.[9]

Seeing the art is one thing; reporting it is quite another, and no cave art, as far as we know, was ever mentioned in print before the nineteenth century.[10] This is quite understandable, since prehistory did not 'exist' until then, and so the pictures had no significance. However, once prehistoric studies got under way and portable art of the Ice Age had been discovered and authenticated, a few scholars at last began to notice what had been staring them in the face for so long.

The first definite suspicions of this type can be attributed to Félix Garrigou, a scholar who investigated scores of caves in the French Pyrenees in the 1860s and who found some pieces of portable art, including the famous bear engraved on a pebble, from the cave of Massat. During a visit to Niaux in 1864, he was taken to the Salon Noir, destined to be recognized as one of the glories of Ice Age cave art. As mentioned above, the Salon had often been visited in the past, and the local guides even called this part of the cave the 'museum'! In his notebook Garrigou wrote that 'there are paintings on the wall. What can they be?'[11] Alas, he did not make the crucial mental leap, and Niaux therefore narrowly missed being the first true discovery of Ice Age parietal (wall) art (its figures were not rediscovered until 1906, although a local schoolteacher called Claustres had mentioned the 'Salle des Bêtes' of Niaux in a note in 1885).

A few years later, in 1870, the archaeologist Jules Ollier de Marichard wrote in his excavation notebook that he had seen 'signs of the zodiac' engraved in a deep gallery of the cave of Ebbou (Ardèche), and in 1873 he added that he had seen animal silhouettes sketched on the walls.[12] But, like Niaux, Ebbou had to wait a considerable time (until 1946!) before being rediscovered. In 1879 de Marichard wrote a letter to Emile Cartailhac stating that some caves in Ardèche had red paintings of 'fantastic animals',[13] but unfortunately he put nothing in print.

One scholar, however, did print something about wall art: in 1878, Léopold Chiron, a schoolteacher, noticed deep engravings in the cave of Chabot (Gard); he not only took photographs and imprints of them, but also published a note about them, although he could not know their date.[14] He mistakenly thought he could see birds and people among the lines; unfortunately, the Chabot engravings are difficult to decipher, and the figures are far from clear. In May 1879, Chiron wrote to the eminent Gabriel de Mortillet to tell him of the discovery of a cave with Palaeolithic flint tools and with engravings on the walls — Chiron had no doubt the drawings were ancient, since they were covered in calcite. De Mortillet, however, who — as we shall see below — was certain that no parietal art could exist in Palaeolithic times, did not deign to reply, or to present the information in the journal he published.[15]

However, a more decisive event occurred in 1878, for it was in that year, at the Paris Universal Exhibition, that Don Marcelino Sanz de Sautuola saw Palaeolithic portable art on show, an experience which led eventually to his discoveries at Altamira. Unfortunately, whereas the pictures in Chabot were too modest to make an impact, those of Altamira were too splendid to be believed.

Altamira

The cave of Altamira, near the north coast of Spain, was found in 1868 by a hunter, Modesto Cubillas, on freeing his dog which had become wedged in some rocks. Sanz de Sautuola, a local gentleman and landowner, first visited the cave in 1875; he noticed some black painted signs on a wall, but thought little of them. His subsequent trip to the Paris Exhibition left him profoundly impressed and excited by the implements and art objects of the Ice Age on show there – we know that he was drawn back again and again to those displays. He was also fortunate enough to meet Edouard Piette, the greatest prehistorian in France, who was in the process of amassing a remarkable and unrivalled collection of Palaeolithic portable art; the two men discussed caves, artifacts and excavation methods.[16]

Back in Spain, Sanz de Sautuola began his own excavations in caves, and eventually returned to Altamira in 1879, when he was forty-eight years old. As is well known, during November that year he was digging in the cave floor, searching for prehistoric tools and portable art of the sort he had seen in Paris, while his little daughter Maria (according to her own account) was 'running about in the cavern and playing about here and there … Suddenly I made out forms and figures on the roof.'[17] She exclaimed, 'Mira, Papa, bueyes!' ('Look, Papa, oxen!')[18] She had seen the cluster of great polychrome bison on the ceiling (Fig.1.5), which had lain concealed in total darkness for 14,000 years, awaiting her.

Her father was, to say the least, astonished. At first he laughed, then he became more interested, particularly when he found that the figures were done with what seemed to be a fatty paste. Finally, 'he was so enthusiastic that he could hardly speak'.[19] He must have noticed at once the close similarity in style between these huge figures and the small portable depictions with which he was familiar, but we do not know exactly when he dared make the deduction that this wall art was of equal age. He certainly knew that it had not been done since the cave's discovery in 1868, because no stranger could spend so much time and effort in there (and to what purpose?) without everyone in the area knowing about it.

In any case, Sanz de Sautuola's greatness lies in the fact that he not only made the crucial deduction, he dared to present his views to the sceptical academic establishment. First, he wrote to Professor Juan Vilanova y Piera, Spain's foremost palaeontologist, in Madrid. Vilanova visited him, dug a little in the cave, and was convinced that Sanz de Sautuola was right about the paintings. As a result of a lecture given about Altamira by Vilanova in Santander, it made front-page news all over Spain, and tremendous crowds of people came to see the cave. Even King Alfonso XII visited Altamira, crawling into the smallest galleries, which were lit by the candles of his entourage of servants, and leaving his name in candle-black on the wall near the entrance!

In 1880, Sanz de Sautuola published a booklet on his discoveries at Altamira, including the Ice Age paintings;[20] it was a cautious piece of work in which he did not affirm the contemporaneity of the painted ceiling and the Palaeolithic deposits, but merely posed the question. In the same year, Vilanova presented the discovery at the International Congress of Anthropology and Prehistoric Archaeology in Lisbon, a gathering of some of Europe's greatest prehistorians – among them Montelius, Pigorini, Virchow, Lubbock, Evans and Cartailhac.

One would think that such a remarkable find, albeit unprecedented and unexpected, would have met with interest – even excitement – at this gathering; that there would have been a stampede of specialists to see the site; that Sanz de Sautuola would have been fêted and congratulated. It is to archaeology's undying shame that the very opposite occurred, and that it helped to bring about his premature death in 1888, a sad and disillusioned man still under suspicion of fraud or naivety, with his discovery rejected by most prehistorians. He had taken the hostility personally, as an attack on his honour and his honesty.[21]

There were a number of reasons for the rejection; for a start, he was a complete unknown, rather than an established prehistorian. Second, nothing remotely similar had been found before; after all, figures scratched on bone were a very different phenomenon from sophisticated polychrome paintings. Moreover, the discovery was made in Spain – at this time, many of the leading prehistorians were French, and south-west France was the region par excellence for Palaeolithic finds. So it must have seemed illogical and suspect for a discovery such as Altamira to occur in a different area.

At the Congress, Vilanova exhibited drawings of the Altamira figures in the halls; but his presentation was met with incredulity, and even an abrupt and contemptuous dismissal of the very idea – Cartailhac, one of the most influential French scholars, had been warned by his virulently anti-clerical friend, Gabriel de Mortillet, that some anti-evolutionist Spanish Jesuits were going to try to make prehistorians look silly.[22] Where Altamira was concerned, therefore, Cartailhac smelt a rat, rather than a bison, and walked out of the conference hall in disgust. When writing to Sanz de Sautuola in 1880 to acknowledge receipt of his booklet, he even informed him that his paintings of 'aurochs' (wild cattle!) were unlike the prehistoric animals, and had differently shaped horns![23]

It was claimed that the paintings were far too good to be so ancient, and that a modern artist (either duping Sanz de Sautuola or conspiring with him) had faked them. Cartailhac referred to the affair as a vulgaire farce de rapin (a dauber's vulgar joke).[24] Yet none of the objectors had seen the original pictures. Vilanova issued formal invitations to the leading prehistorians to visit the cave – but, incredibly, all refused.

In 1881, one Frenchman, the engineer Edouard Harlé, did turn up to examine the art of Altamira; but, as an emissary of Cartailhac and de Mortillet, he had clearly made up his mind beforehand that the paintings had been done in the 1870s, between Sanz de Sautuola's

two visits. His judgement, published in the most influential journal of prehistory of the time,[25] was based on a number of factors: the pictures were too good to have been done by prehistoric savages (this despite the quality of the authenticated portable art!); they were anatomically inaccurate (here he expanded Cartailhac's views about the anatomy of the aurochs, totally irrelevant since the Altamira animals are clearly bison!); the paint looked too fresh to be ancient, it came away on the finger [sic], some of it had been applied with a modern paintbrush; and the cave was too humid and the rock too friable to have preserved art for so

Fig 1.5 Photomontage of the painted ceiling at Altamira (Santander), covering roughly 18 by 9 m. The hind on the left is 2.25 m long, while the curled-up bison are 1.5 to 1.8 m in length. Magdalenian.

long; some of the figures were on top of stalagmite, while others were covered by only a thin layer of it, which need not denote great age (this is perfectly true); in addition, prehistoric artists would have spent long periods of time here, far from daylight, and the soot from their lamps would have blackened the ceiling, which was not the case. This last objection was more reasonable, given our knowledge at the time (no Palaeolithic lamps had yet been authenticated, although a few had already been found);[26] but, as will be seen (p.126), experiments now suggest that animal fat in lamps gives off no soot.

Cartailhac was pleased to publish Harlé's account, since he was irritated at the persistence of the two Spaniards at different meetings;[27] his own report of the Lisbon meeting had a few disdainful words about Altamira, while the official records of the Congress did not mention it at all! Other anti-Altamira articles began to appear in scientific journals and in the newspapers, both in France and in Spain. Two Spanish visitors decided that the paintings were the work of Roman soldiers, while another declared the cave to be a troglodytic temple of the Celts![28] Sanz de Sautuola and Vilanova tried to present their case again at French

Congresses for the Advancement of Science in Algiers and La Rochelle in 1881 and 1882, but in vain: at the latter meeting, neither Cartailhac nor Harlé attended that session, even though solid arguments were being presented, comparing the parietal art with portable art. The booklet and a report were also presented to an International Congress in Berlin in 1882, but again no discussion was held. Finally, Vilanova gave up the fight, and Sanz de Sautuola stood alone.[29] Neither de Mortillet's book *Le Préhistorique* (1883) nor Cartailhac's *Les Ages Préhistoriques de l'Espagne et du Portugal* (1886), nor the International Congress in Paris in 1889 mentioned Altamira at all.

However, it would be wrong to give the impression that all the leading French prehistorians shared the opinions of Gabriel de Mortillet and Emile Cartailhac, the two most virulent and powerful voices against Altamira. For example, Henri Martin (grandfather of the famous prehistorian of the same name) had to decline an invitation to visit the cave in 1880, but said in a letter to Vilanova in October of that year that he saw definite analogies between the drawings and portable art, and thought them of the same age;[30] and Edouard Piette, who had met Sanz de Sautuola in 1878, was in no doubt about Altamira, since he had always thought that portable art must have been accompanied by wall art[31] – indeed he wrote to Cartailhac in February 1887 stating that he firmly believed the paintings were Magdalenian; and in the late 1880s, when he discovered the painted pebbles of Le Mas d'Azil which characterized the 'Azilian culture' at the very end of the Ice Age, he presented them as the oldest known paintings in France, matched or surpassed in age only by Altamira.[32] Needless to say, this discovery too produced howls of rage from the establishment until its validity was proved. Piette was courageous, imaginative and open-minded throughout his career, and stands among the true pioneers of early French prehistory.

It is worth noting that Altamira was also taken seriously by some outside France and Spain. For example, a major German archaeological journal published a note on the cave in 1882 which included a drawing of the whole ceiling taken from a Spanish news magazine, *Ilustración*, of 1880.[33] This in turn led Garrick Mallery, a US army colonel, to give the cave a mention in his book of 1893, the first major work to appear on North American rock art.[34] Nevertheless, these instances were exceptions to the rule. In general, Altamira was ignored or dismissed out of hand.

Acceptance at last

How, then, did cave art change from a source of ridicule to a subject worthy of splendid monographs? It is no coincidence that the metamorphosis took place in south-west France. After the Altamira débâcle, all was quiet on this academic front, but discoveries were still being made: the stratigraphic position of Piette's pebbles, for example, proved that ochre could adhere to rock for millennia.

The real breakthrough – literally – came at the cave of La Mouthe (Dordogne), where in 1895 the owner decided to remove some of the fill and exposed an unknown gallery; a group of four youngsters led by Gaston Berthoumeyrou explored it on 8 April, returning on the 11th with Gaston's father, Edouard; in the course of one of these visits, they spotted a bison engraved on one wall, 100 m inside the cave, as well as other figures; because of Palaeolithic deposits in the blocking fill, it was clear that the pictures must be ancient.[35]

French prehistorian Emile Rivière – who had just visited Altamira – came to the cave and carried out some excavations, in the course of which he found more parietal pictures, and in 1899 he unearthed a Palaeolithic lamp here, the first to be accepted,[36] and still among the finest because of its carved red sandstone and its engraving of an ibex (p.125), which he compared with the engravings on the wall.[37] So La Mouthe provided proof of age, and of a lighting system. Cave art could no longer be denied.

Meanwhile, Chiron had again called attention in 1889 (at the Lyon Anthropological Society) and 1893 to his discovery of engravings in Chabot – but still found no support – while in 1890 he also mentioned engravings in the nearby Grotte du Figuier (Ardèche). Moreover, in 1881, a gentleman named François Daleau had begun digging in the cave of Pair-non-Pair (Heads-or-Tails) near Bordeaux and found bones of Ice Age species;

on 29 December 1883 he first noticed engravings on the cave walls exposed by the removal of occupation layers, but he paid little attention to them, apart from mentioning the fact in his notebook. It was not until thirteen years later that the discoveries at La Mouthe led him to clean the walls with a water-spray from the vineyards and study the pictures – he drew sketches of the animal figures in his notebook, and published an article about them.[38] Since some of them had been covered by Gravettian occupation layers, it was indisputable that the pictures were at least as old as that Upper Palaeolithic culture. Indeed, in 1898 in his last published paper, de Mortillet finally admitted that he had been convinced by Pair-non-Pair, although he still saw the images as 'art for art's sake' (see p.170) and hence no evidence for religion.[39]

Daleau was visited by Piette, and also by Cartailhac who admired the engravings and tried to photograph them. Cartailhac was growing troubled: during a visit to La Mouthe, he himself had removed some Palaeolithic sediment and exposed the legs of a painted animal on the wall.

Further crucial revelations awaited him in the Pyrenees. A clergyman, the Abbé David Cau-Durban, had been digging since 1883 in the Pyrenean cave of Marsoulas, where he had found cultural material and portable art of the Upper Palaeolithic. The cave is small and narrow, and one wall is festooned with engravings and red and black paintings. The paintings, at least, are very visible; Cau-Durban had seen them, but assumed they were modern, like the

Fig.1.6 Photomontage of the painted panel at Marsoulas (Haute-Garonne), showing bison, horses and barbed signs. Probably Magdalenian. Total length: c. 5 m. Note the bison composed of red dots (length: 87 cm).

names and dates on the wall, even though some were partly covered by sediment (Fig.1.6).

After the discoveries at La Mouthe and Pair-non-Pair, Félix Regnault, a local scholar, returned to Marsoulas in 1897 and discovered the animal pictures. But his claims were greeted with hilarity: Cau-Durban mocked him, saying that he was naive and gullible, and that the pictures had been done by children;[40] although Regnault mentioned the paintings at a session of the Société Archéologique du Midi, stressing the analogy between the portable and the wall art in the cave, Cartailhac, cave art's old adversary, refused to publish a note about it in the Society's journal! Regnault got two lines, with a possible misprint that described the paintings as historic rather than prehistoric. Invitations were issued in 1898 to see the paintings, but Cartailhac did not come. Rivière, however, did pay a visit, and found the colours rather fresh – consequently he had doubts about the age of the paintings, though he accepted the engravings. He may have been influenced by Cau-Durban and Cartailhac, both of whom told him they had seen no animal paintings during the excavations. Ironically, as we shall see, it was a visit to Marsoulas in 1902 which finally opened Cartailhac's eyes.

In 1901, one of Rivière's diggers, Pomarel, had found engravings in the cave of Les Combarelles, near Les Eyzies (Dordogne), and told his teacher, Denis Peyrony, about them. The latter visited the cave with prehistorian Louis Capitan and the young Henri Breuil, and found many more pictures. A few days later, Peyrony found the art in the nearby cave of Font de Gaume. These discoveries, together with all that had gone before, finally penetrated Cartailhac's stubbornness. In 1902 he visited Marsoulas and from there he went to Altamira with Breuil and at last saw its art for himself – after which he became one of the most enthusiastic scholars of cave art, even purchasing Marsoulas in order to ensure its protection!

As mentioned earlier, Piette had written to him about Altamira in 1887; once cave art began to be accepted at places such as La Mouthe, he had again reminded Cartailhac about Altamira, comparing it with La Mouthe. Yet in later years, Cartailhac claimed that he himself had re-opened the debate; he boasted

that he discovered many engravings at Marsoulas which Regnault had missed;[41] above all, he published his famous article of 1902, 'Mea Culpa d'un sceptique',[42] which is still cited in France as the epitome of an objective scientist admitting his mistakes. Certainly its title and its intentions are laudable. However, on close inspection, the paper contains only limited contrition: indeed, Cartailhac states that his actions in 1880 were fully justified at the time, and that there is nothing to object to in Harlé's 1881 report. It is clear that Cartailhac changed his mind at the last minute only when the evidence became overwhelming – it was thanks to the new discoveries that 'We no longer have any reason to doubt the antiquity of Altamira.'

Harlé revisited Altamira in 1903, having seen Font de Gaume, and realized his mistake, though somewhat grudgingly; he wrote to Cartailhac: 'it is the Font de Gaume that has made me change my mind. But for that I would still say the Bison Ceiling is a forgery.'[43]

Cartailhac later admitted that he should have expected parietal art to exist, once portable art had been discovered, and he blamed early opposition to Altamira on the fact that 'we were blinded by some dangerous spirit of dogmatism';[44] he even told Sanz de Sautuola's daughter Maria that his profound remorse would only end with his death, while Harlé admitted that his own article was an indelible stain on his scientific record. Cartailhac's Mea Culpa of 1902 was reported by the whole Spanish press with satisfaction.[45] Sanz de Sautuola was vindicated.

On 14 August 1902, a number of prehistorians – including Emile Cartailhac – attending the Montauban Congress of the French Association for the Advancement of Sciences made an excursion to the Les Eyzies area and visited the decorated caves there. This marked the official recognition by science of the existence of cave art. Had more minds been open and willing to accept the unexpected, had Chabot been more splendid and Altamira less so, cave-art studies could have begun two decades earlier. However, they quickly made up for lost time.

THE OLDEST ART
IN THE WORLD

In the last few years it has become self-evident that 'art' (however it may be defined)[1] by no means began with what we call the 'Upper Palaeolithic' or even with what we call 'modern humans' – both of them highly artificial concepts in any case. New examples of pre-Upper Palaeolithic 'art' are accumulating, though many of them are still a source of contention.

Just as the earliest crude stone tools are notoriously difficult to differentiate from natural flaking (the difference between artifacts and 'geofacts'), so it is very easy to make mistakes in recognizing early 'art objects' – many archaeologists in the past have been led astray by apparently engraved lines on bones or shells which were actually natural (rootmarks, etc.),[2] or by perforations in bones and shells which were thought to be humanly made but which, on closer examination, proved to be made by carnivore-gnawing (bones) or oceanic carnivores (shells), or by other factors such as an inherent weakness of an object at a certain point, which therefore frequently breaks in the

Fig. 2.1 Reddish cobble of ironstone, 6 cm across, found at Makapansgat, South Africa, in 1925.

same way in the same place, making the damage seem purposeful. On the other hand, it is a fair bet that natural perforations were not only found useful by early prehistoric people, who thus had ready-made beads available, but were also a source of inspiration for them, leading them to make their own.

The earliest example of some kind of aesthetic sense, or at least evidence for recognition of a likeness, occurs at a very remote period. In 1925 a dark red, water-worn ironstone cobble was found in the South African cave of Makapansgat which Australopithecines (fossil hominids) of about 3 million years ago had brought from at least twenty miles away, presumably because of the extraordinary resemblance to a human face on one side (Fig. 2.1).[3]

Many scholars in the past have pointed out that the symmetry and 'beauty' of many Lower Palaeolithic (Acheulian) handaxes, hundreds of thousands of years old, went far beyond what was technically required of the tool, whatever its uses; and indeed specimens are known from England in which fossil shells, embedded in the flint, were deliberately preserved there as the tool was shaped around them – most famously a handaxe from West Tofts, Norfolk, containing a bivalve mollusc, *Spondylus spinosus*, and another from Swanscombe, Kent, containing a fossil echinoid.[4]

As will also be seen in Chapter 3, there are numerous incidences of very early usage of ochre in many parts of the world, although one cannot be sure in what ways it was used. Pieces of haematite or ochre even appear to have been carried into sites in South Africa up to 800,000 or 900,000 years ago; a small ochre pebble, faceted and with oblique striations, which was found in an Acheulian layer at Hunsgi, southern India, dating to between 200,000 and 300,000 years ago, seems to have been used as a crayon on rock; a Stone Age site at Nooitgedacht near

Kimberley, South Africa, has yielded a ground ochre fragment with an estimated date of over 200,000 years ago, and pieces of pigment have been recovered from Zimbabwean shelter deposits of more than 125,000 years ago. A piece of red mineral, with vertical striations resulting from use, was found in the Acheulian (c. 250,000 BC) rock shelter of Becov, Czech Republic, which had been occupied by Homo erectus, the early human ancestor. The even earlier site of Terra Amata (about 300,000 BC) at Nice, France, produced seventy-five bits of pigment ranging in colour from yellow to brown, red and purple; most of them have traces of artificial abrasion, and were clearly introduced to the site by the occupants, since they do not occur naturally in the vicinity.[5]

Fig.2.2 The Acheulian 'figurine' from Berekhat Ram, Israel, a 35-mm-high fragment of basaltic tuff.

More specific examples of Acheulian art are appearing now that people are becoming more receptive to the idea: a classic example of 'I would not see it if I did not believe it'! For example, there are a number of bones and stones from this period in Europe that bear what seem to be series or patterns of deliberate and non-functional incisions — for example, twelve horse bones from the French rock shelter of Abri Suard (Charente);[6] a fragment of bovine rib from Le Pech de l'Azé (Dordogne), of c. 300,000 BC, bearing an intentional engraving comprising a series of connected double arcs;[7] and several bones from Bilzingsleben (Germany), of about the same date, likewise have series of parallel incisions which seem to have nothing to do with cutting or working.[8] There is also parietal art of this period — two petroglyphs, a large circular cupule and a pecked meandering line, have been found in Auditorium Cave, Bhimbetka (India), which were covered by an Acheulian occupation layer.[9]

However, the real breakthrough in this subject has come about recently through analyses by American researcher Alexander Marshack of new or hitherto neglected pieces of evidence from the Near East: in particular, from the Acheulian site of Berekhat Ram, Israel, a small, shaped piece of volcanic tuff that dates to somewhere between 233,000 and 800,000 years ago. The fragment bears a natural resemblance to a female figurine, and seems to have grooves around its 'neck' and along its 'arms'. Much rested on the question of whether these grooves were natural or humanly made; but Marshack's microscopic analysis has now made it quite clear that humans were responsible. In other words, this was an intentionally enhanced image, and indisputably an 'art object' (Fig.2.2).[10] There are more, varied examples of 'art' from the Middle Palaeolithic. Some have been known for some time, such as the decorated block from the French Mousterian site of La Ferrassie (Dordogne): a Neanderthal child burial lay beneath a large limestone rock with a series of small cupmarks (mostly in pairs) carved in it and apparently placed at random;[11] a circular sandstone pebble from a Mousterian level at the rock shelter of Axlor in the Spanish Basque Country has a central groove and two cupules on it.[12] A piece of flint from the Israeli site of Quneitra, dating to 54,000 years ago, is carefully incised with four semi-circles and other lines (Fig.2.3),[13] and, similarly, among a number of engraved flint flakes and pebbles from the Italian Mousterian shelter of Riparo Tagliente there is one with a double arc incised on it.[14]

There are numerous incised bones from this period: for example, at Prolom II Cave in the Crimea[15] where the Middle Palaeolithic layers have yielded a large series of perforated bones as well as a horse canine marked with five parallel engraved lines and a saiga phalange bearing a fan-like engraved motif. The Mousterian levels of the Spanish Basque cave of Lezetxiki have yielded several perforated and incised bones.[16] The French site of La Quina (Charente) yielded a bovid shoulder-blade engraved with very fine, long parallel lines, while one of the La Ferrassie Neanderthal burials contained a small bone with a series of fine, intentionally incised marks, which reminded the excavator of the notched bones of the later Aurignacian period — similar fragments with regular notches are known from the Mousterian of other European sites such as Cueva Morin, Spain, and several in the Charente.[17] The Bulgarian cave of Bacho Kiro has a Mousterian bone fragment, dating to 47,000 years ago, with a zigzag motif engraved on it.[18]

In the Ukraine, a Mousterian layer at Molodova, more than 40,000 years old, has a mammoth shoulder-blade decorated with little pits, patches of colour, and notches that seem to form complex patterns including cruciform and rectangular forms in which some Soviet scholars saw the outline of an animal,[19] although most of the marks are in fact probably natural and/or cutmarks.

Where ornaments are concerned, these too exist in the Middle Palaeolithic — for example, two bones (a wolf foot-bone and a swan vertebra), with holes bored through the top, from Bocksteinschmiede (Germany), dating to c. 110,000 years ago; a carved and polished segment of mammoth molar, and a fossil nummulite with a line engraved across it (making a cross with a natural crack), from Tata, Hungary, dating to c. 100,000 years ago; a bone fragment from Pech de l'Azé (Dordogne), with a hole

carved in it; a reindeer phalange with a hole bored through its top, and a fox canine with an abandoned attempt at perforation, from La Quina (Charente).[20] Moreover, the rich array of ornaments from the French cave of the Grotte du Renne at Arcy-sur-Cure (Yonne) — including wolf and fox canines made into pendants by incising a groove around the top, at least one sawn reindeer incisor, a bone fragment with a wide carved hole, a sea-fossil with a hole bored through its centre, and a fossil shell with a groove cut around the top — can safely be attributed to Neanderthal craftsmanship, since they come from a Châtelperronian layer (c. 34,000 BC) containing a Neanderthal temporal bone.[21]

Finally, Neanderthals are also known to have collected quantities of non-utilitarian aesthetic items, such as the shells of non-edible species, and heavy pieces of iron pyrites, as well as ochre lumps and 'crayons'; all such behaviour can be seen as artistic expressions, probably related to ritual contexts.[22] And the usage of ochre, already underway for many hundreds of millennia, increased markedly in the Middle Palaeolithic. During the period from about 200,000 to 30,000 BC, such pigments become

Fig.2.3 The engraved flint cortex from Quneitra, Israel, 7.2 cm across, and dating to c. 54,000 years ago.

increasingly frequent and abundant, not only in occupation deposits but also in burials which now occurred for the first time. In France, for example, a single layer at the cave of Pech de l'Azé I yielded 218 blocks of manganese dioxide (black/blue), plus twenty-three of iron oxide (red), scores of which were rounded or polished into a 'crayon' shape, as if they had been used on some soft surface.[23] A Neanderthal skeleton at Le Moustier (Dordogne) was sprinkled with red powder, and red pigment was also found around the head of the famous skeleton of La Chapelle aux Saints, near two skeletons at Qafzeh, Israel, and many others. In addition, there is evidence for actual mining of haematite (iron oxide) in southern Africa from around 45,000 to 50,000 BC onward — indeed, it is estimated that 100 tons were mined at the site of Ngwenya alone, dating to about 43,200 years ago — and in Hungary from 30,000 BC.[24] In Australia's Northern Territory, as will be seen in Chapter 3, used blocks of red and yellow ochre and ground haematite have been found in occupation layers at rock shelters dating to around 60,000 years ago.

Many of the above items, especially the Lower Palaeolithic incised bones and stones, demonstrate that the technique of engraving on these materials was already at a highly developed level long before the Upper Palaeolithic. Although no clearly figurative engraving has yet been found in these periods, there is definitely a first, rudimentary form of graphic expression here. There may appear to be a huge difference between the apparently simple markings of the early periods, and the sophistication of Upper Palaeolithic art. However, we shall never know the thought processes behind the earlier markings, which may have been highly complex — these apparently meaningless markings may have been filled with meanings for the people of the time, meanings so obscure in our eyes that we would find them hard to comprehend. In any case, as with any developmental or evolutionary sequence, such as that of fossil humans themselves, one simply cannot expect to find an unbroken, gradual series of 'missing links' leading from one stage to another. The archaeological record is far too patchy, and besides, the early humans of the Middle and (especially) the Lower Palaeolithic may have had completely different aims and priorities in their graphic expression which did not involve figurative naturalism.

It is often pointed out by those who still cling to the belief that 'art' was first invented by anatomically modern humans (amazingly, some even continue to point to Europe or even the Dordogne as its birthplace!) that the specimens claimed for earlier periods are very few in number and highly disparate. However, the reasons why there is so little solid evidence for 'art' before the Upper Palaeolithic are simple. First, it has been a longstanding dogma among archaeologists and anthropologists alike that no such thing could exist before the Aurignacian — earlier occurrences were simply dismissed as utilitarian marks, contamination from later levels, copying from or imports from Aurignacian neighbours (in the case of Châtelperronian beads), or freak one-offs (such as the La Ferrassie cupmarks) — with the result that many examples in the archaeological record have probably been missed or ignored.[25]

Second, there is the role of taphonomy: 'the severity of taphonomic distortion of archaeological evidence increases with its age';[26] in other words, we should expect much less evidence of artistic activity to have survived from the Middle Palaeolithic than from the Upper, and far less again from the Lower. The earliest abundance of any form of archaeological evidence (especially of perishable kinds, such as the much-touted beads and shells of the Early Upper Palaeolithic) should never be interpreted automatically as the earliest occurrence of a phenomenon. The further back in time we look, the more truncated, distorted and imperceptible will the traces of 'art' appear. But they are there, and now that we are beginning to recognize them, it is certain that many more will surface. As the history of archaeology shows repeatedly (the discoveries of open-air Palaeolithic engravings since 1980 being merely the most recent case in point, see Chapter 9), once a phenomenon is accepted as real, it starts to be looked for and to be found.

3 | A WORLDWIDE PHENOMENON

After the first discoveries (Chapter 1), examples of portable art and new decorated caves continued to be found in Europe (as will be seen below); but for a long time it was believed that, apart from the material in Siberia, Ice Age art did not exist elsewhere, and was an exclusively European phenomenon. Little by little, however, and especially in the last few years, it has become apparent that, towards the end of the Pleistocene, artistic activity was underway all over the world. The technical, naturalistic and aesthetic qualities of European Palaeolithic images remain almost unique for the moment, but it is nevertheless true that, at this period (and sometimes even earlier), in other parts of the world, one can see traces of the same phenomenon.[1]

The New World

Amazingly, the first clue to Pleistocene art in America was found in 1870, only a few years after Lartet's finds in France were authenticated. Unfortunately, the object in question was badly published, and disappeared from 1895 until its rediscovery in 1956! Consequently, very few works on Palaeolithic art mention it. This mineralized sacrum (base of spine) of an extinct fossil camelid was found at Tequixquiaq, in the northern part of the central basin of Mexico. The bone has perhaps been carved, and definitely engraved – two nostrils have been cut into the end – so as to represent the head of a pig-like or dog-like animal; the circumstances of its discovery are unclear, but it is thought to be from a late Pleistocene bone bed, and to be at least 11,000 or 12,000 years old;[2] it is on exhibit in Mexico's National Museum of Anthropology.

Other examples of portable art in the New World are equally vague as to date or authenticity. For example, a bone with an apparent engraving of an animal from Jacob's Cave, Missouri, is thought to be of Pleistocene date;[3] and it has been claimed that an implement made of mammoth ivory, which was found in Florida's Aucilla River, and which bears an incised zigzag design, may be the oldest decorated object yet known from the Western Hemisphere, more than 10,000 years old.[4]

As for parietal art, this is far more difficult to date (see Chapter 5). The New World has an abundance of decorated rocks, shelters and caves in many areas, and early dates for rock art have now been obtained in a number of different countries. In North America, accelerator mass spectrometry (AMS) and cation ratio ages have been obtained from organic material trapped in the rock varnish covering petroglyphs, some of which are in the late Pleistocene.[5] For example, the varnish on a geometric pattern in Arizona gave results around 18,000 years ago, while that on a bighorn sheep petroglyph from California's Coso Range was dated to more than 14,000 years ago – several other petroglyphs in the Coso Range have given cation ratio ages of between 12,600 and 19,000. Moreover, petroglyphs in Wyoming have also produced radiocarbon results of 10,660 and 11,650 years ago.[6] However, all these results remain highly controversial, not simply because they conflict with the orthodox but rapidly disappearing North American view that people were not yet in the New World before about 11,000 or 12,000 years ago, but primarily because of questions of methodology. Even if the dating of the organic material in the varnish is accurate, its source, and therefore its chronological relationship with the petroglyphs beneath, is far from certain, and therefore even the researchers who have produced the dates advise extreme caution and scepticism.

Where paintings are concerned, the

Peruvian cave of Toquepala, which has red figures of camelids, deer and armed hunters on the wall, was found to contain two small 'brushes' of wool impregnated with red ochre stratified in levels dating to almost 10,000 years ago.[7] More direct dates have been obtained for rock paintings in several widely separated parts of Brazil and Argentina. In the latter's north-western province of Jujuy, the deep sandstone rock shelter of Inca Cueva 4 contains some geometric motifs on the heavily exfoliated back wall (Fig. 3.1). They were painted on a preparation of gypsum, and analysis by X-ray diffraction has revealed that exactly the same mineralogical composition of pigments and support is found on fallen, stratified fragments of paint, as well as on artifacts, in an occupation level that has been radiocarbon-dated to 10,620 years ago. In other words, the paintings must predate that time.[8]

At the other end of the country, in Santa Cruz province (Patagonia), two sites have yielded similar evidence. At Los Toldos, excavations in Cave 3 recovered fallen fragments of ceiling bearing red paint, thought to be perhaps from ancient hand stencils, in a layer which indicates an age of around 11,000 years.[9] Farther south, at the site of El Ceibo, as occupation layers were dug away, two dark red paintings of guanacos (wild llamas) appeared on the newly exposed rock wall (Fig. 3.2). Judging by their

Fig. 3.1 Drawings in the rock shelter of Inca Cueva 4, Jujuy province (Argentina), thought to date back to more than 10,600 years ago.

Fig. 3.2 Two dark red paintings of guanacos exposed by excavation at El Ceibo, Santa Cruz province (Argentina). They were covered by a layer of the Casapedrense period, and thus were produced in the Toldense period and must be at least 10,000 years old.

depth and the age of the layer masking them, they must be of similar age to the paintings of Los Toldos.[10] Simple linear petroglyphs covering about 16 square metres of the bedrock floor of Epullán Grande Cave, in Argentina's Neuquén Province (northern Patagonia), are thought to date back to at least 10,000 years ago.[11]

In Brazil, samples of pigment from red paintings in the sandstone Caverna da Pedra Pintada at Monte Alegre on the Lower Amazon have, through Scanning Electron Microscopy, been found to be similar to samples from the hundreds of lumps and drops of red pigment – as well as two small fragments of painted wall – which were stratified in Paleoindian levels radiocarbon-dated to a period from approximately 11,200 to 10,000 years ago.[12] In the arid Piauí region of south-east Brazil, on the other hand, rock paintings at the huge sandstone rock shelter of Boqueirão da Pedra Furada have been dated to the same period. One fallen fragment, bearing a clear red human stick-figure, has been

found in occupation layer XII; from its position in relation to layers above and below, dated by charcoal, this level has been assigned to about 10,000 or 12,000 years ago (Fig. 3.4), while another bearing two red lines, very probably the legs of a human or animal, came from a layer dating to around 17,000 years ago. These are therefore minimum ages for the fallen art.[13]

At the nearby shelter of Toca do Baixão do Perna I, direct dating of organic carbon in a pigment ball fashioned by humans and apparently worn as an ornament gave a result of 15,250 years ago.[14] Likewise at Perna – as at El Ceibo, mentioned above – a panel of small red figures was exposed by excavation of the layers which had covered it, and though faded, the images survived burial amazingly well (Fig. 3.5). One fragment of charcoal still adhering to the panel gave a radiocarbon date of 9650 years ago, while charcoal from the layer touching the bottom of the panel has been dated to 10,530 years ago – hence, unless one envisages artists painting at nose level while lying on the floor, the

Fig. 3.3 Hand stencils in the rock shelter of Los Toldos 2, Santa Cruz province (Argentina). Excavations in shelter 3 found what may be fragments of hand stencils in layer 10 (Early Toldense), dating to c. 11,000 years ago.

Fig. 3.4 A sandstone spall from Pedra Furada, Brazil, bearing a red human stick-figure, from an occupation layer of 10,000–12,000 years ago.

Fig. 3.5 Small red human stick-figures exposed on the wall of Perna rock shelter, Brazil, by excavation. Over 10,500 years old.

panel must be somewhat older than this date.[15] Its figures correspond perfectly in size and type to that of similar age at Pedra Furada. Together with art at many other sites in this region, they have been attributed to the 'Serra da Capibara' style, which is thus thought to date back to at least 12,000 and probably at least 17,000 years ago.

Finally, it has been claimed that petroglyphs in the same area (most notably at Caiçaras and Racho Santo) are probably of Pleistocene age on the basis of their heavy patination and their

non-figurative motifs,[16] while paintings in central Brazil (Bahia region) may perhaps depict extinct species such as giant sloths and the hippo-like Toxodonts.[17]

Africa

It is likely that some of Africa's rich abundance of rock art is of Late Pleistocene age. In Zimbabwe, what seems to be a paint palette has been excavated from a layer dating back more than 40,000 years at Nswatugi Cave. In addition, at Pomongwe Cave fragments of painted stone are known from layers dating to between 13,000 and 35,000 years old, and the same site has yielded granite spalls from the walls, bearing traces of paint, in deposits dating back at least 10,000 years and perhaps even to the twelfth millennium BC.[18] Moreover, as we saw in Chapter 2, pieces of pigment have been recovered from Zimbabwean shelter deposits of more than 125,000 years ago, while a Stone Age site at Nooitgedacht near Kimberley has yielded a ground ochre fragment with an estimated date of over 200,000 years ago, and pieces of haematite or ochre even appear to have been carried into sites in South Africa up to 800,000 or 900,000 years ago. At some decorated sites in Tanzania, levels have been excavated containing ochre fragments, ochre pencils and stained 'palettes' from about 29,000 years ago onward.[19]

Portable Palaeolithic art has been well authenticated in Namibia, where seven fragments of stone found by Eric Wendt in the Apollo 11 Cave have paint on them, including four or five recognizable animal figures (Fig. 3.6) such as a black rhino and two possible zebras; they display a use of two colours, and

Fig. 3.6 Charcoal drawing of an unidentified animal on a stone plaquette from Apollo 11 Cave, Namibia, c. 25,000 years old. The two fragments are 11 by 8 cm and 9 by 5 cm.

were associated with charcoal which has provided a radiocarbon date of at least 19,000 and perhaps even 26,000 years ago.[20] The site also yielded some notched bones of similar or greater age.

Engraved pieces of wood and bone (including a baboon fibula with twenty-nine parallel, incised notches), have been recovered from Border Cave, Kwazulu, and dated to between 35,000 and 37,500 years ago – the baboon fibula has been compared with similar calendar sticks of wood which are still in use by some Bushman clans in south-west Africa.[21] More recently, this site has also yielded several bone fragments 'decorated' with multiple incised parallel lines in levels dating to more than 100,000 years ago.[22] Similarly, a Middle Stone Age layer at the Klasies River Mouth Caves, South Africa, has yielded some notched and grooved bones dating to more than 100,000 years ago.[23]

A bored stone decorated with incisions has been found in a layer of Matupi Cave, Zaïre,

dating to about 20,000 years ago.[24] One should also mention the ceramic fragment with incised decoration, dating to about 20,000 years ago, from Tamar Hat Cave, Algeria,[25] especially as the Upper Palaeolithic levels (more than 13,000 years ago) of the site of Afalou-Bou Rhummel, in eastern Algeria, have yielded dozens of small terracotta figurines (Fig. 3.7), mostly of horned animal-heads, made of a red clayey paste.[26]

The Middle East, Central Asia and India
Anatolia (Turkey) contains a few pieces of Pleistocene portable art,[27] but no definite parietal art as yet – figures in Kara'in Cave, near Antalya, have been attributed to the Palaeolithic, but they remain undated.[28]

Israel has recently begun to produce a few pieces of art from the Pleistocene. Among the youngest are the limestone pebble from the Epi-Palaeolithic (c. 19,000–14,500 years ago) site of Urkan e-Rub which has engravings of 'ladder' motifs and parallel lines on it[29] and a

Fig. 3.7 Fragment of a terracotta animal figurine from Afalou-Bou Rhummel, Algeria, found in a hearth in 1989. Made of fine red clay, it is 32 mm long, 21 mm high. Iberomaurusian period (more than 13,000 years old).

number of figurines from the Natufian of El-Wad Cave, dating to 12,800–10,300 years ago.[30] There is also a crude engraving of a possible animal on a limestone slab from the Aurignacian of Hayonim Cave.[31] One should also mention the faint possibility that some linear rock engravings in a number of caves at Mount Carmel are Palaeolithic.[32]

However, the two most important finds from Israel are the Quneitra flint plaque and the Berekhat Ram figurine, mentioned earlier (see pp. 24–5).

In Syria, recent excavations at Jerf el Ahmar, on the left bank of the Euphrates, have uncovered a series of stones, 10,000 years old, which bear series of pictograms – combinations of lines, arrows and animal outlines – which have been seen by some researchers as an intermediate stage in an evolution from Palaeolithic art to true writing, which arose about 5000 years later in the form of Sumerian cuneiform.[33]

It has been claimed that the oldest rock art in central Arabia dates back some 14,000 years, though the only evidence for this is the apparently Pleistocene fauna depicted.[34] The same situation exists in India, where the hundreds of caves and rock shelters around Bhimbetka, near Bhopal, contain parietal paintings spanning a long period. Claims have been put forward that the earliest are Upper Palaeolithic in age,[35] especially since engraved ostrich eggshells from excavated layers here are said to have been dated to between 25,000 and 40,000 years ago.[36] One particularly fine fragment from Patne has a criss-cross pattern engraved between two parallel lines (Fig. 3.8).

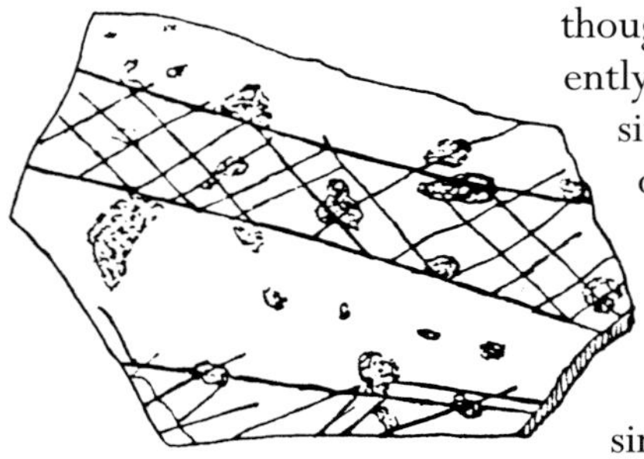

As mentioned earlier (see p. 24), Bhimbetka also has the oldest reported petroglyphs in the

Fig. 3.8 Tracing of an engraved fragment of ostrich eggshell from Patne, India, dating to at least 25,000 years ago. (After Kumar et al.)

world, and it has been claimed that nine other cupmarks in the same cave are of similar age.[37] Recently, Daraki-Chattan ('fractured rock'), a deep cavity between two sandstone rocks at Indragarh in India's Chambal Valley, has been found to contain 498 cupmarks of very archaic appearance, which have been linked with Lower and Middle Palaeolithic artifacts on the cave floor.[38] As we saw in Chapter 2, a small ochre pebble from Hunsgi (a site in southern India, dating to between 200,000 and 300,000 years ago, and containing abundant pieces of ochre) seems to have been used as a crayon on rock.

In Afghanistan, the excavations by Louis Dupree and his associates in the early 1960s at Aq Kupruq produced two small engraved or carved stone artifacts which comprise the only examples of intentionally worked art objects known from the Upper Palaeolithic or pre-Neolithic of Afghanistan. The date of the artifacts is the subject of some confusion, early reports claiming c. 20,000 BC and later reports c. 8000 BC. One object is an apparently humanoid head carved and engraved on a small, soft limestone pebble, 6 cm high and 4 cm wide.[39] The eyes are engraved as circles, but the 'nose' forms a depression rather than a prominence, while the 'mouth' is equally bizarrely rendered, and what may be an ear is engraved at one side. Unfortunately, this intriguing object has disappeared through the repeated looting of Kabul Museum in recent years.

The Far East
China's Palaeolithic art was limited for many years to the 120 beads and other decorative objects from the upper cave at Zhoukoudian. In recent years, some hand stencils found in caves have been claimed to be of Palaeolithic age, through simple analogy with those dated to the Pleistocene in France and Tasmania, but proof is still lacking for the moment. However, a definite piece of Palaeolithic art from China did emerge a few years ago – an engraved 14-centimetre piece of antler from Longgu Cave (Hebei province) radiocarbon-dated to 13,065 years ago (Fig. 3.9).[40] The site of Shiyu (Shanxi province) has yielded half of a perforated stone disc, 8 cm in diameter, which is about 28,000

years old, and thus one of the world's oldest known perforated stone objects.

In Korea, a number of claims have been made for portable art in the Middle Palaeolithic: these take the form of bones supposedly modified to depict animals,[41] but serious doubts exist about all of them. Similarly, primitive pecked rock-carvings in a number of sites are claimed to be Palaeolithic, partly because they are thought to depict extinct reindeer and grey deer, and partly because they display 'Palaeolithic mentality',[42] but, once again, proof is lacking.

In Japan, on the other hand, there are some extremely interesting engraved pebbles from the cave of Kamikuroiwa. Layer IX (Initial Jomon) has been dated to 12,165 years ago, and contained several little pebbles with engravings on them, some of which seem to represent breasts and 'skirts' (Fig. 3.10).[43] It should also be noted, in the light of early examples of work in terracotta (see p. 31), that pottery vessels in Japan – the oldest known in the world – currently date back to around 16,000 years ago, while fragments of a badly fired clay vessel have been found in the site of Gasja, on the lower Amur River in eastern Siberia, associated with charcoal dated to 12,960 years ago.[44]

A cave in the Mongolian Altai, Khoist-Tsenker-Agui or Chojt-Cencherijn-aguj (Cave of the Blue River), contains paintings of animals in niches. Their subject matter – horses, goats, bulls, ostrich-like birds, camels, and elephants or mammoths – has led to their being ascribed to the Palaeolithic, especially as there are no human figures and no recognizable compositions; however, there is absolutely no proof of their age, and it has been pointed out that the site's gypsum rock is probably too unstable to preserve pictures from that period.[45]

Australia

It is in Australia, with its incredible wealth of rock art, that one finds most of the non-European examples of Pleistocene parietal art.[46] The first site where its existence was authenticated was Koonalda Cave, in southern Australia, which was found to contain abundant 'digital flutings' (lines made with fingers) on the ceiling and walls, in total darkness, hundreds of metres inside; they seemed to be asso-

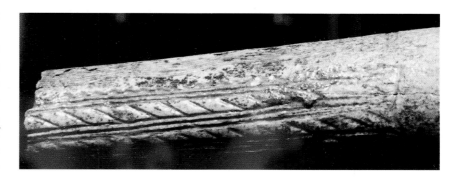

ciated with the extraction of flint.[47] The site's archaeology showed that the mining activity took place at least 15,000 to 30,000 years ago, and so the marks are probably of similar age. These finger flutings are identical to those known in several of the European Palaeolithic caves.

Like those of Koonalda, finger flutings in the Snowy River Cave, Victoria, can be dated only by assuming that they are contemporaneous with archaeological deposits at the cave entrance, which are about 20,000 years old.

The first direct proof of the antiquity of Australian art came from the Early Man shelter in Queensland, where very weathered and patinated engravings (circles, grids, and intertwined lines) covered the back wall and disappeared into the archaeological layer (Fig. 3.11). As the layer yielded a radiocarbon date of 13,200 years ago,[48] it is clear that the engravings must be at least this old: it will be recalled that it was precisely this sort of proof, at La Mouthe, which finally clinched the authenticity of Palaeolithic parietal art in Europe.

Engravings resembling those of the Early Man shelter exist in Tasmania; as that region became separated from the continent by a rise in sea-level around 12,000 years ago, and since its art contains no engraved dingo prints (the dingo having arrived in Australia after that date), some scholars believe that the Tasmanian engravings are older than the separation; others think they are much younger. Since 1986, twenty-three red hand stencils have been found 20 m from daylight inside Ballawinne Cave in the Maxwell Valley, south-west Tasmania. They are undated, but, from a comparison of the cave's archaeological material with that of similar sites in the region, it has been estimated that they may be at least 14,000 years old.[49]

Fig. 3.9 Decorated fragment of deer antler, 13.4 cm long, from Longgu, China, dating to about 13,000 years ago.

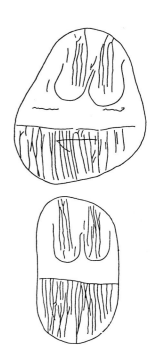

Fig. 3.10 Tracing of pebbles from Kamikuroiwa, c. 5 cm long. (After Aikens and Higuchi)

Since 1987, another twenty-three hand stencils, a roughly drawn circle and extensive areas of wall smeared with deep red pigment have been found in another large limestone cave in south-west Tasmania, Wargata Mina (Judds Cavern), mostly in a high alcove about 35 m from the entrance, at the limit of light penetration.[50] The red pigment has been found to contain human blood, and AMS radiocarbon dating of samples of it from the cave has produced results of 9240 and 10,730 years ago.[51] In 1988, a third cave in south-west Tasmania, TASI 3614, was also found to contain three red hand stencils, situated 10 m and 50 m from the entrance.

There is evidence of even older rock paintings in other parts of the country. What may be a painting on the wall of the Sandy Creek 2 rock shelter in Queensland has produced a date of 24,600 years ago;[52] while excavations in Sandy Creek 1 produced two striated fragments of red pigment dating to around 32,000, and two yellow fragments date to 28,000 and 25,900 (this site also yielded a fragment of sandstone, with part of a deeply pecked motif, in a layer dating to 14,400 years ago).[53] On a boulder at Walkunder Arch Cave in the Chillagoe region of north Queensland, a crust sample containing layers of paint from three different painting episodes has produced AMS radiocarbon estimates of 16,100, 25,800 and 28,100 years ago.[54]

At Carpenter's Gap rock shelter in the Kimberley, what seems to be a fallen fragment

Fig. 3.11 Engravings in the Early Man shelter, Queensland.

of painted wall was found in a layer containing charcoal dated to more than 39,000 years ago, suggesting a minimum age for the painting;[55] while in Arnhem Land, Northern Australia, used blocks of red and yellow ochre and ground haematite have been found in occupation layers at the shelters of Malakunanja II and Nauwalabila I, dating to around 60,000 years ago. Most notable, in the latter site, is a piece of high-quality haematite weighing a kilogram, which was brought in from some distance and whose facets and striations are clear signs of use.[56] Many ochre 'pencils' with traces of wear have also been found in this region's shelters throughout the Pleistocene layers. Although much of this colouring material may have been used for body-painting or some other purpose, it is likely that at least some of it was used to produce wall art. Some of the oldest paintings in the region are covered by a thin siliceous film, which is deposited only in very arid conditions, and the last such period here occurred

18,000 years ago. Moreover, among these apparently ancient figures are animals which some researchers interpret as species extinct in Australia for at least that length of time (e.g. the marsupial tapir Palorchestes);[57] other scholars disagree with these interpretations and the proposed chronology,[58] but everyone agrees that some of the art (the Boomerang/Dynamic/ early Mimi period style) is 'pre-estuarine' and thus perhaps thousands of years earlier than 9000 years ago.

Samples of weathered dark pigment taken from sandstone rock shelters along Laurie Creek in the Northern Territory have been claimed to contain human blood, and radio-carbon dating produced a date of 20,320 from them, but serious doubts have since been expressed as to whether the detection of blood was accurate and, therefore, whether the pigment was indeed humanly made at all.[59]

Australia has produced little portable art from the Pleistocene so far. In the cave of

Fig. 3.12 The oval petroglyph, *c.* 15 cm wide, from Wharton Hill, Olary Province (South Australia), varnish on which yielded a date of more than 42,000 years.

Devil's Lair, Western Australia, three perforated bone beads have been found in a layer that is 12,000 to 15,000 years old, together with a possible stone pendant and some slabs of stone with possible engraved lines on them, although it is by no means certain that the lines are artificial.[60] More recently, the upper incisor of a *Diprotodon* from a megafaunal assemblage at Spring Creek, Victoria, dating to the eighteenth millennium BC, has been found to bear a series of twenty-eight incisions or grooves which observation and experimentation suggest strongly were made by humans.[61] Finally, a necklace of twenty-two small Conus shells deliberately modified as beads, dating to about 32,000 years ago, has been discovered at Mandu Mandu Creek rock shelter, Western Australia.[62]

In north-west Australia, in the Pilbara, there are thousands of petroglyphs in the open air, and many researchers are beginning to think that the patina on some proves that they are extremely ancient, probably some 10,000 to 15,000 years old. At Gum Tree Valley, on the Burrup Peninsula (Western Australia), some

visibly very old engravings are closely associated with sea-shells that have been dated to 18,510 years ago.[63] At Sturts Meadows, in New South Wales, a compact carbonate overlying a desert varnish on petroglyphs has given AMS results of 10,250 and 10,410 years ago, suggesting that most of this huge site's figures are probably of at least that age.[64]

As in North America (see above), the dating of desert varnish covering petroglyphs remains highly controversial and of uncertain validity, but has produced the earliest direct dates for rock art in the world so far: in the Olary region of South Australia the cation ratio technique first produced a date of over 30,000 years,[65] while more recently dating of varnish by both AMS radiocarbon and cation ratio has produced results of more than 42,000 years for an oval motif at Wharton Hill (Fig. 3.12) and 43,140 years for a curvilinear motif at Panaramitee North.[66]

Most recently, claims have been made for rock art dating back at least 75,000 years at Jinmium in the Northern Territory, where some enormous sandstone boulders are cov-

Fig. 3.13 Cupmarks on sandstone boulder at Jinmium, Australia, which are certainly prehistoric, but of uncertain antiquity – perhaps up to 20,000 years old, possibly even 75,000.

Figs. 3.14 and 3.15 Large, deeply carved circle with internal lozenge lattice in entrance of Paroong Cave, Australia, and deeply engraved circles (diameter between 15 and 45 cm) on the opposite wall of the entrance.

ered with thousands of pecked cupmarks — around 3500 on one rock, and 3200 on another nearby (Fig. 3.13). Some of these were masked by sediments dated by the thermoluminescence technique to between 58,000 and 75,000 years ago, thus suggesting a minimum age for the art. Lower layers of sediment nearby, containing ochre and stone artifacts, have likewise been dated by the same method to 116,000 years, and even up to 176,000.[67]

The marks on the rocks are definitely humanly made, and it has been estimated that each cupule would have taken at least an hour to make. They were probably produced by a large number of people or accumulated over a long period. They were first noticed above ground on boulders and rock outcrops, and as earth was dug away it was found that they continued underground. It is the earth covering the lowest markings, two cupules on a fallen rock

Fig. 3.16 Digital tracings in Karlie-ngoinpool Cave, South Australia.

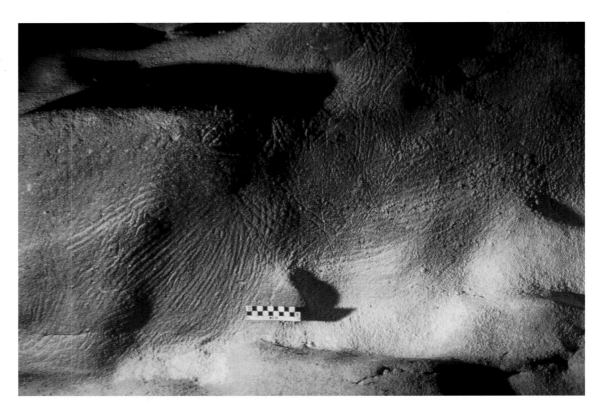

fragment more than a metre below the present ground level, which produced the early date of up to 75,000 years. However, contamination may have affected these dates.

There are not many caves in Australia, and those which do exist tend to be shunned by modern Aborigines, who are wary of the dark and of caverns. Consequently, any art found inside a cave is unlikely to be of recent origin. The discovery of digital flutings at Koonalda was followed by similar though lesser finds in the small Orchestra Shell Cave (Western Australia)[68] and in two other caves in the south of the country. Recently, however, a whole series of decorated caves has been discovered in South Australia, near Mount Gambier. This region contains hundreds of caves in its Tertiary limestone karst. Up to now, 140 have been examined; almost all contain traces of animals (clawmarks on the walls, and so forth), and thirty-five of them also contain non-figurative marks made by humans on the walls and ceilings – indeed this seems to be a remarkable concentration of non-figurative parietal art.

These caves are not like the well-known examples in France and Spain; many of them have vertical entrances, at ground level in fields. Farmers have always used them as garbage dumps, throwing in scrap metal, machines and all kinds of refuse, to prevent livestock falling in; consequently, the descent into the depths can be dangerous, as one negotiates a tottering mountain of rusting barbed wire and old refrigerators. This state of affairs, however, has helped to protect the caves and their art: very few potholers or explorers have entered them, especially as they are on private land and often near farmhouses. These caves are being cleared of rubbish and locked, one by one: until then, their exact location must remain a secret.

It has been established[69] that they contain three distinct traditions of petroglyphs, which are sometimes superimposed on the same wall, separated by thin layers of carbonate. It is hoped that radiometric dating (the Uranium/Thorium method) of these deposits will one day provide an approximate chronology for the traditions.

The earliest phase comprises the digital flutings; all thirty-five caves have them, and they always lie beneath the engraved lines and seem to follow or emphasize the topography of the walls. As with those of Koonalda, Orchestra

Shell Cave and others, there is a possibility that they may be at least 20,000 years old. If that is so, then, at that time, this tradition of finger-marking extended along the entire southern part of the continent, a distance of 3000 km!

The second phase, apparently also very ancient, is called Karake (after the cave where such traces were first seen); it is characterized by deeply engraved and weathered circles – simple, concentric, or divided up by gouged lines – which resemble the archaic petroglyphs of Early Man shelter. Finally, the terminal phase, comprising shallow engravings, is probably much more recent, dating to the Holocene period.

It is noteworthy that several of these caves, like Koonalda, served as chert mines at the time of their decoration; the region around Mount Gambier was already known to archaeologists for the great abundance of Pleistocene stone tools and knapping-waste found there, around the good-quality surface sources of stone. The decorated cave-mines are thus only the nucleus of a considerable activity area.

Two of the caves in particular are outstanding. Paroong has an important collection of Karake-type petroglyphs, especially at its entrance: the vertical walls here are literally covered with a profusion of deeply carved circles, one next to another, up to a height of 4 m above the present floor (Fig. 3.14–15).

The cave of Karlie-ngoinpool (which means 'many' in the old local indigenous language) contains one of the greatest known concentrations of non-figurative parietal art. Its well-preserved finger-flutings cover 75 square m

Fig. 3.17 Deeply engraved circles associated with extraction of flint nodules, Karlie-ngoinpool Cave, south Australia.

(Fig. 3.16). There are also quantities of deeply engraved circles and other Karake motifs, and some of them (covering a 10-metre width of wall) remain hidden behind the mountain of rubbish in the entrance. Many of these circles are closely associated with the extraction of nodules of chert of very poor quality (Fig. 3.17). The cave ends in a little 'sanctuary', a tiny chamber in total darkness and covered in engravings, especially circles, which appears to have been of particular significance for the people of this perhaps remote period.

Clearly, therefore, the map of Pleistocene artistic activity is rapidly filling up, with every continent – and especially Australia and South America – providing well-dated examples. It is probable that there are innumerable similar surprises in store for us. Nevertheless, Europe remains supreme, for the moment, in the quantity and quality of its surviving Palaeolithic art.

Europe

The Palaeolithic pictures of Europe are generally treated as two distinct entities, the portable and the parietal – whereas in reality these are merely the two ends of a continuous range; in other words, there is an overlap between the two categories, comprising cases in which it is impossible to decide whether detached frag-

ments of wall were decorated before or after falling, and blocks which could be moved but were too large to be carried around (Breuil called them parois mobiles, or movable walls): at the French rock shelter of La Marche, for example, the size of the engraved slabs varies from specimens a few cm long to some over a square metre in area and weighing tens of kilograms. However, for the sake of convenience, the usual artificial division will be retained here.

Portable

The two categories have somewhat different distributions within Europe. Portable art is found from the Iberian Peninsula and North Africa to Siberia,[70] and has notable concentrations in Central and Eastern Europe; occasional specimens also turn up in countries around the fringes of Europe, such as those of Israel (see above) or the handful of Upper Palaeolithic engraved bones from England.[71]

It is hard to quantify portable art objects, since many are broken, and unknown numbers of them remain unpublished, whether in private or clandestine collections, or lying forgotten and unstudied in museums around the world.[72] One estimate in 1980[73] that there are

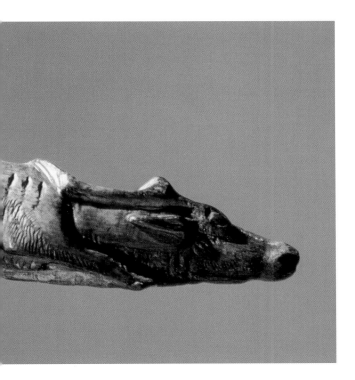

Fig. 3.20 Painted plaquette from Parpalló, Spain, dating to the Solutreo-Gravettian period, and showing a horse and some short red, parallel marks. It is 37.5 cm long, 27 wide, and 4.5 thick.

OPPOSITE: Fig. 3.18 Antler spear-thrower carved in the form of a mammoth, from Bruniquel (Tarn-et-Garonne). The trunk joins the legs while the tusks jut forwards. Probably Magdalenian. Length: 12.4 cm.

LEFT: Fig. 3.19 Ivory carving of two reindeer, perhaps a male following a female, from Bruniquel. Probably Magdalenian. Length: 20.7 cm.

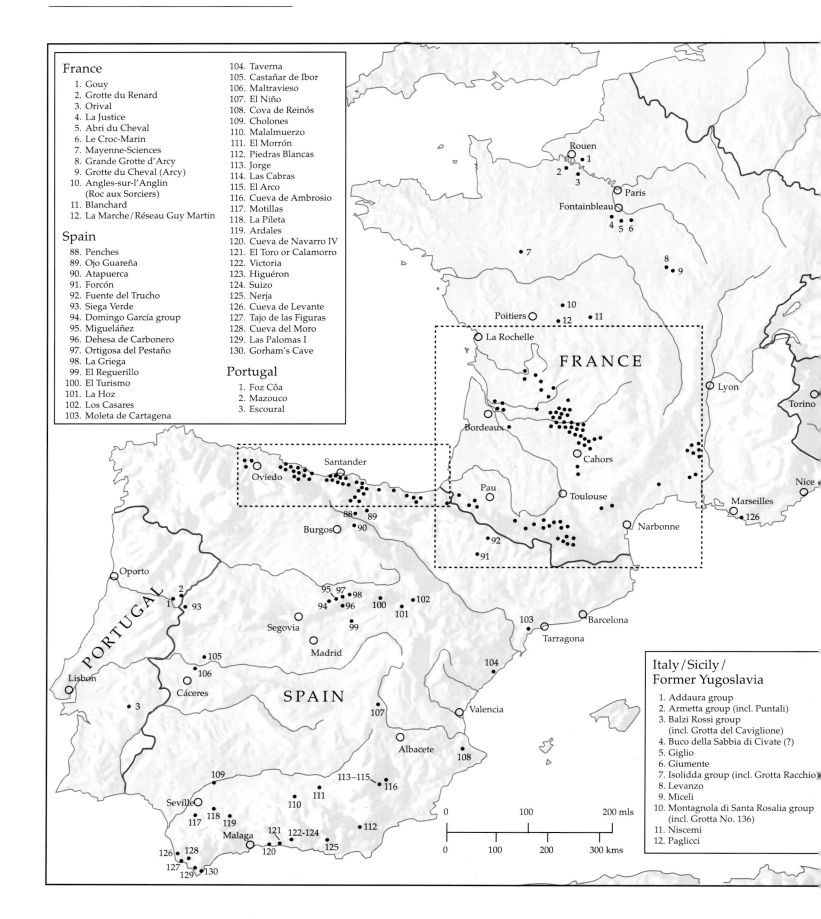

France

1. Gouy
2. Grotte du Renard
3. Orival
4. La Justice
5. Abri du Cheval
6. Le Croc-Marin
7. Mayenne-Sciences
8. Grande Grotte d'Arcy
9. Grotte du Cheval (Arcy)
10. Angles-sur-l'Anglin (Roc aux Sorciers)
11. Blanchard
12. La Marche/Réseau Guy Martin

Spain

88. Penches
89. Ojo Guareña
90. Atapuerca
91. Forcón
92. Fuente del Trucho
93. Siega Verde
94. Domingo García group
95. Migueláñez
96. Dehesa de Carbonero
97. Ortigosa del Pestaño
98. La Griega
99. El Reguerillo
100. El Turismo
101. La Hoz
102. Los Casares
103. Moleta de Cartagena

104. Taverna
105. Castañar de Ibor
106. Maltravieso
107. El Niño
108. Cova de Reinós
109. Cholones
110. Malalmuerzo
111. El Morrón
112. Piedras Blancas
113. Jorge
114. Las Cabras
115. El Arco
116. Cueva de Ambrosio
117. Motillas
118. La Pileta
119. Ardales
120. Cueva de Navarro IV
121. El Toro or Calamorro
122. Victoria
123. Higuéron
124. Suizo
125. Nerja
126. Cueva de Levante
127. Tajo de las Figuras
128. Cueva del Moro
129. Las Palomas I
130. Gorham's Cave

Portugal

1. Foz Côa
2. Mazouco
3. Escoural

Italy/Sicily/Former Yugoslavia

1. Addaura group
2. Armetta group (incl. Puntali)
3. Balzi Rossi group (incl. Grotta del Caviglione)
4. Buco della Sabbia di Civate (?)
5. Giglio
6. Giumente
7. Isolidda group (incl. Grotta Racchio)
8. Levanzo
9. Miceli
10. Montagnola di Santa Rosalia group (incl. Grotta No. 136)
11. Niscemi
12. Paglicci

13. Perciata
14. Porto Badisco
15. Pizzo Muletta
16. Rocca Rumena group
17. Romanelli
18. Grotta del Romito
19. Sallinella
20. San Teodoro
21. Santa Maria di Agnano
22. Vaccari
23. Vieste group
24. Villabruna A
25. Za Minica I / II
26. Badanj

well over 10,000 pieces of Palaeolithic portable art in Western Europe is certainly a minimal figure. A 1965 study[74] of the portable art from sites in the Périgord region of France produced a total of 2329 objects – once again a minimal figure, since there are many more in foreign or private museums and collections.

The scattering of these objects around the museums of the world poses great problems for their study, which will be alleviated only when good casts of the 'absentees' can be housed in some European research centres, and when full details of every specimen are available from the computerized databanks which are now being compiled.[75]

On the other hand, the uneven distribution of the objects among archaeological sites is a stimulating puzzle for the archaeologist. Many Palaeolithic sites in Europe have no decorated objects at all, while others have only one or a few. Some Magdalenian 'supersites', however, have hundreds: the cave of Parpalló, in eastern Spain, produced 5034 plaques of engraved and painted stone (with 6245 decorated faces), together with scores of decorated bones, from its long Upper Palaeolithic occupation, and especially the Magdalenian.[76] The site of La Marche, already mentioned above, has at least 1512 engraved slabs of stone (1268 after some fragments were fitted together), weighing a total of 4 tons![77] The open-air site of Gönnersdorf in West Germany has about 500 engraved slates.[78] The cave of Enlène, in the French Pyrenees, has recently yielded over 1100 engraved stones, in addition to its wealth of decorated bone and antler[79] – in some areas of the cave floor, over 100 engraved slabs have been found in less than a square metre!

Clearly, therefore, some sites are infinitely more important than others in terms of decorated objects. This becomes particularly clear from the above-mentioned survey of Périgord specimens: the 2329 items were spread over seventy-two sites; but two rock shelters alone, La Madeleine (582 objects) and Laugerie-Basse (560), account for 1142, or 48 per cent of the total; if one adds Limeuil (176) and Rochereil (132), then only four sites account for 1450 objects, or 66 per cent of the total. Similar figures would emerge from studies of this kind in the Pyrenees or Spain and, as we shall now see,

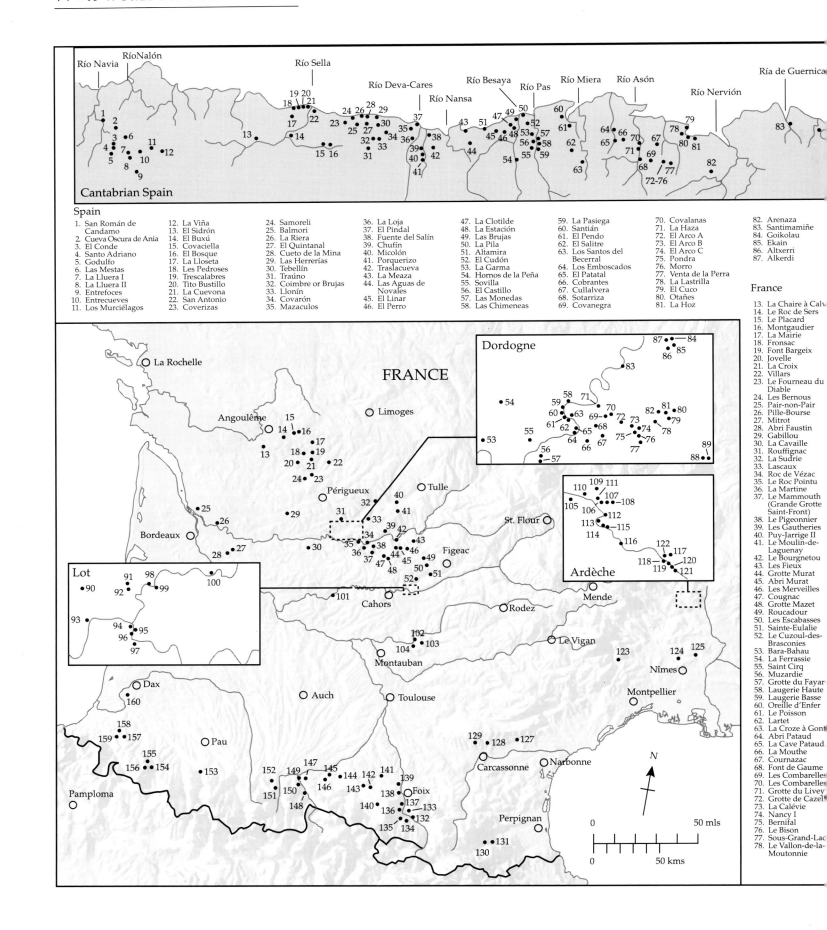

Río Navia RíoNalón Río Sella Río Deva-Cares Río Nansa Río Besaya Río Pas Río Miera Río Asón Río Nervión Ría de Guernica

Cantabrian Spain

Spain

1. San Román de Candamo
2. Cueva Oscura de Ania
3. El Conde
4. Santo Adriano
5. Godulfo
6. Las Mestas
7. La Lluera I
8. La Lluera II
9. Entrefoces
10. Entrecueves
11. Los Murciélagos
12. La Viña
13. El Sidrón
14. El Buxú
15. Covaciella
16. El Bosque
17. La Lloseta
18. Les Pedroses
19. Trescalabres
20. Tito Bustillo
21. La Cuevona
22. San Antonio
23. Coverizas
24. Samoreli
25. Balmori
26. La Riera
27. El Quintanal
28. Cueto de la Mina
29. Las Herrerías
30. Tebellín
31. Traúno
32. Coimbre or Brujas
33. Llonín
34. Covarón
35. Mazaculos
36. La Loja
37. El Pindal
38. Fuente del Salín
39. Chufín
40. Micolón
41. Porquerizo
42. Traslacueva
43. La Meaza
44. Las Aguas de Novales
45. El Linar
46. El Perro
47. La Clotilde
48. La Estación
49. Las Brujas
50. La Pila
51. Altamira
52. El Cudón
53. La Garma
54. Hornos de la Peña
55. Sovilla
56. El Castillo
57. Las Monedas
58. Las Chimeneas
59. La Pasiega
60. Santián
61. El Pendo
62. El Salitre
63. Los Santos del Becerral
64. Los Emboscados
65. El Patatal
66. Cobrantes
67. Cullalvera
68. Sotarriza
69. Covanegra
70. Covalanas
71. La Haza
72. El Arco A
73. El Arco B
74. El Arco C
75. Pondra
76. Morro
77. Venta de la Perra
78. La Lastrilla
79. El Cuco
80. Otañes
81. La Hoz
82. Arenaza
83. Santimamiñe
84. Goikolau
85. Ekain
86. Altxerri
87. Alkerdi

France

13. La Chaire à Calv[in]
14. Le Roc de Sers
15. Le Placard
16. Montgaudier
17. La Mairie
18. Fronsac
19. Font Bargeix
20. Jovelle
21. La Croix
22. Villars
23. Le Fourneau du Diable
24. Les Bernous
25. Pair-non-Pair
26. Pille-Bourse
27. Mitrot
28. Abri Faustin
29. Gabillou
30. La Cavaille
31. Rouffignac
32. La Sudrie
33. Lascaux
34. Roc de Vézac
35. Le Roc Pointu
36. La Martine
37. Le Mammouth (Grande Grotte Saint-Front)
38. Le Pigeonnier
39. Les Gautheries
40. Puy-Jarrige II
41. Le Moulin-de-Laguenay
42. Le Bourgnetou
43. Les Fieux
44. Grotte Murat
45. Abri Murat
46. Les Merveilles
47. Cougnac
48. Grotte Mazet
49. Roucadour
50. Les Escabasses
51. Sainte-Eulalie
52. Le Cuzoul-des-Brasconies
53. Bara-Bahau
54. La Ferrassie
55. Saint Cirq
56. Muzardie
57. Grotte du Fayar[d]
58. Laugerie Haute
59. Laugerie Basse
60. Oreille d'Enfer
61. Le Poisson
62. Lartet
63. La Croze à Gont[...]
64. Abri Pataud
65. La Cave Pataud
66. La Mouthe
67. Cournazac
68. Font de Gaume
69. Les Combarelles
70. Les Combarelles
71. Grotte du Livey
72. Grotte de Cazel[le]
73. La Calévie
74. Nancy I
75. Bernifal
76. Le Bison
77. Sous-Grand-Lac
78. Le Vallon-de-la-Moutonnie

FRANCE

Dordogne

Ardèche

Lot

La Rochelle · Angoulême · Limoges · Périgueux · Tulle · St. Flour · Bordeaux · Figeac · Mende · Cahors · Rodez · Le Vigan · Montauban · Nîmes · Montpellier · Dax · Auch · Toulouse · Pau · Carcassonne · Narbonne · Pamplona · Foix · Perpignan

N

0   50 mls

0   50 kms

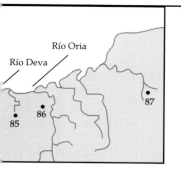

parietal art's distribution is equally patchy.

Parietal

Cave art is difficult to quantify, both in numbers of caves and in numbers of pictures. The most striking fact, however, is that its apparent geographical distribution in Europe is very different from that of portable art, although it is most abundant in areas which are also rich in decorated objects: the Périgord, the French Pyrenees and Cantabrian Spain.

Palaeolithic decorated caves are found from southern Portugal and the very south of Spain up to the north of France: the most northerly known at present is that of Gouy, just southeast of Rouen.[80] None has yet been found in Belgium or Britain,[81] though that may change in the future. They extend to the east of France along the Mediterranean through Sicily and Italy to western former Yugoslavia.

It has recently been discovered that the cave of Geissenklösterle in south-west Germany has painting on its wall, in the form of a black triangle, while excavation of its Aurignacian layer yielded a fallen block bearing traces of black, red and yellow paint; and the cave of Höhle Fels, near Schelklingen, in the same region, likewise yielded ten exfoliated fragments of limestone wall, highly polished by the repeated passing of cave-bears, lying in Gravettian occupation layers dating to 22,000 – 23,000 years ago. Many are covered in fine, criss-cross engravings, and clearly fell off owing to the destruction of the walls by intense freeze-thaw actions.[82]

This underlines the fact that the surviving sample of cave art is determined (and highly distorted) by geomorphological factors – i.e. by the type of rock on which it was made – rather than by 'cultural refugia' (see p. 200). Indeed, the apparent importance of decorated caves in the Upper Palaeolithic record is almost certainly illusory. Just as cave mouths and rock shelters were merely the sites where early archaeologists could most easily discover the accumulated residues of Ice Age occupation, so these were also the places where parietal art of the period had survived successfully. Neither cave dwelling nor cave decoration is truly characteristic of the Upper Palaeolithic. Seen against the 30,000-year span of the period, the

roughly 300 decorated sites known at present are remarkably few. We have no idea what percentage of the original total has survived, let alone what percentage has been discovered.

As will be seen below (Chapter 9), a great deal of art was produced in the open air; and it seems probable that the decoration of rock shelters was also far more common than we suppose, even in the earliest phases of the Upper Palaeolithic, but that most of the evidence has weathered away over the millennia. This is exemplified by the Abri Pataud (Dordogne), where every layer of occupation contains fallen fragments of what seem to have been a permanently painted and engraved wall and ceiling; and similarly, the collapsed Aurignacian rock shelter of Blanchard (Dordogne) has yielded several fragments of wall or ceiling bearing bichrome painting.[83]

Claims for Palaeolithic cave art arise sporadically in Central Europe – for example in Hungary where some possibly Palaeolithic engraved motifs have been found on the wall of the cave of Jenö Hillebrand[84] – but all are lacking in proof at present. For a while, an engraving of a stag in the Bavarian cave of Schulerloch was thought to be Palaeolithic,[85] but, together with another German candidate, it is now thought to be of more recent date.[86] Similarly, claims for Palaeolithic engravings in Austria seem to be highly dubious.[87] In former Czechoslovakia, some 'V' signs in red ochre, covered in calcite, on the wall of Mladec Cave are likely to be ancient,[88] while a recently discovered drawing inside Byci Skála cave in Moravia may also be a well-founded claimant.[89]

The first problem in quantifying decorated caves is that they cannot usually be dated directly (see below, p. 60) and therefore many examples remain doubtful, especially where they contain only non-figurative designs. For example, the former USSR has a number of these dubious cases, from Georgia to Siberia, in caves, rock shelters and on open-air cliff-faces; they comprise both engravings and paintings, and both realistic and stylized animals and geometric motifs.[90] However, there are at least two caves in the southern Ural Mountains which contain what are almost certainly Palaeolithic depictions. In 1959, paintings were discovered in the great cave of Kapova (or Kapovaya, or

Shulgan-Tash); the red figures include seven mammoths, a rhino and three horses.[91] Excavations in the cave have uncovered an occupation layer dating to about 14,000 years ago, and containing beads of brown and green serpentine, the base of a fired clay vessel with traces of pigment, and a fallen fragment of cave wall bearing part of a red painting of what may be a mammoth.[92]

The cave of Ignatiev, 100 km distant, whose art was found in the 1980s, contains a number of parietal figures, including signs, horses, several mammoths, and a remarkable stylized bovid and a stylized human female with twenty-eight dots between her legs.[93] Some Ice Age occupation, dated to around 14,000 years ago, has been found in the cave, but that does not necessarily date all of the parietal figures, since more recent periods are also present in the occupation layers. It has also been claimed that some red markings in the nearby cave of Serpievka-2 are also of Palaeolithic age.

The fact that these caves are 4000 km away from the main clusters of decorated caves in Western Europe serves to highlight the difference in distribution between portable art and surviving parietal art. No decorated caves have yet been found through most of Central Europe (apart from those few mentioned above), between north-east France and the Urals, despite an abundance of portable art across this void. Of course, cave art can occur only in areas with caves (although, as will be shown in Chapter 9, examples of open-air art are now being found); but Central Europe has plenty of suitable sites. Some theories about the discrepancy will be examined later.

It should not be thought, however, that all decorated caves in Western Europe are well authenticated; apart from occasional suspected fakes (also a problem with portable art – see Chapter 6), there are many cases in which reported examples of cave art have never been rediscovered; have been proved to be natural shapes, cracks or mineral colours in the rock (lusus naturae); have been deemed post-Palaeolithic; or simply remain doubtful or totally undated.[94] This means that no two tallies of decorated caves are likely to be identical: in a recent detailed and authoritative survey of French cave art, no fewer than forty-two examples were considered erroneous or very doubtful in the Aquitaine region alone.[95] Another problem with quantification is that some caves which are now separate may have formed a single system in the Upper Palaeolithic; while other caves have been given several names which become confused in the literature, so that the same site acquires two or three different entries in the list!

At the time of the most recent survey (1993), there were 147 accepted or authenticated examples of Palaeolithic parietal art in France,[96] although at least thirteen more have been discovered since then; there are 130 examples in Spain,[97] up to twenty-five caves/groups in Sicily/Italy,[98] three sites in Portugal (Escoural, Mazouco and the Côa Valley),[99] one in former Yugoslavia (Badanj),[100] and one in Romania (Cuciulat).[101] As has already been mentioned, their distribution within countries is very uneven, with important clusters occurring both in the Périgord (fifty-six sites), in the French Pyrenees (thirty sites) and in Cantabrian Spain (eighty-seven sites from Asturias to Navarre).[102]

Moreover, just as portable art varies markedly in quantity from site to site, the same is true of parietal images. As mentioned above, it is difficult to quantify the drawings: this is because of superimpositions, deterioration and, above all, a lack of consensus about how to count 'signs' – should a series of dots be considered separately or as a unit? And how can one quantify a mass of meandering finger-marks? Consequently, different authors often have widely divergent totals of figures for the same cave.

Nevertheless, it is clear that there are some sites with few figures and others with hundreds. For example, the Grotte du Roc at Saint-André-d'Allas (Dordogne), or the Cueva de los Murciélagos at Portazgo and the Abrigo de Godulfo at Bercio (both in Asturias) have one figure apiece;[103] others, like the Grotte de Pradières (Ariège) or the Cova Bastera (Pyrénées-Orientales) have only a few red marks on their walls;[104] while the Grotte du Cheval at Foix (Ariège) or the Abri du Poisson (Dordogne) have a mere handful of figures.[105] On the other hand, Lascaux (Dordogne) has about 600 paintings and nearly 1500 engravings;[106] the Grotte des Trois Frères (Ariège) has over 1100 parietal figures,[107] more than any other Pyrenean site – indeed, since that figure was calculated, whole new areas of engraving have been found by careful removal of clay from the walls, and more remain covered for the moment;[108] some Spanish caves, such as Castillo and Altamira, are equally dominant in their region. It is very hard to see how caves with one figure and supersites like these could be equivalent in any way; similarly, some caves are huge while others are tiny, though size does not always equate with numbers of figures – there are large caves with few, and small caves with many.

In an exercise similar to that on the Périgord's portable art, the 'graphic units' of the parietal sites in Ariège (and part of Haute-Garonne) have been quantified: the total of over 2600 'graphic units' is spread over eleven caves, but four of them (Trois Frères, Niaux, Fontanet, Marsoulas) account for 2216 units, or 84.9 per cent,[109] with Trois Frères predominant of course (over 50 per cent).

Both types of art

Thus we have some sites with huge quantities of portable art, and some with similar concentrations of parietal images. Do the two phenomena ever coincide? Surprisingly, in those regions of Europe where the two types of art are found, there are relatively few sites which have both (and many which have neither form!): a 1983 study of this topic, focusing on figurative art, found only twenty-seven sites in France, nine in Spain and three in Italy[110] – and some of these had only one portable item, while others had over 100. Some local areas rich in one form are noticeably poor in the other – for example, in the Quercy region of France, areas rich in portable art are poor in parietal art.

It seems reasonable to suppose that caves which have both art-types, with one or both in large quantities, were very special places in the last Ice Age: one can certainly include the great river-tunnel of Le Mas d'Azil (Ariège) and Isturitz with its three decorated tunnel-caves – these are the true 'supersites' of the Pyrenees, with their huge concentrations of tool-production and artistic activity; they must have served as storehouses, meeting-places, ritual foci and socio-economic centres not only for local groups but also for a far wider area, as is confirmed by their artifactual links with far-flung sites and with each other.[111] The same is true, though on a lesser scale, of Altamira and Castillo.[112]

However, perhaps the most dramatic example can be seen in the Volp caves: Enlène (which, as we have seen, has huge amounts of portable art) is connected by a passage to Trois Frères, the richest parietal art site in the whole region. This cave-system, therefore, was clearly of the greatest importance in the Magdalenian period, an importance further underlined by the immediate proximity of the remarkable Tuc d'Audoubert (see p.110–11).

Enlène was inhabited; it has hearths, but almost no wall art. Trois Frères is profusely decorated, but has almost no trace of habitation. These caves thus suggested to early scholars of Palaeolithic art that living sites were largely undecorated and 'sanctuaries' were largely uninhabited. Their view seemed to be further strengthened by the case of Niaux and La Vache, two caves facing each other across a narrow valley in Ariège. Niaux is rich in wall art but, as far as we know, was not inhabited; La Vache is a living site, rich in portable art but with no Palaeolithic wall decoration. However, it is now known that some decorated sites were inhabited (e.g. Lascaux, Fontanet, Marsoulas), though in some cases it may still be true that a decorated site was visited rarely, or only once, for whatever rites were carried out there.

As will be discussed later (see p. 127), there are clear regional and chronological differences in the techniques and content of both portable and parietal art in Europe; taken as a whole, however, this phenomenon now comprises about 300 decorated sites and many thousands of objects. More are being found every year – the objects mostly by archaeologists, and the caves (an average of at least one per year) primarily by speleologists and quarrymen. It may seem a lot, but seen against the 30,000 years of the Upper Palaeolithic, or even just the few millennia of the Magdalenian, the known quantity of Palaeolithic art actually remains quite small. One might say that as a medium it is rare, but often well done …

4 MAKING A RECORD

Since the first discoveries and the acceptance of Palaeolithic art, a great deal of effort has gone into copying and recording it in one way or another. This is primarily to make reproductions available for scholars to work with (many examples need to be studied for research or synthesis), and to present the material to the public at large.

Tracing and copying

Copying has the additional advantage of reducing the number of occasions on which an original object might need to be handled or a cave visited – and in a few cases even specialists cannot visit the caves at present: the Grotte Cosquer's entrance is 35 m below the sea, and comprises a 150-m tunnel which is extremely dangerous (three divers have already drowned in it), so that only experienced professional divers can enter this cave. The Réseau Clastres at Niaux is closed off by an underground lake and kept sealed in this way so that instruments inside it can send out data on the cave's undisturbed 'climate'; while Erberua (Pyrénées-Atlantiques) lies at the end of a long 'siphon', and can thus be visited only by divers with suitable equipment.

Finally, of course, the making of copies ensures that any example of Palaeolithic art which might deteriorate or be lost, stolen or damaged will at least survive in reproduction: for example, one of the Laussel 'Venus' bas-reliefs is thought to have been destroyed in Berlin during the last war, and survives only in the form of casts and pictures.[1] Photographs and tracings made over the last 100 years also enable us to monitor changes to the original images through time.[2]

Breuil

The doyen of recorders of the art was the Abbé Henri Breuil, who was a young man at the turn of the century when the first decorated caves were authenticated (Fig.4.1): he was a talented artist, excelling in animal figures, and had been employed by Edouard Piette to draw his remarkable collection of Palaeolithic portable art.[3] He therefore had the great good fortune to be present – young and vigorous, with time on his hands and possessing the necessary skills – at the very time when the copying of cave art needed to begin. His first attempt entailed tracing a few figures in La Mouthe in 1900;[4] subsequently, he did Altamira in 1902, Marsoulas, Font de Gaume, Combarelles, and all the other early finds – indeed, he himself found many caves or figures.

His working methods were very crude by modern standards, though he was hampered by the limited materials then in existence. Sheets of florist's or rice paper (the most transparent available, though merely translucent at best) were held on to the cave wall, often by assistants, and the Palaeolithic lines beneath were traced with pencil or crayon. Carbide lamps had to be held, often at arm's length for long periods, by helpers; Breuil was quick-tempered, and might administer a sharp slap if a boy's fatigue or cramp caused the light to move;[5] he was utterly absorbed in the work, hardly speaking for hours on end, simply moving the arms of his 'human candelabras' when necessary.[6]

Later the tracings would be redrawn for publication, though not always immediately – Breuil did so much copying, and had so many other commitments, that publication sometimes occurred many years later (twenty years in the case of Trois Frères, twenty-four for the Galerie Vidal at Bédeilhac, twenty-six for Pair-non-Pair), and the redrawing may have been done by a collaborator (e.g. by Michaut in the case of the Galerie Vidal), and then published unchecked. This inevitably led to mistakes: for example, his published tracings of Pair-non-Pair contain numerous meanders and circular images, but recent investigation has shown that these are all natural accidents in the rock-face: fissures, bumps and hollows; Breuil undoubtedly noted them as such on his original tracings, but when the time came for publication the meaning of his graphic conventions had been forgotten.[7]

The 'direct tracing' method had its disadvantages, since contact with the wall inevitably damaged the art very slightly in some places (as can be seen in some modern macrophotographs), and Breuil was even known on rare occasions to dust surfaces with a few flicks of his handkerchief to make some lines clearer![8] Even fragile engravings in the clay of cave floors, such as those at Niaux, were traced by the direct method, running a pencil without pressure on to soft paper in contact with the clay.[9]

At Altamira, his working conditions in 1902 were appalling: since the ceiling figures were done in a very pasty paint, his paper could not be placed on them, as this would detach pigment. Consequently, he had to copy the animals while lying supine on sacks filled with ferns, using the imperfect yellow light of candles which spattered his clothing with wax, and undergoing constant fatigue and strain. His method in this cave was to make rough sketches, then measure the originals, and finally make his copies. He had brought a box of water-colour paints with him, but the cave's atmosphere was too humid for the paper to dry, so he used pastels instead; having no black (a colour of great importance in the cave), he had to improvise with burnt wood and crushed charcoal mixed with water, but the results were unsatisfactory; he returned to Altamira in 1932 to do the job properly, redrawing a few figures and retouching the old copies of the rest.[10] Not only was electric light now available; the floor had been lowered, which meant that he could be comfortably seated and could see the figures with less distortion and foreshortening than had been the case in 1902. It is worth noting that a few figures had already shown marked deterioration in the thirty intervening years; and he later noted that they had paled even further between 1932 and his last visit in 1952.[11]

Fig. 4.1 Henri Breuil (1877–1961).

Students of Palaeolithic art owe a huge debt to Breuil for his patient toil in the caves — by his own reckoning he spent over 700 days deciphering and copying their art during the course of his long life. There are a number of cases where paintings have faded since he copied them (e.g. the Galerie Vidal in the cave of Bédeilhac), so that only the Breuil versions give one any idea as to how the originals may have looked.[12] Indeed, he himself wrote that 'after me, nobody will really have a true, precise knowledge of the content of these caves except through my tracings.'[13] But a major problem — which has come to the fore only in the last few decades — is that all his copies are indeed a 'Breuil version'. His colossal output (including his drawings of hundreds of portable items), his influence,[14] and his sheer dominance (not only in Palaeolithic art but in all prehistory) until his death in 1961, have ensured that Breuil copies are to be found in most textbooks and in most works on cave art; in many cases they are far more familiar to us than the originals, and this is a very dangerous state of affairs because, like every artist, Breuil had a style of his own and made mistakes. We are therefore seeing Palaeolithic figures that have passed through a standard 'Breuil process': they are subjective copies, not faithful facsimiles. It is probable that a scholar of Ice Age art, shown six versions of

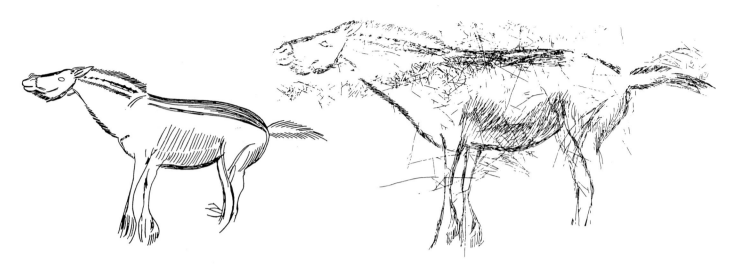

Figs. 4.2 & 4.3 'Spot the
Breuil': his version and a
more recent tracing (by
V. Féruglio) of a horse
engraving from Trois
Frères (Ariège): 27.5 cm
long.

Figs. 4.4 & 4.5 'Spot the
Breuil': his version and a
more recent tracing (by
Bernaldo de Quirós and
Cabrera) of an ibex
engraving from Castillo
(Santander): about 40 cm
long.

the same · figure by different prehistorians, could 'spot the Breuil' at a glance (Figs.4.2–5).

There are three important consequences of this situation. First, the Breuil style has inevitably made a wide range of Palaeolithic figures seem somewhat similar to our eyes; it may therefore exaggerate our view of the art's unity through space and time (and, conversely, comparison of figures produced by two different copiers may exaggerate the differences, since their own styles enhance the contrast).[15]

Second, Breuil's tracings are.so well known and so firmly established that they can hamper

new readings of a figure, since they influence what the researcher sees.[16]

Third, Breuil's copies reflect his own preoccupations, his own notions about the art. As will be seen (p. 171), he believed in ideas of hunting magic: this was a hunting art, and he therefore went all out to decipher animal outlines, sometimes filling in missing elements. According to some accounts, he was fairly objective in his copying: while working in Trois Frères one day, he was asked the meaning of the particular line he was tracing; 'I don't know,' he replied, 'I copy what there is, we'll explain it

later … if it's possible.'[17]

Nevertheless, it is clear from his books and articles that, despite this attitude to his task, he was in fact extremely selective in the lines which he chose to copy or at least 'extract' for publication. Where lines seemed to have no relevance to the animal figures, he often ignored them, dismissing them as *traits parasites*, unlike his contemporary, Jean Bouyssonie, whose copies were often fuller and more faithful. As a result, we have lots of nice clean tracings of individual animal figures, pulled out of their context, copied from book to book, and giving the impression that Palaeolithic art contains nothing else; whereas in fact there is a vast amount of non-figurative, abstract or geometric marking in the caves which may have been just as important, if not more important, to prehistoric people.

In recent years, as Breuil's work has been checked, and as objects or figures have been re-studied, it has become very apparent that his work is by no means always as reliable as had been thought. For example, at Les Combarelles, it has become clear that he sometimes 'let his pencil go' and that some figures do not correspond at all to how he drew them;[18] at Altamira itself, re-analysis of the Final Gallery has shown that some of his copies are grossly in error or simply unrecognizable, and must have been done as rough freehand sketches, not as tracings – he seems to have had little patience for, or interest in, the cave's engravings, some of which he ignored completely;[19] while some of his copies of portable art, especially of engraved plaquettes, have turned out to be even less precise than his parietal copies (see p. 185–6).[20]

Modern methods of copying

Since Breuil was the first to do this kind of work, he had to learn as he went along; and in view of this lack of precedent, and the materials available to him, it is scarcely surprising that he made mistakes in his methods and his results; indeed the wonder is that he did not make far more, and modern scholars are still filled with admiration for the quality and perception of much of his work.

Today, conditions have greatly improved, not only in terms of light-sources, but also of the truly transparent, supple plastics and acetates available to draw on, and new types of pens and markers. Nevertheless, direct tracing is now (theoretically) taboo, since we are more aware of the damage it can cause. Instead, 'tracing at a distance' is done, with the sheet set up in front of the wall; and some new techniques are also available: in the past, moulds made of parietal engravings (as at Pair-non-Pair) had to be of plaster or clay, and the risk to the originals was enormous. Rubber latex was a significant improvement, but today casts can be made relatively safely of engravings on hard surfaces using elastomer silicones and polyesters which not only cause virtually no damage but are also quick and easy to apply, and produce far more precise and resistant results.[21]

Indeed, the results are entirely faithful, exact replicas of the original in size and volume. Moulds have the added advantage that they turn parietal art into portable; they can be removed for study elsewhere under different lighting systems, and can eventually be displayed to the public. Casts can even be treated so that the engraved lines stand out far more clearly than on the wall: water mixed with ink is spread over the cast; when it is wiped away, some remains inside the engraved lines. This extra clarity has enabled researchers to find no fewer than twenty-five new engraved figures on a cast from Trois Frères Cave.[22]

A similar technique has been applied to portable engravings such as those of La Marche and La Colombière. Plasticine or silicone imprints of the engraved surface turn the incisions into raised lines, and in some cases one can see in which order they were made. The imprints are also easier on the eye, since they do not have the distracting nuances of colour and texture of the originals.[23]

The study of portable art has also made great strides through use of the microscope, particularly in the pioneering work of Alexander Marshack;[24] this new methodology has revealed hitherto unnoticed details of content and composition, and, as we shall see (p. 89), can even suggest exactly how the different marks were made. A recent approach, combining casts and microscopes, involves making varnish replicas of the engraved surfaces of pebbles; these replicas, unlike the stones them-

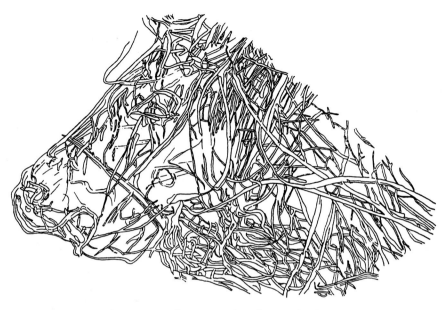

Fig. 4.6 Tracing of the plasticine imprint of an engraved horse head from La Marche (Vienne). Magdalenian. Length from muzzle to ear: 6.3 cm. Note the probable halter. (After Pales and de St-Péreuse)

selves, can then be studied under the Scanning Electron Microscope, at enormous magnifications; this method not only provides data on the engraving technique (see p. 90), but can also show in what order superimposed lines have been engraved.[25]

Photography

All the new techniques are used in close association with photography. Thanks to the dominance of Breuil's tracings, little attention was paid to photographs in the field of Palaeolithic art until comparatively recently, despite some excellent early efforts: pictures were taken of the Altamira figures in 1880 using electric light; from October 1912 onwards, Max Bégouën was achieving extraordinary results in the Volp caves. He took the voluminous family camera to the farthest depths of the caverns; its glass negatives required poses of an hour, or quantities of magnesium, and were developed on the spot with water from the cave! He worked on them to produce maximum contrast, and the results thus give the false impression that the engravings have lost visibility since that time.[26]

Nowadays, thanks to the major advances in cameras, films and lighting, a wide range of techniques has been applied in the caves. For example, photographs of panels or figures are enlarged to full size; tracings are then made from these, though it is always necessary to check the results against the original too — in fact, it is preferable to do the work in the cave,

in front of the original. This technique is superior to other tracing methods, since the wall need never be touched, and one avoids the difficulties of installing the paper parallel to the panel, and of reflections and shadows caused by irregularities in the wall's surface.

Enlargements to greater than actual size, or macro-photographs of small areas, allow one to see tiny details or superimpositions more clearly. Photographs of paintings, of course, may be published as they are, but engravings often require tracings to help the reader decipher what is there: thus photography and drawing are complementary, not substitutes for one another.

By and large, photographs are taken perpendicular to the wall; for large figures or panels, or for those in narrow places where one cannot stand far enough back, a number of overlapping pictures are taken and then amalgamated into a photo-montage. The same technique is also invaluable for parietal figures which cannot be seen all at once (such as the horse 'falling' around a rock at Lascaux, p. 176) or for portable objects such as those where the composition is engraved around a cylindrical baton.

A single picture of a parietal figure is far from sufficient today: a whole series is now taken, using different films (both monochrome and colour), light-sources (lamps and electronic flash) from different angles, lenses (normal, wide-angle, macro, etc.), filters and degrees of contrast. Multiple flashes can bring out the relief of a sculpture, and even the traces of modelling clay by hand, as on the Tuc d'Audoubert bison (p. 111).[27] The last few years have seen the introduction of computers for the enhancement of photos and the storage of digitized copies. Computer-scanning of portable engravings with a series of different lightings has also been introduced, enabling a perfectly objective and exact image to be stored and manipulated.[28]

Another new technique, pioneered in the caves by Alexander Marshack and adopted by Jean Vertut and others, involves the use of infrared and ultra-violet lamps. Ultra-violet radiation makes certain materials fluoresce: any calcite and living organisms on the cave walls do so, but the ochres and manganese used as pigments do not; consequently one can assess any damage to the figures caused by growths or calcite-flows, while ultra-violet can also show up any paint beneath the calcite and thus 'restore' fading detail, as shown by Marshack on some of the figures at Niaux.[29] Consequently, ultra-violet lamps were used for both tracing and photography during the systematic recording of this cave's art.[30]

On the other hand, infra-red light or film makes red ochres appear transparent, so that one can see other pigments beneath them; moreover, any impurities in the ochres remain visible, and so different mixes of paint, with different impurities, can be detected. By this method, Marshack claims to be able to assess in what order the famous 'spotted horses' panel at Pech Merle (Lot) was built up.[31] Recently, infra-red photography of the newly discovered parietal images in the Grande Grotte of Arcy-sur-Cure (Yonne) has led to the finding of new figures, and has made others far easier to see.[32]

Infra-red can sometimes make the original composition clearer by 'removing' the thin trickles where pigment ran: Jean Vertut's picture of a 'tectiform' sign at Bernifal reveals that it was not painted in continuous lines, as some had thought, but as a series of dots from which the colour had run (Fig.4.9).[33] The use of infra-red also counteracts the effects of changes in humidity and wall conditions, which can make certain painted lines visible on some days and not on others – for parietal paintings 'live' in accordance with atmospheric conditions.[34]

This phenomenon of changes in figures through the year has been studied in depth by Michel Lorblanchet at some caves in the Quercy. During eighteen months (in 1984–6) in Pech Merle and Cougnac he took photographs every two weeks of the same details, at the same distance, and with the same film and lighting. The total of more than 600 slides revealed that contrast was at a maximum in the autumn (September to December), weakest in the winter, increased in the spring, and oscillated in the summer. In short, while some details (such as the tail on the central horse in Pech Merle's Frise Noire, see p.122–3, or the hump of a Cougnac megaloceros) were barely visible for much of the year, and disappeared in the winter, they were most visible in October and November.[35] This needs to be borne in mind

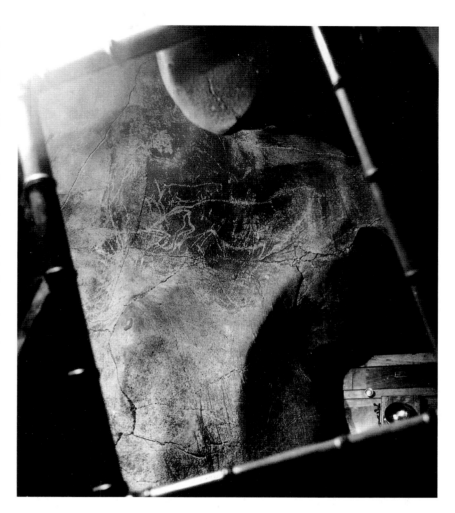

when tracings are to be made. One cannot simply take cave panels at face value: the conditions of conservation are of paramount importance, and profound long-term studies of each cave are required.

In addition to the traditional tracings, therefore, it has become essential for archives to include complete coverage in slide form (Jean Vertut took about 1200 slides in the Volp caves), and in some cases also on video and film – as, for example, in the 'Corpus Lascaux', a cinematic archive of that cave's decoration.[36]

However, for art of this kind, the ideal is to have a record in three dimensions; so the very expensive technique of photogrammetry (the method by which contour maps are made from aerial photographs) is increasingly being used; in Palaeolithic art, by means of stereo-photographs and some key measurements, it is possible to produce a detailed 'contour map' of an object or panel, which can then be used to

Fig. 4.7 Pioneering photography in Trois Frères (Ariège) in the late 1920s. The cramped conditions in this gallery led to the use of the mirror. Note the camera's reflection and the horse engravings.

Fig. 4.8 Ultra-violet photo of red 'claviform' sign in Trois Frères (Ariège) painted on a surface prepared by scraping. Probably Magdalenian. The claviform is 76 cm in length.

make an accurate 3-D copy – in other words, making a cast without even touching the original! This has been done, for example, for the clay bison of the Tuc d'Audoubert, which could not be copied by any casting technique owing to their fragility[37] – Jean Vertut's expertise was invaluable in making this project a success.

The best-known 3-D copy at present is the excellent Lascaux II, completed in 1983, which reproduces part of that cave and its paintings;[38] but the technology which produced it is already almost obsolete, and even more accurate copies of decorated caves are already being made (such as the 5 by 7-metre facsimile of the Altamira ceiling in Japan)[39] or planned – of Cosquer, and perhaps Chauvet. In a different approach, Niaux's Salon Noir has been recreated in a building in the Prehistoric Park at Tarascon-sur-Ariège, not as it looks today but with the animals drawn as they would have appeared originally, before the calcite-flows and the damage from water-flows.[40]

A different kind of 3-D copy of Lascaux now exists at Disney's EPCOT Center in Florida, where it is possible to visit 'Virtual Lascaux'. Artists and computer experts have combined to recreate the cave in three dimensions using 'texture maps' – photos of the paintings are scanned in from books, and computer programmes vary the shadow and light on the walls to make them look like the real thing when one is wearing virtual-reality goggles. One can 'move' through the cave with the help of a joystick.[41]

The quest for objectivity

Every tracing or copy of a Palaeolithic figure is inevitably subjective to some degree; it is also a distortion of reality since – except in the case of accurate casts – three dimensions are being reduced to two. Every tracing is a personal piece of work – different people see different things – and it is impossible to remove subjectivity completely.[42] A copy is only as good as the copier, and all copiers make mistakes, the number and extent of which depend on the method they use and their experience and personality: good examples include a series of eighteen published versions of an engraved human head from the Grotte du Placard (Charente), none of which proved to be completely reliable;[43] five different versions of an engraved reindeer from Les Combarelles;[44] nine of a reindeer from Saut-du-Perron and seven of a rhino from La Colombière;[45] eight of the chamois on a disc from Laugerie-Basse;[46] and about fifty of the famous mammoth of La Madeleine, found in 1864.[47] Some of them were clearly done from the original, others were simply copied from book to book, with the distortions increasing each time. If the original were destroyed or lost, which of the versions should one 'believe'?

There are even a few cases where a truly appalling copy has been published, then republished by others who neglected to compare it with the original object (or even with a cast or photograph), although a glance would have sufficed to reveal the numerous errors.[48]

A great deal of emphasis is now placed on the 'depersonalization' of copying; the pendulum has swung away from Breuil's technique of putting some spirit into tracings (the 'artistic approach'); we now have teams producing 'collaborative copies', or individuals making accurate but often lifeless versions of the Ice Age images (the 'cartographic approach'). Both methods have their merits and their disadvantages.

Deciphering or copying images on a cave wall is rather like an excavation, except that the 'site' is not destroyed in the process; the pictures are 'artifacts' as well as art and, if superimposed, they even have a stratigraphy.[49] Moreover, instead of selecting and completing animal figures from the mass of marks, like early archaeologists seeking, keeping and publishing only the belles pièces and ignoring the 'waste flakes', the aim for the last thirty years has been to copy everything. This helps to reduce psychological effects akin to identifying shapes in clouds or ink-blots: faced with a mass of digital flutings or engraved lines, the mind tends to find what it wants to find, in accordance with its preconceptions, and often detects figurative images which are really not there.[50] In addition, one needs to counteract the psychological effect whereby the eye is drawn to the deeper lines (although these may have been of secondary importance) and to lines in concave areas which are generally better preserved than those on convex areas which are more exposed to wear and rubbing.[51]

To eliminate lines we do not understand is an insult to the artist, who did not put them there for nothing; where there are so many lines that it is difficult to 'isolate' anything, however, it is still necessary to 'pull out' any definite figures which exist hidden in the complex mass (this is also far less of a strain on the eyes), though one should still try to publish the mass, leaving the reader free to make a different choice (Figs.4.10, 4.11).[52] In his herculean twenty-five-year study of the 1512 slabs from La Marche with their terrible confusion of engraved lines, Léon Pales isolated and published only those figures which his expert knowledge of human and animal anatomy revealed to his eye: but he estimated that only one line in 1000 has been deciphered on these stones.[53] Unfortunately, there are very few scholars with similar skills in deciphering and reproducing Palaeolithic engravings.

Fig. 4.9 Infra-red photo of painted tectiform at Bernifal (Dordogne). Probably Magdalenian. Width: c. 30 cm.

Fig. 4.10 & 4.11
Exhaustive tracing of engraved stone from Gönnersdorf (Germany), and a horse figure 'extracted' from the mass (see also p. 159). Magdalenian. The plaquette is 20 cm long, 18 cm wide, and the horse 8.5 cm high. (After Bosinski and Fischer)

Breuil's notion of traits parasites has been completely abandoned: Palaeolithic images are not restricted to the figurative but also include non-figurative, abstract and geometric marks, some of them of enormous antiquity, going by the new evidence from other parts of the world (see above, p.24).

Just as artifacts are no longer dug up for their own sake, but for what their manufacture, context and associations can tell us, so Palaeolithic images are now studied with the same aims. We shall examine techniques of execution in a later section (Chapters 7 and 8), while associations of figures with one another and with differently shaped panels loom large in recent interpretations of cave art (Chapter 11).

As for context, it is obviously necessary for cave plans to be as accurate as possible; however, as with tracings of figures, versions have sometimes been published which are so divergent that four plans of La Baume-Latrone, for example, looked like different caves![54] Old cave maps are often highly inaccurate, and mistakes have also been made in many cases concerning the part of the cave or the particular wall on which figures occur, and many figures are often missed out altogether.[55] All of this requires checking and published corrections. For example, the only parietal figure in the cave of El Quintanal (Asturias) was thought to be lost, but it turned out that it was actually located on the wall opposite the one marked on the published plan![56]

We have already seen that copying now includes as much as possible — and it is no longer always restricted to artificial marks, figurative or non-figurative, since it can often be difficult (and occasionally impossible) to tell whether lines are incisions or cracks,[57] or to differentiate between faded spots of paint and blobs of natural colour in the rock.

Photographs, of course, play a major role in studies of context, since a normal lens sees roughly the same area as an eyeball, and a photo shows everything at once: artificial lines, natural marks and fissures, the grain of the rock, and the overall aspect of the panel. In recent publications of some decorated panels, the whole rock-face has been drawn, though few specialists have the ability or the time to do this (Fig. 4.12).[58] A different technique is 'morphometric cartography': i.e. the making of a detailed contour map, such as that of the panel with the black painted frieze at Pech Merle.[59]

There is also a new emphasis on the physico-chemical interactions of images and wall:[60] close study of the changing conditions in the caves (see above, p.53), and of the natural phenomena affecting the images, helps to assess what has happened to them through time, and provides crucial information for conservation.

In short, since the time of Breuil, the task of copying Palaeolithic art has not only acquired many improved techniques, it has also adopted new aims.[61] It is no longer a question of simply accumulating collections of animal figures for publication. All copies, whether tracings or photographs, are seen as tools for further research — as starting-points, not as ends in themselves. Tracings are no longer seen as faithful reproductions, but as explanations and interpretations of the images, incorporating an inevitable degree of subjectivity, distortion and choice. Consequently, no copy can ever be definitive, no cave art can ever be entirely known.

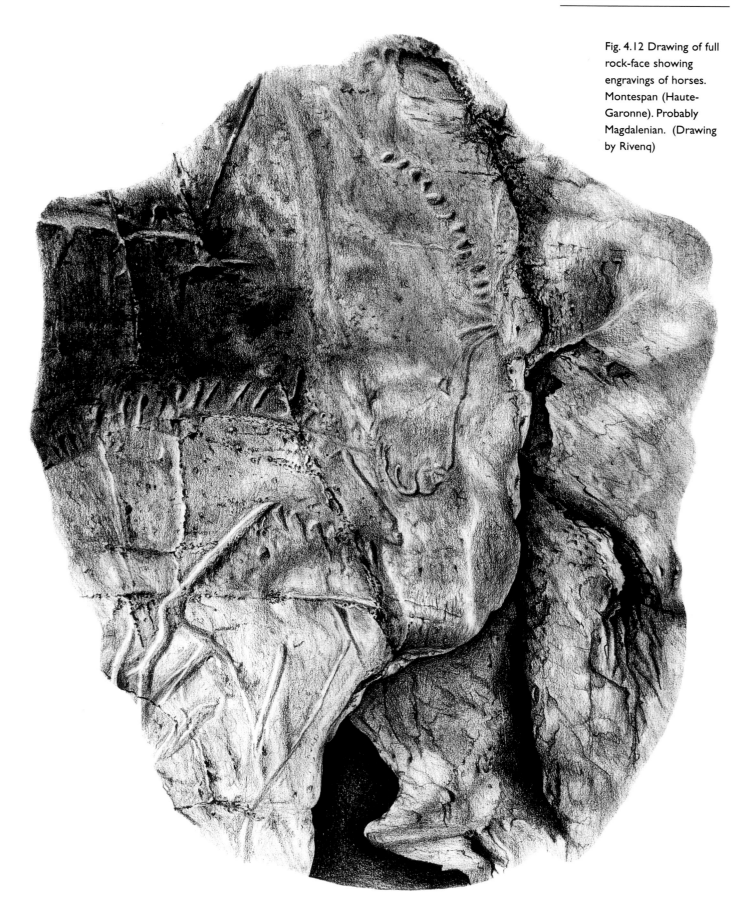

Fig. 4.12 Drawing of full rock-face showing engravings of horses. Montespan (Haute-Garonne). Probably Magdalenian. (Drawing by Rivenq)

5 | How Old is the Art?

How do we know all of this art was made in the Ice Age, and during which particular phase of the Upper Palaeolithic? Until recently, the dating methods had changed little since the nineteenth century, and much still remains uncertain.

Portable art

Where portable objects are concerned, their position in the stratigraphy of a site, together with the associated bone and stone tools, give one a pretty clear idea of the cultural phase involved; of course, the development of radio-carbon dating since the last war has led to fair-ly accurate measurement of the age of some of these levels. Measurement of the radiocarbon content of an actual piece of portable art is rarely done because of the destruction involved (in any case, this dates the death of the animal whose bone or antler was used, rather than the artistic activity), but in recent years nineteen AMS dates have been obtained from portable art on antler from five Magdalenian sites in northern Spain, with results ranging from 14,600 to 10,310 years ago.[1]

Far more common is the dating of organic material from layers containing the portable art. Examples include the date of 14,280 years before the present for the layer of La Marche containing the engraved stones, or that of 11,900 BC for the layer at Enlène which yielded an antler baton with a salmon in champlevé (see below, p. 93), or dates of 10,430 and 10,710 BC for the Gönnersdorf habitations containing the engravings.[2]

There are a number of snags. First, many of the best-known pieces of portable art were found in the last century, and they were dug up like potatoes; only a few people like Piette noted the particular layer from which the objects came: early excavators such as Lartet and Christy, or the Marquis de Vibraye, paid comparatively little attention to exact prove-nance, being carried away by the search for more Palaeolithic 'nuggets'.[3] Consequently, lots of famous specimens can be given no more definite attribution than 'probably Magdalenian'. Second, a great many objects have no details of provenance at all — either because they come from clandestine digs, or because their sites were atrociously excavated or never published — and many may even be fakes (see Chapter 6). As we shall see, very few of the 'Venus figurines' of Western Europe have any stratigraphic context whatsoever. At Lourdes (Hautes-Pyrénées), over 2000 cubic metres of deposits were removed from the cave of Les Esplugues (so that it could be turned into a chapel), and scattered over nearby land — the many items of Palaeolithic portable art subse-quently discovered in these discarded sedi-ments thus have no stratigraphic provenance.[4]

Third, while a layer may accurately date art objects such as plaquettes which seem to have been made and discarded quickly, there are many portable objects, notably statuettes, that seem to have been much handled (a point brought out clearly in analyses by Alexander Marshack, see below p. 206) and which may have been carried around and used for genera-tions: consequently, their final resting-place may be very far from their place of manufac-ture, and the layer in which they are found is merely an indication of when they were lost or discarded, not necessarily of when they were made — they may in fact be considerably older.[5]

Finally, there are inevitably occasional dis-agreements about the precise attribution of the cultural material associated with art objects: for example, the industry with the engraved peb-bles of La Colombière was thought to be an atypical Gravettian, but is now reckoned to be an atypical Magdalenian, which agrees better with its radiocarbon dates around the thirteenth millennium BC.[6] Much portable art without good stratigraphic context has been assigned to different phases on the basis of style, but this involves subjective judgements and relies on 'type fossils' which often have limited validity – for example, one can find the same motifs on objects from different periods, such as the Middle and Late Magdalenian.[7] As will be seen, all the Western 'Venus figurines' have unjustifiably been lumped into two groups, the Gravettian and the Magdalenian. It is a tautology to date an object by its resemblance to standard examples, and then to use that resemblance as proof of the date.

Nevertheless, in a few cases, the resemblance is so striking that there is a strong probability that the same period is involved, and at times even the same artist or group of artists; this is shown in examples where both objects being compared are securely dated, such as the spearthrowers of Le Mas d'Azil and Bédeilhac,[8] or the portable engravings from the Grotte des Deux Avens (Ardèche) which are very similar in style and execution (especially of fish) to those of La Vache (Ariège): both have been dated to the mid-eleventh millennium BC.[9]

Chronological schemes tracing the development of portable art have been devised, but tend to be simplistic: one of the earliest was that of Piette, who thought that sculpture in the round came first, to be followed by relief sculptures, contours découpés and finally engraving.[10] Unfortunately, he ignored or forgot the fact that engravings, for example, already existed in

Fig. 5.1 Engravings in the Chauvet Cave, showing a horse on the left, and two superimposed mammoths on the right. Note the animal claw-marks below some figures.

abundance in his pre-engraving phases! As we have seen (Chapter 2), a variety of techniques were already in use before the Upper Palaeolithic, while some (such as the contours découpés) seem to characterize certain regions and periods, and cannot therefore be taken as fixed stages in the development of portable art as a whole.

Parietal art

Most wall art is extremely difficult to date; as mentioned earlier, attempts are underway to develop means of dating the various types of varnish or patina which form over rocks and their engravings in different parts of the world, or, in Australian caves, the layers of carbonate between different phases of engraving on a wall; and since the discovery that many black drawings in the European caves were done not with manganese but with charcoal (see Chapter 8) it has become possible to obtain radiocarbon dates from this organic material (see below). As with portable art, this technique dates the death of the tree rather than the artistic activity, but the two events are unlikely to be very dissimilar in age in most situations. In all other cases, however, different methods are required.[11] Proof of Palaeolithic age can take a number of forms: for example, the depiction of animals which are now extinct (such as the mammoth, which had disappeared from mainland Eurasia by the end of the Pleistocene),[12] or which were only present during the Ice Age (figures of reindeer in southern France and northern Spain), is a solid argument which, it will be recalled, first convinced the scientific world of the reality of Palaeolithic portable art.

More recently, attempts have been made to date Palaeolithic parietal art by the species depicted: it has been assumed that the pictures reflect the faunal assemblage outside, and that therefore the cave of Las Monedas, which has mostly horses (42 per cent) with some reindeer (13 per cent), represents a cold phase, whereas the neighbouring cave of Las Chimeneas, which has 42 per cent cervids, few horses and no reindeer, represents an older, warmer phase[13] — yet both sets of figures seem to be of the same style, and the direct dates they have recently produced point to a much more complex situation.

However, whereas Monedas and Chimeneas once seemed to be simple, homogeneous 'sanctuaries', matters become far more complicated for heterogeneous collections of animal figures, such as the 155 of Castillo which have been assigned to the Aurignacian (twenty-seven), Gravettian (eight), Solutrean (twenty-five), Lower Magdalenian (eighty-eight) and Upper Magdalenian (seven); the ten species represented are not particularly indicative of a cold or warm climate; and if the chronological attributions are correct, then there was a massive presence of red deer in the Solutrean, whereas only bison existed in the Upper Magdalenian![14]

Clearly, there are many other possible reasons for these differences in depicted species; they should not be taken simply as a tally of what was available outside. As will be seen in Chapter 11, species percentages in depictions rarely correspond to those in animal bones; both assemblages are a conscious selection of what was available. While a picture of a reindeer may prove that the animal was present, an absence of reindeer pictures does not prove that the animal was no longer around; and in any case, parietal and portable art of the same period, and even in the same site, tend to depict different species — thus, at the Tuc d'Audoubert, there are close analogies between both art-forms in style and details of execution, and the two are clearly contemporaneous. The portable material has been dated to 12,400 BC. Yet the bison dominates on the walls, and felines and reindeer are present, together with a hundred claviform signs (see Chapter 10), but there are no fish or anthropomorphs; the portable art, on the other hand, has no claviforms, felines or reindeer, but it does have fish and anthropomorphs, and it is dominated by horses![15] Other examples of differences in the overall content of the two art-forms will be given in Chapters 7 and 8.

In the past, it was often assumed that a formation of stalagmite or calcite on top of a painting or engraving was proof of great age, but nowadays we are more cautious, since such layers can form very quickly under certain conditions, and may merely prove that the figures are not recent fakes. Very different rates of stalagmite accumulation can occur in the different microclimates of a single cave.[16] For instance, in the Grotte du Cheval at Arcy-sur-Cure (Yonne), photographs taken of an engraved mammoth in 1946 have been compared with others taken in the 1950s and today, and show a significant increase in the size of calcite growths nearby, and a stalagmite-flow has formed on top of the clay spoilheap from excavations in the 1950s, whereas other parts of the engraved wall in the same cave do not seem to have altered in 15,000 years![17]

Quite a few caves seem to have been blocked during or just after the Ice Age — Cosquer's entrance, for example, was drowned by the rise in sea-level, while Fontanet's entrance collapsed during the Magdalenian, sometime after 11,860 BC (a date obtained from some charcoal in the cave); and it was occupation deposits of the Palaeolithic which blocked and masked the decorated gallery at La Mouthe — consequently any art located behind blockages of this type has to be of Palaeolithic date.

Similar proof occurs in cases where all or part of the decorated walls themselves are covered by Palaeolithic deposits (datable through their bone and stone tools, and sometimes by the radiocarbon method). The classic example of this type is Pair-non-Pair, whose engravings appeared only when the Gravettian occupation layers were removed; in the same way, the parietal engravings of Ste Eulalie (Lot) were covered by a Late Magdalenian level, the top of which was dated to 10,830 years ago. They are at hand-height when one stands on the Magdalenian III level, but are too high above the Solutrean occupation layer. It is therefore likely that the engravings date back to the Magdalenian III period, about 15,000 years ago. The sculpted friezes of Cap Blanc, La Chaire à Calvin and Angles-sur-l'Anglin were masked to some extent by Magdalenian deposits; and it will be recalled that the engravings of the Early Man rock shelter in Queensland disappeared into a layer dating to more than 13,000 years ago. Similar cases are known from Brazil and Argentina (see Chapter 3).

There are also a few cases where fragments of decorated wall have fallen and become stratified in the archaeological layers; at Teyjat (Dordogne), for example, a block of stalagmite with engravings on it was partially covered by late Magdalenian deposits, and a detached portion bearing a bison engraving was found in the lower layer ('Magdalenian V'). At the Abri du Colombier (Ardèche), a fallen rock spall, bearing a finely engraved ibex on it which closely resembles others still on the wall, was found in a Magdalenian layer above levels dating to the twelfth/eleventh millennia BC. And at the Grotte du Placard (Charente) the parietal art has been dated to the Early Magdalenian and, especially, the Final Solutrean since several engraved blocks, fallen from the wall, are covered by Solutrean layers, and are therefore more than 20,000 years old.[18] We have seen (p. 45) that fallen decorated fragments may constitute a large part of the art from Aurignacian and Gravettian layers: for instance, the Early Aurignacian layer 11 at the Abri Pataud contained a piece of ceiling with a ring in it, and a painted fragment, while the Aurignacian of the abri du Poisson also contains fragments of engraved ceiling.[19]

All the above circumstances, however, merely provide a minimum age for the parietal art. The Pair-non-Pair engravings have usually been attributed to the Gravettian, but they could just as easily be Aurignacian (they were above the level of the Aurignacian occupation layers). The same applies to the Laussel bas-reliefs, which were partly covered by Gravettian layers, but could be older. Similarly, the Isturitz engravings were above the Solutrean level and covered by Magdalenian layers — yet they have often been assigned to the Magdalenian.

A piece of wall may fall to the ground years, centuries or even millennia after it was decorated. Its stratigraphic position records its destruction, not its execution. Only in exceptional cases does the opposite occur: at the Abri du Poisson, one of the excavated layers contained thin pieces of limestone which had crumbled from the ceiling; since the sculpted fish itself bore no signs of any such deterioration, it must have been carved after the crumbling (dating to 'Perigordian IV', a Gravettian phase);[20] here, therefore, we have a rare example of a maximum possible age, rather than a minimum.

A different type of maximum age occurs in

Fig. 5.2 One of the first examples of well-dated parietal art: the painted panel at Tête-du-Lion (Ardèche), showing aurochs (70 cm long). Solutrean.

some high valleys, such as that of Vicdessos (Ariège), where excavation has shown that caves were not occupied before the Middle Magdalenian because of glacial activity; therefore the art in caves in or near that valley, such as Niaux, Fontanet or Les Eglises, cannot be older than that period. But how can one obtain a more precise date?

One way is to find out if and when the decorated cave or shelter was occupied. Some sites have no known Palaeolithic occupation (for example Monedas, Chimeneas, Pindal, etc.); in others the deposits were removed, unexcavated, to make visits easier for tourists; and other caves were visited on many occasions in the Upper Palaeolithic (Castillo, Font de Gaume, etc.). In some cases, however, there is only one brief period of occupation, and it is therefore likely – though by no means certain –

that the artistic activity coincided with this occupation. For instance, the decoration of Font-Bargeix (Dordogne) was probably done towards the end of the Magdalenian, since the only known Palaeolithic occupation belongs to 'Magdalenian VI'.[21] The cave of Gabillou had atypical Early Magdalenian material at its entrance,[22] and it is thought extremely likely that its very homogeneous collection of parietal figures can be attributed to that occupation, and perhaps even to one person.

Some caves – notably in Cantabria (La Pasiega, Chufín, El Buxú, etc.)[23] and the Rhône Valley (e.g. Chabot) – only have Solutrean deposits. At the Tête-du-Lion, the charcoal fragments of the *foyer d'éclairage* (see below, p. 125) were next to some spots of red ochre on the ground, analysis of which proved them to be of exactly the same composition as the bovid painting on the wall – the radiocarbon date of 19,700 BC therefore dates this cave's decoration to the Early Solutrean, like much of the art in this region.[24] Similarly, at Fuente del Salín (Santander), a hearth which is assumed to be related to the production of the cave's twenty hand stencils and two positive hands has produced a radiocarbon date of 22,340 years ago.[25] Inside the Grande Grotte of Arcy-sur-Cure (Yonne), charcoal associated with burnt bone and traces of ochre has produced dates from 30,160 to 24,660 years ago, providing a likely time-range for the art on the walls.[26] In the cave of Gargas (Hautes-Pyrénées), a bone fragment stuck into a fissure close to some hand stencils, and thus assumed to be somehow linked with them, has produced a radiocarbon age of 26,860 years ago, remarkably close to results from hand stencils in the Cosquer Cave (see Table, p. 75).[27]

Since some caves were decorated but never inhabited, the two phenomena were clearly sometimes separate, so that a single occupation layer, while suggestive, may be much younger or older than the art. The situation improves if evidence of artistic activity can be found in a site's occupation layers – especially colouring materials in a painted site, as at Tête-du-Lion. This is the case at Altamira, Lascaux and Tito Bustillo, for example. However, one has to be careful how occupation layers are dated: taking radiocarbon dates from charcoal in a cave's hearth, as has been pointed out, is rather like dating a church by analysing the residue from its candles! The study of pollen from occupation layers may also help to assess the date.[28]

A different type of artistic evidence from the occupation levels is portable art; and at some sites which have not only parietal art but also well-stratified portable art, one can see clear analogies between the two in technique and style; this method has been in use since the beginning of the century – indeed, the first monograph on Altamira devoted an entire chapter to the shared engraving techniques of the two art-forms.[29] In some cases it is quite probable that the same artist was responsible, and at any rate the method can provide a fairly reliable date for the wall art: at Gargas, there are some resemblances between one or two Gravettian portable engravings and those on the walls. Examples abound in the Magdalenian – the Tuc d'Audoubert has already been mentioned, and in the same way there are close analogies between the portable art of Enlène and, beyond its passage, the parietal figures of Trois Frères, so that many of the latter can probably be dated to the twelfth millennium BC; engravings on plaquettes at Labastide which are in the same style as its parietal figures[30] come from a layer dated to 12,310 BC.

At Angles-sur-l'Anglin, a 'Magdalenian III' occupation was established on bedrock, its upper surface eventually touching the base of the sculpted frieze, while the 'Final Magdalenian' layer covered the sculptures completely. The artists clearly belonged to the Magdalenian III phase, especially as it featured massive picks, lumps of pigment, ochre crayons, grinding-stones and spatulas; in addition, a portrait of a man, engraved on a plaquette from this layer, bears a close resemblance to the big polychrome specimen detached from the site's parietal art,[31] and this phase has been dated to 14,160 years before the present (as mentioned above, the Magdalenian III engraved stones of La Marche, only 30 km away, have been dated to 14,280 before the present, indicating that the two sites were roughly contemporaneous, and there was clearly contact between them. Angles, like La Marche, has limestone plaquettes with animals engraved on them).

The best-known examples of identical portable and parietal figures are the engraved hind heads found on deer shoulder-blades at Altamira and Castillo, and on the walls of both caves. The respective excavators claimed that the Altamira specimens came from a final Solutrean layer and the Castillo ones from an initial Magdalenian, and since then there has been debate about the reliability of these observations, especially as most scholars were keen to include these figures in the Magdalenian.[32] The culprit, as is often the case, was the artificial nomenclature created by prehistorians: it has become clear from more recent excavations that there was continuity between the two phases, and in fact they are probably contemporaneous in Cantabria.[33] It is more sensible to refer to this period by its date than to attach cultural labels to it, and direct radiocarbon

Fig. 5.3 The painted panel at Tito Bustillo (Asturias), showing horses and a reindeer, both engraved and painted. The reindeer's muzzle is superimposed on the black horse. The deer is just over 1 m long. Magdalenian.

dating of one engraved shoulder-blade from Altamira has assigned it to 14,480 years ago.[34]

Stratigraphy of figures

Although the Altamira ceiling has sometimes been taken as a single accumulated composition, it actually comprises a series of superimpositions: Francisco Jordá distinguished five separate phases of decoration, beginning with some continuous-line engravings, followed by figures in red flat-wash, then some multiple-trace engravings, some black figures, and finally the famous polychrome paintings.[35] Since the multiple-trace engravings (mostly heads of hinds, and a few other animals) are identical to the portable specimens from the cave, it is clear that the two earlier phases of ceiling-decoration are probably more than 14,480 years old, while the black figures and polychromes are younger.

Some of the radiocarbon dates recently obtained for the polychrome bison confirm this, while others seem to contradict it (see Table, p. 75).

It was the cave of Marsoulas which first inspired in Cartailhac the idea for dating by superimposition – he noted the different styles present, and thought he could distinguish at least three layers: black animal figures, polychromes, and finally red figures.[36] Subsequently Breuil adopted and extended this approach.

Nevertheless, it is tricky to use superimposition as a chronological guide (quite apart from the problem of establishing the order in which layers were applied, which, as will be seen below, p. 122, can be very difficult). Theoretically, of course, all the layers could have been produced within a very short space of time – a superimposition could represent half an hour! – but it is perhaps more likely that they span a few years, at least, and the timespan could be decades, centuries, or even millennia.

A problem similar to the Altamira ceiling is posed by the great painted panel of Tito Bustillo, where nine superimposed phases of engraving and painting have been differentiated, culminating in the famous polychrome horses and reindeer. The last five phases are thought to date to the thirteenth millennium BC, on the basis of comparisons with engraved plaquettes from the cave and from the occupation at its entrance, and dates obtained from the layers containing residues of artistic activity at the foot of the panel (stone tools with traces of pigment, etc.);[37] but the first four phases are more difficult – the excavators believe that they are probably not much older than the rest, but since Phase IV comprises engravings resembling the multiple-trace type, it is possible that it too dates to the thirteenth millennium BC as at Altamira, with only the other three phases being older. On the other hand, it is becoming apparent that the 'multiple-line engravings' may not be such a reliable chronological indicator after all, since they constitute a very poorly defined category.

Even more problematical is Lascaux which, in recent years, has been treated as a homogeneous collection of figures, all produced within a maximum of about 500 years; but Breuil said he could discern twenty-two different episodes of decoration in the cave, while the Abbé Glory saw six in the Hall of the Bulls alone.[38] There is certainly some heterogeneity of style in the cave, and a great deal of superimposition. A single date of 17,190 years ago for the cave came from charcoal in the Passage, while later dates (16,000 and 15,516) were obtained from charcoal down the 'well-shaft' (probably the end of a separate cave rather than a sanctuary within Lascaux). Little help can be obtained from the simple geometric signs carved on some objects (a lamp, and spearpoints) which are similar to those engraved or painted on the walls, or from the pigments and undiagnostic, worn flints in the thin archaeological layer.[39]

There is no doubt that some of Lascaux's art can be attributed to the traditionally accepted date of 17,190 years ago, the Early Magdalenian; but this certainly does not prove that the whole thing is a coherent entity spanning only a few centuries. For example, the cave is dominated by its score of great aurochs figures, yet in the whole of south-west France bones of the aurochs are not found between the Gravettian and the Final Magdalenian.[40] It is also worth bearing in mind that there is another cluster of dates from the cave in the ninth millennium before the present, and that the remarkable stylistic resemblances between some of its figures (notably the bulls and some deer) and those of some Spanish Levantine art sites such as Minateda and Cogul, which Breuil noticed immediately and mentioned in print,[41] have yet to be satisfactorily explained. It is therefore extremely probable that Lascaux is by no means a homogeneous whole, but belongs to a number of different periods, perhaps even in part to the Early Holocene. When the time comes to date organic material in its paintings (e.g. charcoal has already been detected in one bull, see p. 114), one can confidently predict that there will be some surprises, and that Lascaux will be revealed as a highly complex accumulation of compositions spanning a far longer period than has been supposed.

Dating by style

If a decorated site was unoccupied, and has no portable art of its own or any organic material in its parietal images, it becomes necessary to seek stylistic comparison with material from other sites and even other regions. As with undated portable art, one inevitably encounters all the problems of subjectivity, of 'type-fossils', and of over-simplistic schemes of development. All stylistic arguments are based on an assumption that figures which appear similar in style or technique were roughly contemporaneous in their execution. But there are other factors involved – for example, much depends on the technique used by the artist and the type of rock: in Gargas, the most delicately detailed figures are in the zones with smooth, clear, fine-grained walls, which conserve the technical details of the engraving, while the more 'archaic' figures are in the galleries with more irregular or less compact walls, or those with soft, chalky or clayey walls. In other words, there is a clear influence of wall-type on the distribution and 'style' of the figures. The same phenomenon can be seen in the dramatic difference between two Gravettian horses from Labattut, the finer being on a hard pebble, the cruder on a limestone block.[42]

Visibility can also play a role here: for example, during a study of the Grotte Carriot (Lot), a big red painting of a deer looked very simple and archaic in June, but as the year progressed and the condition of the wall altered (see above, p. 53) more details became visible, so that finally, by December, this 'archaic' figure was completely visible and revealed itself to be of Final Magdalenian style![43]

Suggested sequences of appearance of different forms of representation tend to be highly subjective, and sometimes twist the facts in order to make them fit – for example, Stoliar's scheme, based on only a handful of sites, progresses from the exhibition of part of an actual animal body in the Lower and Middle Palaeolithic, through lifesize dummies of various types to sculpture, bas-relief and finally engraving in clay. It requires the Montespan clay statues to be early Aurignacian or even Châtelperronian in date, though there is not the slightest evidence for this view, and the figures are almost certainly Magdalenian like the rest of the cave's art.[44]

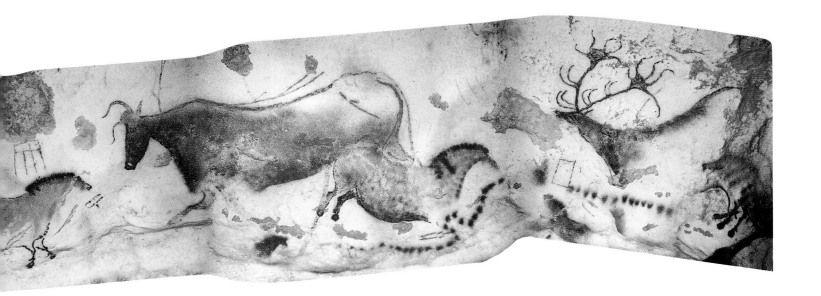

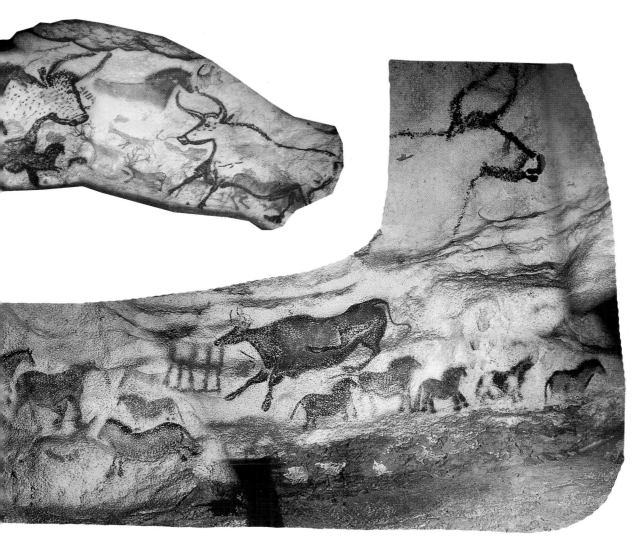

Figs. 5.4, 5.5 & 5.6 Three photomontages of parts of Lascaux (Dordogne). Two are from the 'Diverticule Axial', while the central one is part of the 'Hall of the Bulls'. The bottom one includes a falling or jumping aurochs cow (1.7 m long) with a bull's head above, as well as ibex, quadrangular signs, and small horses. The top one includes deer, aurochs, dots and quadrangular signs, and the three 'Chinese horses'. The second horse is 1.5 m long, the large cow 2.8 m, and the deer on the right 1.5 m. The middle one shows the 'unicorn' on the left (1.65 m long), as well as the great aurochs, horses, and some small deer figures. Probably Magdalenian.

Breuil's cycles

Breuil based much of his chronological scheme on the presence or absence of 'twisted perspective', a feature which he considered primitive, and which means that an animal figure in profile still has its horns, antlers, tusks or hooves facing the front (rather like Mickey Mouse's ears, which always face forward no matter what the position of his body, since this makes animation easier).

It can be argued that this graphic convention is not necessarily primitive, since it enhances the impact and beauty of features like horns and antlers,[45] and can help to suggest depth; indeed, it occurs frequently in the art of later cultures in various parts of the world, up to and including Western modern art.[46] Moreover, in the Ice Age figures it is possible that hooves, at least, were drawn in twisted perspective so that they resembled the animal tracks which must have been of fundamental importance to the hunters and those to whom they had to teach their skills – a side view of a hoof shows nothing.[47] Indeed, it has been said that if one were to wipe out the bison on the Altamira ceiling and leave only their feet, a professional hunter would at once recognize them as a good representation of a bison's spoor.[48]

For Breuil, however, twisted perspective was a decisive indicator of archaism: he had observed that, at Gargas, the Gravettian engraved figures had horns seen from the front; the bison on the wall of La Grèze was engraved with horns in twisted perspective, and since this cave had Late Gravettian and Solutrean deposits, he thought the bison was relatively early; but in Magdalenian figures, hooves and horns were drawn in proper perspective. He therefore developed an 'evolution of perspective' which included a 'semi-twisted' stage at Lascaux.[49]

Nevertheless, like Piette's, his scheme was inconsistent, since twisted perspective is also known in the Magdalenian, as in the hooves of the Altamira bison mentioned above. It is also dominant in the cave of Gabillou, which, as we have seen, is probably attributable to the Early Magdalenian. The Magdalenian parietal male portrait from Angles-sur-l'Anglin, mentioned above, has a 'full-frontal' eye, as in Egyptian and Cretan art.[50] Breuil himself placed the portable engraving from Laugerie-Basse known as the 'Femme au Renne' firmly in the Middle Magdalenian (its stratigraphic position was uncertain), but the animal profile's hooves face the front! Yet he claimed that a bovid head with twisted perspective from Isturitz, found in a Final Magdalenian layer, had been brought up from a Gravettian level![51] On the other hand, true perspective is sometimes found in early phases, as in some ibex at Gargas, the deer of Labattut,[52] or the numerous different methods of showing perspective in Chauvet Cave, more than 30,000 years ago.[53] Twisted perspective, therefore, is very far from being a reliable chronological marker.

Breuil began by proposing a four-stage scheme, but eventually he conceived a development in two cycles, the 'Aurignaco-Perigordian' and the 'Solutreo-Magdalenian', which were largely inspired by the art in the cave of Castillo, and are now only of historical interest, especially as they virtually ignore portable art. With the benefit of hindsight, it seems too neat and symmetrical that Palaeolithic art should have developed as two essentially similar but independent cycles of evolution: each of them progressed from simple to complex forms in engraving, sculpture and painting; each started with 'primitive' or archaic figures, and led on to more complex and detailed images. The second cycle was derived from, and built on, the first; overall, he saw a progression from schematic to naturalistic, and finally to degenerate forms.[54]

The 'Aurignaco-Perigordian' cycle featured hand stencils, macaroni finger-markings, simple outline animals with legs omitted, and large red animals in flat-wash (as at Altamira), eventually leading to bichrome and giant figures (as at Lascaux, which Breuil thought Gravettian) with an improvement in perspective from twisted to semi-twisted. He thought that finger-markings led to engravings, and thence to bas-reliefs like those of Laussel.

The 'Solutreo-Magdalenian' cycle again began with simple outline animals, then black figures with flat-wash, followed by infill and hatching, and finally the bichromes and polychromes, as at Altamira and Font de Gaume. The sculpture of the Solutrean (such as the carved blocks of Roc de Sers, found in a late

Solutrean level) led to the delicate, detailed engravings of the Magdalenian. The cycle ended with the abstract designs of the Azilian. He believed that his system was applicable almost unchanged to all regions with Palaeolithic decorated caves.[55]

Clearly, this system posed problems; we have already seen that a rigid application of the twisted perspective criterion led him to inconsistencies and errors. Another puzzle was why the bas-reliefs of the Gravettian should be separated in this way from those of the Solutrean, when it was far more likely that they were a continuum.

One solution was proposed by Annette Laming-Emperaire;[56] her approach was somewhat different from Breuil's, since she believed that superimpositions were actually purposeful compositions representing a group or a scene. She simplified the evolution of Palaeolithic art into three basic stages: her 'archaic phase' was roughly equivalent to Breuil's first cycle, and included the simple outline animals, for example; her middle phase covered the material where Breuil's two cycles overlapped — i.e. the bas-relief sculptures, flat-wash paintings, and so forth; and her final phase included the polychromes and other characteristic Magdalenian figures. A division into three 'cycles' was proposed by Francisco Jordá, primarily for Cantabria: they covered the Aurignaco-Gravettian, the Solutrean and Lower Magdalenian, and the Magdaleno-Azilian.[57]

Leroi-Gourhan's styles

All these schemes were soon superseded, however, by that of Leroi-Gourhan,[58] which was based primarily on the characteristics of what seemed to be securely dated figures (both portable and parietal), and which reigned supreme until the advent of direct dating. Like Laming-Emperaire, he believed that superimpositions were deliberate compositions, spanning a very short period, and therefore he treated the Altamira ceiling, for example, as a coherent entity — this is probably valid for some of its polychrome figures but, as mentioned above, there are four earlier phases of decoration beneath them, and it is very risky to assume that they are all contemporaneous, especially as more than one occupation is represented in the archaeological layers of the cave.

Leroi-Gourhan proposed a series of four 'styles', the most recent pair being further subdivided into 'early' and 'late' phases. Unlike Breuil's cycles, they were seen as an unbroken development, with a series of 'pushes' separated by long periods of transition.[59] Following a 'pre-figurative' phase covering some of the Mousterian material mentioned in Chapter 2, Style I comprised material from the Aurignacian and Early Gravettian, and notably the carved blocks from those periods, and the fallen fragments of what may have been decorated walls, together with the very rare pieces of portable art which he attributed to this phase, such as a few plaquettes from Isturitz. The motifs included deep incisions, figures with stiff contours (the Belcayre herbivore being the only complete specimen), an apparent obsession with vulvas (see below, Chapter 11), an absence of decorated utilitarian objects, and parietal art only in daylight areas of caves and shelters.[60]

Style II corresponded to the rest of the Gravettian and part of the Solutrean; it featured good animal profiles, with a sinuous neck/back line, often an elongated head, an oval eye, and twisted perspective, but with the extremities rarely depicted — the bison of La Grèze characterized this phase. Style II included far more portable art than parietal — partly because so many 'Venus' figurines had been lumped into this phase, and partly because there are so few portable animal figures of the period with which parietal figures could be compared — in the Franco-Cantabrian region as a whole, it is reckoned that only about 20 per cent of all portable art predates the Middle Magdalenian! Indeed, in Cantabria, all but one or two pieces are attributable to the Late Solutrean or Magdalenian.[61] What little parietal art there is in this phase was still thought to be restricted to daylight zones.

Style III, covering the rest of the Solutrean and the Early Magdalenian, took as its prototypes the animal figures in caves such as Lascaux and Pech Merle, with their undersized heads and limbs, and semi-twisted perspective dominating: the figures in the caves of Domme were attributed to Style III because of their striking resemblance to depictions at Pech

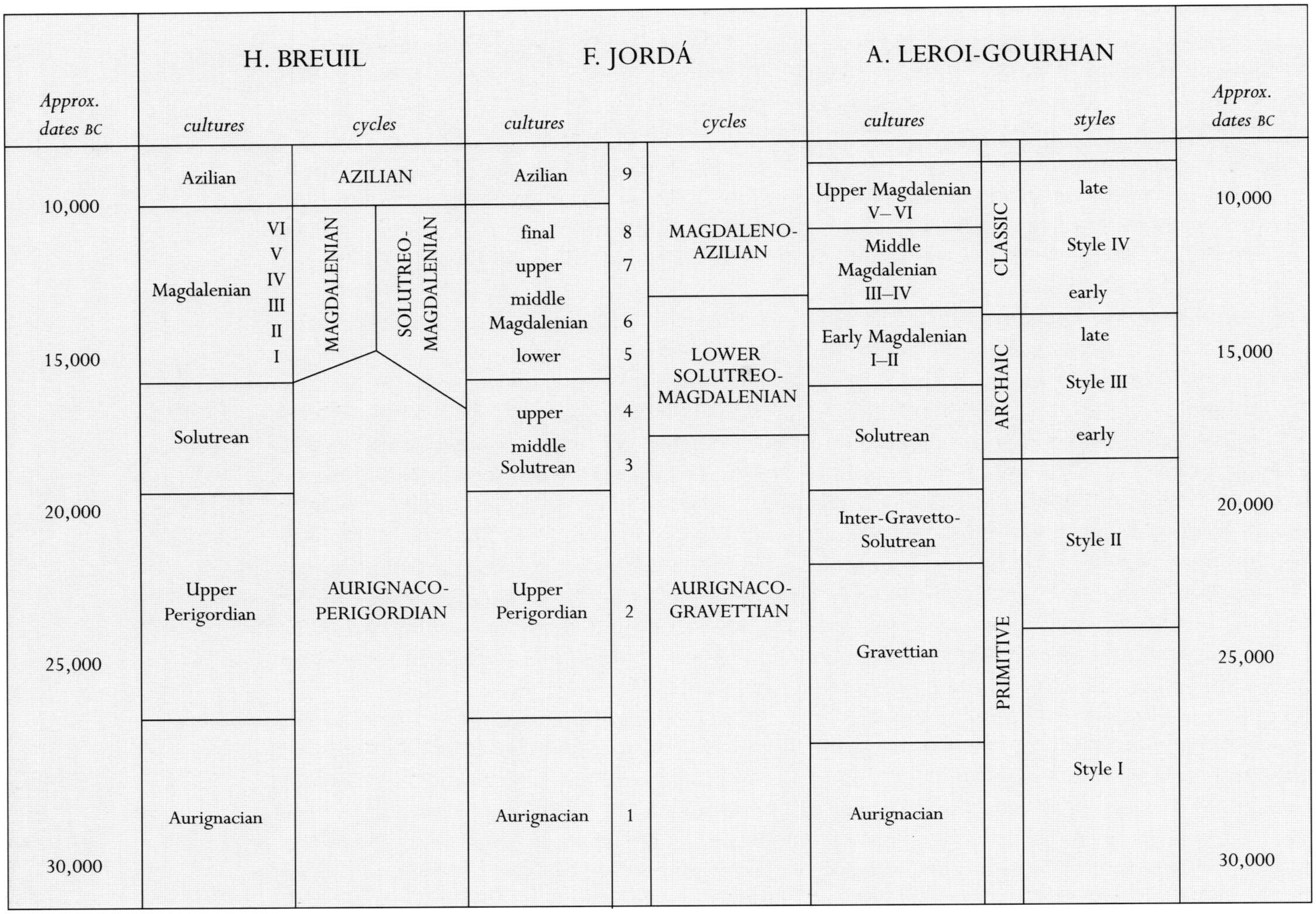

Approx. dates BC	H. BREUIL cultures	H. BREUIL cycles	F. JORDÁ cultures		F. JORDÁ cycles	A. LEROI-GOURHAN cultures		A. LEROI-GOURHAN styles	Approx. dates BC
10,000	Azilian	AZILIAN	Azilian	9	MAGDALENO-AZILIAN	Upper Magdalenian V–VI	CLASSIC	late	10,000
	Magdalenian VI V IV III II I	MAGDALENIAN / SOLUTREO-MAGDALENIAN	final	8		Middle Magdalenian III–IV	CLASSIC	Style IV	
			upper	7				early	
			middle Magdalenian	6		Early Magdalenian I–II	ARCHAIC	late	15,000
15,000			lower	5	LOWER SOLUTREO-MAGDALENIAN			Style III	
	Solutrean		upper	4		Solutrean	ARCHAIC	early	
			middle Solutrean	3					
20,000	Upper Perigordian	AURIGNACO-PERIGORDIAN	Upper Perigordian	2	AURIGNACO-GRAVETTIAN	Inter-Gravetto-Solutrean	PRIMITIVE	Style II	20,000
25,000						Gravettian	PRIMITIVE		25,000
	Aurignacian		Aurignacian	1		Aurignacian	PRIMITIVE	Style I	
30,000									30,000

Fig. 5.7 Comparison of the dating systems (based on style) of Breuil, Jordá and Leroi-Gourhan. (After Naber et al.)

Merle — particularly the volume of the front half of bison-figures, and their elongated bodies.[62] It was in this phase that decoration of the dark parts of caves seemed to have begun.

Finally, Style IV included all the wonders of Middle and Late Magdalenian art, and the decoration of really deep galleries, sometimes marked to their furthest accessible points. The early part was closer in spirit to Style III, with little movement in the figures, which are simply 'suspended in mid-air'; but the more recent phase featured more supple animals. In the later part of Style IV, Leroi-Gourhan believed that the decoration of caves gave way to that of plaquettes and similar objects. Contrary to popular belief, the art of the Azilian is not limited to spots on pebbles, but also has naturalistic animal figures in some areas (for example, in portable engravings at Pont d'Ambon, the Abri Morin, and the Abri Murat).

Leroi-Gourhan's scheme, therefore, was fundamentally like Breuil's in that, with the outlook of modern people, it saw an overall progression from simple, archaic forms to complex, detailed, accurate figures of animals, while signs developed from simple and naturalistic to abstract and stylized forms. It treated Palaeolithic art as an essentially uniform phenomenon. Diversity was played down in favour of standardization, and the development was greatly oversimplified.[63]

It should be stressed, however, that Leroi-Gourhan, like Breuil before him, was fully aware of the very tentative nature of his scheme, and that his Styles had extremely blurred boundaries. He was able to draw on more securely dated examples than were available to Breuil, but since, as mentioned above, little early parietal art could be dated, he still had to fall back on some of the same criteria as

Breuil, such as twisted perspective. Both found it very hard to apply their schemes to painting. The difference between their two schemes was therefore not so much one of principle as of results – Breuil had the art spread throughout the Upper Palaeolithic (he saw the parietal art of Altamira as dating from the Aurignacian to the Late Magdalenian), but with a very low output in the Solutrean, whereas Leroi-Gourhan compressed the majority of the art into his last two Styles, and especially IV – in other words, their absolute chronologies may have differed, but their relative chronologies were similar.

In recent years, with the benefit of hindsight and of new discoveries and dates, the profound problems with both these schemes have become readily apparent. Even the most ardent followers of Leroi-Gourhan, for example, found it difficult to distinguish between styles I and II;[64] the neck/back line was of little help as a criterion, since a 'Style II' figure from La Mouthe was just like Style III figures in Spain, while some Style IV figures from Teyjat and Les Combarelles were similar to Style III images at Lascaux.[65] There were also numerous contradictions and inconsistencies in the stylistic criteria used to differentiate the Middle and Late Magdalenian within Style IV.[66] The discovery of Chauvet Cave, of course, has not only revealed unexpected sophistication of parietal art at the very start of the Upper Palaeolithic, but also shows that the dark depths of some caves were indeed decorated right from the start, contrary to Leroi-Gourhan's scheme.

The ladder and the bush

Palaeolithic art did not have a single beginning and a single climax; there must have been many of both, varying from region to region and from period to period. It is self-evident that within those 20,000 or 25,000 years there must have been periods of stagnation, improvement and even regression, with different influences, innovations, experiments and discoveries coming into play. The same is true of the cultures of the Upper Palaeolithic, which come and go during this period, each with its own specialities (such as Solutrean flintwork or Magdalenian bonework), and constituting a complex interweaving of traditions, with continuities in some aspects of culture and sharp

breaks in others.[67]

The development of Palaeolithic art was probably akin to evolution itself: not a straight line or a ladder, but a much more circuitous path – a complex growth like a bush, with parallel shoots and a mass of offshoots; not a slow, gradual change but a 'punctuated equilibrium', with occasional flashes of brilliance. One must never forget that art is produced by individual artists, and the sporadic appearance of genius during this timespan cannot really be fitted into a general scheme. Each period of the Upper Palaeolithic almost certainly saw the coexistence and fluctuating importance of a number of styles and techniques (both realistic and schematic),[68] as well as a wide range of talent and ability (not forgetting the different styles and degrees of skill through which any Palaeolithic Picasso will have passed in a lifetime). It is also naive to assume that all Palaeolithic images are purposeful masterpieces – much of what we call stylization may simply be compensation for a lack of skill; there must be a certain percentage of meaningless scribble, limited ability, or simply crude attempts by children or beginners, and so we should not assume that everything had complex symbolic meaning.[69] Moreover, there must have been different developments at different times in different regions, and similar styles in two separate regions are not necessarily contemporaneous.[70]

Consequently, not every apparently 'primitive' or 'archaic' figure is necessarily old (Leroi-Gourhan fully admitted this point), and some of the earliest art will look quite sophisticated – as Chauvet Cave has proved. Who, for example, would have assigned the Vogelherd animals, the Hohlenstein-Stadel statuette, the Galgenberg figurine or the Brassempouy head to the Aurignacian if they had not been found in layers from that period? Leroi-Gourhan preferred to put the Vogelherd animals into his Style II, thus denying their actual provenance, and ensuring that his scheme was totally unprepared for the shock of Chauvet Cave.

In some problems, both Breuil and Leroi-Gourhan seem to have been wrong: at Altamira, for example, both of them attributed the polychrome paintings of the ceiling to the late Magdalenian, but excavation only revealed layers of the Final Solutrean and the Middle Magdalenian, which were rich in portable engravings and in colouring materials, as we have seen; so it seems likely that all of Altamira's decoration can be attributed to those periods (see Table, p. 75), and that it was blocked during or after the Middle Magdalenian (there is likewise no trace of Aurignacian occupation, contrary to Breuil's theory).[71]

There are many examples where Leroi-Gourhan's scheme proved more satisfactory than Breuil's: for example, at Trois Frères, Breuil thought several parietal bison in the 'Sanctuary' were Gravettian, but Leroi-Gourhan thought them Magdalenian, and showed they were identical to other bison which Breuil considered Magdalenian! His view has been confirmed by the discovery of a Magdalenian pebble from Enlène with a bison and a horse engraved on it: a broken line next to one figure is very similar to those on seven of the parietal bison.[72]

Other problems remain open, until such time as excavation of the sites in question or direct dating of paints may help to resolve them: for example, the parietal art of Altxerri seems to be homogeneous, and has been attributed to the Late Magdalenian on the basis of Leroi-Gourhan's stylistic criteria, through comparison with portable art from Isturitz, and because the reindeer is depicted; but other scholars have thought it Solutrean and/or Lower Magdalenian, and point to supposed resemblances in technique and style between the Altxerri engravings and the 'multiple-trace' engravings of the Late Solutrean/Early Magdalenian of sites like Altamira and Castillo (and in any case, reindeer bones have been found in Solutrean levels in northern Spain).[73]

It is simply impossible to fit everything into a rigid scheme which minimizes the variability of representations in any phase: in Palaeolithic art every rule has exceptions – for example, where all polychrome paintings were automatically thought contemporaneous, it is now clear that some of those of Altamira are earlier than those of Tito Bustillo and Castillo. Some criteria are unreliable chronological indicators – such as hoof-shape, since the hooves on an Aurignacian horse from Blanchard are identical

to those on an Early Magdalenian horse from Solvieux, 20,000 years younger! – while other features seem to belong to a single period: frontal views of animals (mostly ibex, but also a horse on a piece of portable art from Las Caldas, Asturias, p. 135) are, on present evidence, limited to the Late Magdalenian.[74]

Nevertheless, there are no real 'type fossils', and styles are very hard to define and separate. Rather than attempt to define styles, and then fit the images to them, it seems more sensible to take the archaeological and dating evidence at face value, to make an inventory of the different works in each period and their predominant features, and then develop an overall 'evolutionary scheme' (if such a thing is desirable) which incorporates that evidence without distorting it. We must adopt a more flexible system of dating by style, which avoids the problems of rigid systems like those above. A classic example is provided by the cave of Parpalló, Spain, whose numerous pieces of well-stratified portable art (see above, p. 43), spanning many phases of the Upper Palaeolithic, were neglected by Leroi-Gourhan, perhaps because they displayed a number of features which contradict his scheme. Work in recent years, particularly in Spain (at Parpalló and other sites), the Rhône Valley and the Quercy region, has established that, far from being the artistic desert of Breuil's view, the Solutrean witnessed a tremendous amount of aesthetic activity.[75]

The uncertainties in dating by style alone mean that all known parietal sites are featured on the maps, (see p. 42–4), with no attempt at dividing them into chronological slices: it is certainly unsatisfactory to compress 25,000 years of wall decoration – two-thirds of art history! – into one map, but this solution is more honest than a subjective division into stylistic periods.

Nevertheless, for the moment, style remains the only means we have of dating a great many sites. What, then, is the way forward? Rather than rely blindly on a fixed scheme, or on any particular feature such as twisted perspective, it is wiser to establish assemblages of features: in the same way, archaeological layers are no longer given a cultural attribution because of a single 'type-fossil', but rather on the basis of the whole stone and bone industry present, on the associations, interrelations and presence/absence of different features.

One example is provided by a recent study of a piece of Gravettian portable art, in which the authors drew up a list of definite or extremely probable specimens of portable and parietal art of the period (e.g. from Parpalló, Isturitz, etc.), and then looked for relevant criteria by which they could be recognized.[76] It was found difficult to produce firm criteria for Leroi-Gourhan's styles – as already mentioned, twisted perspective was known in the Magdalenian, while an oval eye is not exclusive to Style II either. However, the piece under study had an association of features (elongated horse head, oval eye, hoof-style) which placed it firmly within that phase.

Direct dating

Radiocarbon dating could not be applied to cave art in the past, partly because the pigments were not thought to contain organic material (see Chapter 8) and partly because, even if they did, the amount required to produce a date would have removed entire figures. Scientific advances have now shown that organic material (notably charcoal) was used far more often than had been thought, and AMS can obtain a date from a mere pinprick of pigment.

The application of direct dating to Palaeolithic parietal decoration has only just begun, and so far has produced results at only eleven sites (see Table, p. 75), although many more dates are already in the pipeline. Naturally, the technique is in its infancy, and each date must be considered only tentative and imperfect.[77] Nevertheless, one message has

Fig. 5.8 The Horse Panel in the Chauvet Cave, showing aurochs heads at left, horse heads in the centre, and a number of rhinoceros depictions including, at bottom centre, the fighting rhinos which have been directly dated.

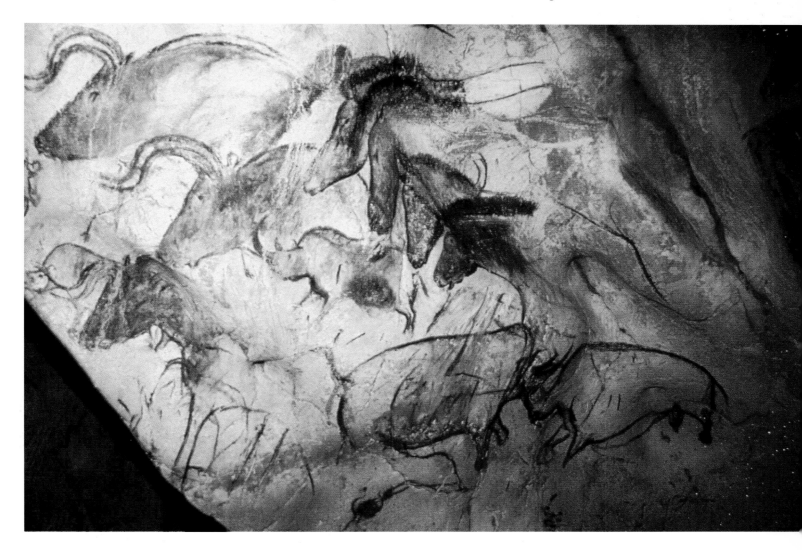

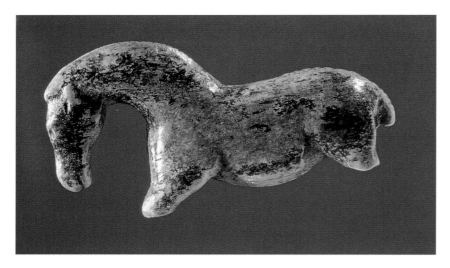

Fig. 5.9 The ivory horse from Vogelherd, Germany. Aurignacian, less than 5 cm in length.

already emerged loud and clear from the sparse and preliminary results obtained: the execution of the decoration in these caves was far more complex and episodic than had hitherto been supposed. It seems that neither Breuil, who saw cave art simply as an accumulation of figures, nor Leroi-Gourhan, who saw each cave essentially as a homogeneous composition, was correct: as usual in archaeology, the truth lies somewhere between the two extremes, and the decoration of caves can be seen as an accumulation of different compositions scattered through time.

By and large the radiocarbon results have broadly confirmed the ages which had been estimated for the caves on the basis of style and archaeological material, but some surprises have also turned up — most notably at Chauvet (where initial estimates were around 20,000 but dates came out at more than 30,000 by AMS),[78] though even here the early dates tally well with the sophisticated portable carvings known from this same period in south-west Germany and elsewhere (Vogelherd, Geissenklösterle, Hohlenstein-Stadel, Galgenberg). It should be noted that the Chauvet dates may in fact represent only the last of the cave's artistic episodes, since older red traces, largely erased, are clearly visible beneath the animal figures.[79]

Cosquer seems to have at least two phases, although two adjacent bison of the same style have produced dates of about 26,500 and 18,500 years ago, suggesting that, unless some contamination has crept in, either the same style lasted for 8,000 years, or both bison were

drawn at the same time, but some ancient charcoal from the cave's earlier occupation was used for one of them.[80]

The Salon Noir of Niaux (Ariège), previously thought to be extremely homogeneous, also has at least two phases.[81] In the past, the whole of Niaux's decoration was assigned on stylistic grounds to about 14,000 years ago; however, charcoal from two bison figures in the Salon Noir has now been radiocarbon-dated and produced strikingly different results: one bison was dated to 13,850 years ago, as expected, but the other produced a date of 12,890 years ago: in other words, Niaux's decoration seems to have been built up in at least two separate phases, while Cougnac — whose famous Megaloceros panel was confidently thought to belong to a single phase — may have at least three or four episodes spanning many millennia, with even its adjacent Megaloceros figures producing markedly different ages.[82] Stylistic studies had assigned Cougnac's art to about 18,000 to 16,000 years ago, but the charcoal in dots on the wall points to a later period, while dates for the cave's figures of the extinct giant deer are several millennia earlier, from 19,500 to 25,120 years ago.

In Spain, charcoal was discovered some time ago in several of the bison paintings from the cave of Altamira: three of them have been dated, and produced complex results.[83] It appears that two big bison were produced around 14,800 to 14,600 years ago, whereas a small one was added much later, about 13,570 to 13,130 years ago. The famous painted ceiling may not, therefore, be a single homogeneous composition after all. Dates from black motifs in the cave's terminal gallery are somewhat earlier than the polychromes. The superficially similar polychrome bison from the nearby cave of El Castillo, traditionally linked with those of Altamira by style and technique, have been dated to about 13,500 to 13,000 years ago, a millennium later than expected, but presumably from the same phase as Altamira's small bison.[84] The dates of the 'classic Magdalenian' bison in the cave of Covaciella, found in 1994, came out much as expected, around 14,000 years ago; those of Las Chimeneas (Santander) were much younger than anticipated from their style, those from

CHRONOLOGICAL CHART OF CAVE DRAWINGS
DIRECTLY DATED BY RADIOCARBON, IN YEARS BEFORE PRESENT:

Chauvet (Ardèche): right-hand confronted rhino:	32,410 + 720 (Gif A 95132)
	30,790 + 600 (Gif A 95133)
Chauvet (Ardèche): left-hand confronted rhino:	30,940 + 610 (Gif A 95126)
Chauvet (Ardèche): bison:	30,340 + 570 (Gif A 95128)
Cosquer (Bouches-du-Rhône): oval sign:	28,370 + 440 (Gif A 96074)
Cosquer (Bouches-du-Rhône): hand stencil No. 19:	27,740 + 410 (Gif A 96073)
Cosquer (Bouches-du-Rhône): bison No. 2:	27,350 + 430 (Gif A 95195)
	26,250 + 350 (Gif A 96069)
Cosquer (Bouches-du Rhône): hand stencil:	27,110 + 390 (Gif A 92409)
	27,110 + 350 (Gif A 92491)
Cougnac (Lot): female megaloceros:	25,120 + 390 (Gif A 92425)
Cosquer (Bouches-du-Rhône): black hand stencil No. 12:	24,840 + 340 (Gif A 95358)
Cosquer (Bouches-du-Rhône): horse No. 5:	24,730 + 300 (Gif A 96072)
Pech Merle (Lot): right-hand spotted horse:	24,640 + 390 (Gif A 95357)
Cougnac (Lot): male megaloceros:	23,610 + 350 (Gif A 91183)
	22,750 + 390 (Gif A 92426)
Cougnac (Lot): female megaloceros:	19,500 + 270 (Gif A 91324)
Cosquer (Bouches-du-Rhône): megaloceros:	19,340 + 200 (Gif A 95135)
Cosquer (Bouches-du-Rhône): feline:	19,200 + 220 (Gif A 92418)
Cosquer (Bouches-du-Rhône): horse:	18,840 + 240 (Gif A 92416)
	18,820 + 310 (Gif A 92417)
Cosquer (Bouches-du Rhône): bison No. 2:	18,530 + 180 (Gif A 92492)
	18,010 + 190 (Gif A 92419)
Cosquer (Bouches-du-Rhône): star-shaped sign:	17,800 + 160 (Gif A 96075)
Altamira (Santander): black marks, terminal gallery:	16,480 + 210 (Gif A 96061)
Altamira (Santander): tectiform sign, terminal gallery	15,440 + 200 (Gif A 91185)
Las Chimeneas (Santander): black stag No. 20:	15,070 + 140 (Gif A 95194)
Altamira (Santander): black hind in 'la Hoya' gallery:	15,050 + 180 (Gif A 96062)
Altamira (Santander): large bison XXXIII, facing right:	14,820 + 130 (Gif A 96071)
	14,330 + 190 (Gif A 91181)
Altamira (Santander): large bison XXXVI, facing left:	14,800 + 150 (Gif A 96060)
	13,940 + 170 (Gif A 91179)
Altamira (Santander): line beneath engraved hind:	14,650 + 140 (Gif A 96059)
Cougnac (Lot): black dot:	14,290 + 180 (Gif A 89250)
Covaciella (Asturias): bison:	14,260 + 130 (Gif A 95364)
Covaciella (Asturias): bison:	14,060 + 140 (Gif A 95281)
Cosquer (Bouches-du-Rhône): 'jellyfish' sign:	14,050 + 180 (Gif A 96101)
Las Chimeneas (Santander): black traces on sign panel 14:	13,940 + 140 (Gif A 95230)
Niaux (Ariège): bison, panel 5:	13,850 + 150 (Gif A 92501)
Cougnac (Lot): black fingermark:	13,810 + 210 (Gif A 92500)
Altamira (Santander): small bison XLIV, facing left:	13,570 + 190 (Gif A 91178)
	13,130 + 120 (Gif A 96067)
El Castillo (Santander): bison 19:	13,570 + 130 (Gif A 95108)
	13,520 + 120 (Gif A 95109)
El Castillo (Santander): large bison 18a, facing right:	13,060 + 200 (Gif A 91004)
Niaux (Ariège): black line, panel 1:	13,060 + 200 (Gif A 92499)
El Castillo (Santander): large bison 18b, facing right:	12,910 + 180 (Gif A 91172)
Niaux (Ariège): bison, panel 1:	12,890 + 160 (Gif A 91319)
Le Portel (Ariège): Niaux-type horse:	12,180 + 125 (AA 9465)
Las Monedas (Santander): black ibex 16:	12,170 + 110 (Gif A 95203)
	11,630 + 120 (Gif A 95284)
Las Monedas (Santander): black horse 20:	11,950 + 120 (Gif A 95360)
Le Portel (Ariège): big horse:	11,600 + 150 (AA 9766)

After Clottes, Courtin & Valladas 1996; Clottes et al. 1992; Clottes et al. 1995; Igler et al. 1994; Lorblanchet 1993a; Lorblanchet, Cachier & Valladas 1995; Moure et al. 1996; Valladas et al. 1992.

Las Monedas (Santander) and Le Portel (Ariège) were slightly younger than expected,[85] and the date from Pech Merle slightly older.[86]

A number of important points need to be stressed in relation to all the direct dates obtained so far from Palaeolithic cave art. First, what has been dated is the death of the tree that produced the charcoal, which is not necessarily the same as the time when the charcoal was used to produce the figure: in most cases the two events are probably not very far apart, but it is theoretically possible that people could have entered a cave and used charcoal from an ancient hearth to draw on the walls (as suggested above for Cosquer), so the charcoal's age represents merely a maximum age for the art. Second, the figures produced by the laboratories — even if they are all accurate and free of contamination, which is a considerable assumption since even minute contamination can produce great distortion — are uncalibrated radiocarbon ages, not precise calendar dates. Third, a single date is of little use or reliability, and it is unfortunate that some of these dated figures can never be redated owing to lack of usable organic material. If just one date of 30,000 had been obtained for the Grotte Chauvet, nobody would have believed the result; it was the series of dates, together with the later figures (c. 26,500 years ago) obtained for torch marks on top of the calcite that covered the art, which convinced everyone despite the universal amazement.

Nevertheless, these uncertainties and caveats concerning radiocarbon dates are — and always have been — equally applicable to the rest of the archaeological record. Because there is only a 68 per cent chance that the true age lies within the span including the plus/minus figure, it is obvious that at least a third of radiocarbon dates may be faulty; some are recognized as such immediately, because they are so incongruous, but others that are currently accepted as correct are probably wrong. The study of the archaeological record by means of the radiocarbon method is still being improved and refined. In the meantime we have to make do with the results obtained, however imperfect, and whatever their limitations; despite the growing list, the dated figures are, of course, only a tiny fraction of the cave art corpus. However, it is certain that in the next few years many more dates will be obtained for images in cave art which will help specialists to fine-tune their knowledge of how, when and perhaps even why these sites were decorated in this way.

Despite the variety of the forms which it encompasses, Palaeolithic art does constitute a recognizable episode in art history. We have seen that, for a number of reasons, almost all of it can safely be attributed to the twenty or twenty-five millennia of the Upper Palaeolithic, and most of it to the last ten. More precise attributions are difficult at present for the majority of sites, and where possible one should therefore try to restrict discussion to examples of portable and parietal art which are securely dated, and thus hope to duck accusations of subjectivity and tautology! A 1995 survey concluded that about forty Palaeolithic parietal sites had been reasonably well dated so far, using one or several of the methods outlined in this chapter — six to the Aurignacian, ten to the Gravettian, six Solutrean, one Early Magdalenian, fourteen Middle Magdalenian and eleven Late Magdalenian.[87] In other words, out of the roughly 300 decorated sites ascribed to the period, barely 13 per cent have been objectively dated, accounting for perhaps 3000 motifs out of an estimated 25,000 or 30,000 figures. However, this does not mean that the other 87 per cent are totally undated — merely that they are less well dated for the time being. Some will never be datable to everyone's satisfaction, but as more radiocarbon estimates emerge, and as other solid criteria are brought to bear, one can confidently expect that the greater part of the corpus will eventually become better established. Nevertheless, in view of the scarcity of organic material in the paint, and the high cost of AMS dating, it will always be necessary to resort to subjective stylistic comparisons in order to assess the age of the vast majority of Palaeolithic imagery.

6 FAKES AND FORGERIES

Human nature being what it is, it was inevitable that, as soon as Ice Age portable art was discovered and accepted in the 1860s, fakes began to spring up. Deliberately manufactured fake 'antiquities' have long been a plague to unwary tourists and museum curators alike, and the history of archaeology is filled with well-known examples, from the work of Flint Jack in nineteenth-century England to the Piltdown hoax, and the task of unmasking them still continues today.

Portable art

Where Ice Age art is concerned, it is virtually certain that some examples of portable and parietal art are deliberate fakes, and the database will not be reliable until these red herrings have been exposed and eliminated.[1]

The French prehistorian Vayson de Pradenne set out the problem very clearly over sixty years ago,[2] stressing that south-west France was already a fertile area for fakers of antiquities as far back as the 1860s. The interest aroused by the first discoveries of Ice Age art objects, together with the high prices they fetched from the start, made fraud a very tempting prospect. And it was also made easy by the circumstances – the caves and rock shelters of the region were filled with bone and antler of the reindeer which, having become soft with age, were very easy to carve or engrave. He concluded that numerous fakes were produced, and that not all have been re-cognized as such whereas, conversely, doubts have probably been cast on authentic pieces (e.g. at the time when Cartailhac was rejecting all cave art – see Chapter 1 – the German pre-historian Ludwig Lindenschmidt was claiming that all Ice Age portable art was fake too!).

Lindenschmidt's attitude had come about in part through his role in exposing two indis-putable fakes from the famous Swiss cave of Kesslerloch (Thayngen). This site's famous (and authentic) engraving of a reindeer (see p. 140) was discovered by Konrad Merk in 1874. However, in his site report, and against his bet-ter judgement, he also included two engravings on bone fragments which a workman named Stamm claimed to have found in the spoilheap. They depicted a bear and a fox, both seated. It was obvious, and noted in the report, that they were far less well drawn than the reindeer and other Ice Age engravings; but it was Lindenschmidt, in 1876, who spotted that the two figures were clearly copied from drawings in a children's book of 1868 on zoo animals![3]

Vayson de Pradenne specified that one famous piece was probably a fake;[4] he had been told by the prehistorian Adrien de Mortillet that the famous antler baton from Gorge d'Enfer (Dordogne), carved into the form of a double phallus, was manufactured as a special commission by an old collector of obscene objects; and he had heard from other well-informed sources that all was not well with cer-tain other classic objects. Unfortunately, no sys-tematic revision had been undertaken, in part no doubt because of the potential embarrass-ment to major museums and the world of scholarship if some of the 'icons' of the past were revealed to be forgeries.[5] More than sixty years later, very little has been done about this problem, and it remains shocking even to express doubts about famous pieces.

For example, a recent study has dared to cast doubt on the little head of Brassempouy (Landes), carved in mammoth ivory, and one of the most famous pieces of Ice Age art in the world, reproduced in countless books as a 'type-fossil' of the period. Yet no image of dubi-ous provenance, no matter what its level of general acceptance, should be immune from investigation – and, like many other pieces of prehistoric art from early excavations, the

Fig. 6.1 The small ivory human head from Brassempouy (Landes), only 3.6 cm high.

Brassempouy head has no real provenance. It was found in a period when workers were often paid by the find, and when excavators (in this case, Edouard Piette) were seldom present on site to supervise the labourers. It is admitted that the position of the Brassempouy statuettes has never been established exactly,[6] and Piette found masses of raw fossil ivory in the site.

A sculptor, Ulrich Niedhorn, has recently questioned the authenticity of the head and the whole series of svelte figurines from the site (he accepts the obese figurines from Brassempouy, which formed a separate series). His arguments[7] – the lustre of freshly polished ivory and lack of surface corrosion on the head, in contrast to other images from the site with a normally decayed surface; the excessive verticality and symmetry in the 'fakes' – may or may not prove to be valid. The fact remains that the doubts he raises about these Brassempouy figures, as well as about the enigmatic Grimaldi figurines (so similar to each other, but not to other images) and even the famous 'Venus' of Willendorf, are legitimate. All early finds without well-documented provenance should be treated with great circumspection until some way can be found to establish or disprove their authenticity. They should not be accepted on the basis of style alone, and certainly not because they have appeared in all the standard works on archaeology by specialists.

This problem has become acute in the last few years through the reappearance of some 'missing' female figurines from Grimaldi in an antique shop in Canada. Serious doubts have always justifiably plagued that site's art objects, owing to their total lack of provenance and uncertainties about when and where their discoverer – or, rather, his labourers – unearthed them. Gabriel de Mortillet attacked their authenticity in 1898 and, in the ensuing discussion, Emile Rivière stated that he had actually seen, in 1892, the sale of numerous, recently made pieces, presented as prehistoric, at the entrance of the Grimaldi caves.[8] Rigorous tests will therefore have to be performed on the 'new' finds, many of them highly bizarre, before they can be accepted as authentic.[9]

In some cases, doubts arise from stories that are handed down. For example, the bas-relief 'Venus' on a block at the Abri Pataud (Dordogne) has been the subject of numerous tales concerning student pranks, although at least one of the original excavators of the site with American prehistorian Hallam Movius believes that there has been some confusion between the 'real' Venus and a second one which was indeed made as a prank. Certainly Movius himself believed sufficiently in the carving to place it on the cover of all the site's monographs.[10] But the doubts remain so strong – some researchers vehemently reject the carving – that a number of specialists feel simply unable to judge whether it is authentic or not; and since the stone-working cannot be dated, it will be very difficult to establish the truth in this case, unless traces of metal tools can be found. The same applies to the Brassempouy head, undoubtedly made of 'genuine' fossil ivory.

However, in other cases, science can be brought to bear: for example, an ivory carving of a male human head (Fig.6.2), said to come from near the Czech Upper Palaeolithic site of Dolní Vestonice. It was not excavated by an archaeologist or discovered in situ, but found in the possession of a Czech family living in Australia, who claimed it had been discovered in a field near the site in the 1890s. It therefore has no definite provenance, stratigraphy, context or date, and it was extremely rash and foolish for a prestigious publication like National Geographic to put the carving on its October 1988 cover, and proclaim it to be the portrait of an Ice Age ancestor; the last-minute addition of a question mark at the insistence of the author involved did little to diminish its impact.[11]

The style of the head is wrong; it neither looks nor 'feels' Palaeolithic; it is far too detailed, having hair, beard, eyebrows, eyeballs (with pinpoint holes in the irises), cheekbones, nostrils, etc. Apart from a painted parietal engraving of a bearded profile at Angles-sur-l'Anglin, there is absolutely nothing else in Palaeolithic art like this head, although the 26,000-year-old 'female' ivory head from Dolní Vestonice, found in the 1920s, does have nostrils. Uniqueness in itself is no guide to fakery – if it were, a huge number of accepted prehistoric images would be rejected – but this male head breaks all the known canons of Ice Age art. What it resembles most is the cruder pair of heads from the French site of Glozel, 'dug' in the 1920s, part of an amazing hodgepodge of hideous pseudo-Ice Age art, pseudo-Neolithic pots and pseudo-Near Eastern inscribed tablets that almost all archaeologists dismiss as ridiculous and clumsy fakes. Like them, this male head suggests an ill-informed forger's idea of what an Ice Age man should look like.

The problem in proving this head a fake (like other objects mentioned above) is that its raw material is certainly ancient – fossil ivory is plentiful in the area. It is the working of the material which needs to be dated, and this may be impossible with present techniques of analysis. However, it has been found that the ivory is so enormously ancient that the carving has to be a fake.[12]

As soon as Azilian pebbles – small flat stones from the end of the Ice Age, decorated with red dots and lines – were discovered in the French Pyrenees in the 1880s, a trade in fakes sprang into existence: for not only were these objects extremely easy to make (all it needed was a pebble and some red colouring) but collectors

– especially those representing American museums in the early twentieth century – were willing to pay large amounts of money for these appealing objects. By 1929, a veritable commerce in fake pebbles was underway, and hundreds of forgeries were produced. A systematic attempt by French researcher Claude Couraud eventually established some criteria for weeding out most of the fakes – genuine pebbles tend to be of a certain shape and size, and to have a limited range of motifs of dots or lines, so any unusual shapes or motifs are inherently suspect. Fakes often look new and glossy, while authentic pebbles are dull; the red used on fakes may not match any natural earth colouring (which was the only red available to the Azilians), and in some cases the marks left by a crayon's 'lead' can be seen under a magnifying glass. And any incomplete pebble with paint on the broken area is certainly a fake, since the Azilian artists had an endless supply of perfect pebbles to choose from. Applying these criteria has exposed a large number of fakes in museums in France, Britain and America.[13]

Occasionally, direct dating of the raw material can expose the forgery: for example, a unique engraving of a mammoth on a pendant carved from a whelk-shell from Holly Oak, Delaware (Fig. 6.3). Its 'discoverer', Hilborne Cresson, claimed that he had found it in late Pleistocene deposits in 1864, but he did not actually bring it forth till 1889 (he had been in France from the 1870s till the 1880s), and

Fig. 6.2 The male ivory head allegedly from Dolní Vestonice (Czech Republic), 8 cm tall. (After Schuster/ Carpenter)

Fig. 6.3 Tracing of the mammoth engraved on a piece of large whelk-shell, from Holly Oak (Delaware). (After Kraft and Thomas)

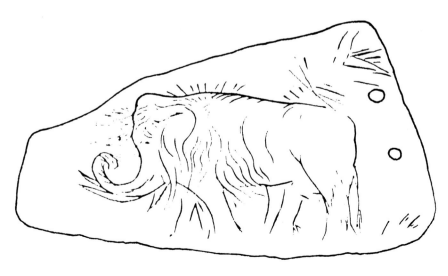

never mentioned the object in print, although he published work on Palaeolithic people in Delaware. In 1894 he committed suicide, believing himself to be under suspicion of counterfeiting money.

Re-examination of the engraving in the 1970s suggested that the incisions showed the same stages of weathering as the shell surface itself; but the problem of the shell's date remained. No such pendant was known from the late Pleistocene, being characteristic of the late prehistoric period of the Ohio Valley, and the assemblage of objects supposedly found at the same time as the engraving came from very different periods, none of them Pleistocene.

Cresson's contemporaries had not taken the engraving too seriously, since many hoaxes and forgeries occurred around this time, aimed at proving the existence of an American Palaeolithic of equal antiquity to the European. Some of the suspicion arose from the fact that the Holly Oak engraving bore a superficial resemblance to a poor copy of the unquestionably authentic mammoth engraving found by Lartet and Christy at La Madeleine, France, in 1864 (see p. 15).

Nevertheless, despite these factors and the engraving's uniqueness in New World prehistory, it was not possible to dismiss it completely, and the Holly Oak mammoth was eventually resurrected from relative obscurity by its supporters, even being placed on the cover of Science, although argument continued to rage over its authenticity.[14]

The issue was settled by radiocarbon dating of the shell, which produced a result of only 1530 years ago.[15] Since mammoths had become extinct in America about 9000 years earlier, this was clearly a fake. It seems pretty obvious that Cresson had taken the whelk shell from an archaeological site of the ninth century AD, and made the engraving on it in imitation of the La Madeleine mammoth, probably in the 1880s rather than in 1864 as he claimed.

Similarly, the famous supposedly Ice Age engraving of a horse head on a rib-bone (Fig. 6.4), 'found' by two schoolboys in a quarry at Sherborne, Dorset, in 1911, was always considered of doubtful authenticity, and probably based on the horse engraving from Creswell Crags. Recent microscopic analysis revealed that the engraved lines did not have the same patina as the bone surface, and displayed none of the features normally visible in experimental lines produced by stone tools on fresh bone, such as sharp edges and multiple parallel striations. It seems clear that the engraving took place on an already weathered bone, and a fragment of the piece was eventually radiocarbon-dated to about 610 years ago, the fourteenth century AD, proving the object to be a fake.[16]

Cave art

Fakes of cave art are — as far as we know — much rarer than those of portable art. It is northern Spain which has produced the best-known examples, such as the Cantabrian cave of Las Brujas, where fake paintings were detected by Breuil and Carballo in 1909, and destroyed in 1960:[17] or, in Asturias, the cave of Cueto de

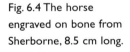

Fig. 6.4 The horse engraved on bone from Sherborne, 8.5 cm long.

Fig. 6.5 Panel in Zubialde Cave, Spain; the dot and plantlike sign near the dreadful hand stencil appeared after the first photographs were taken.

Lledias or de Cardín where some figures, including a bison, were produced in the 1960s.[18] One incidental but nevertheless important contribution of the new scientific techniques described in Chapters 5 and 8 is that we are now better able to investigate potential fakes in parietal art and either expose or authenticate them: for example, a doubtful 'vulva' figure in the French cave of Niaux has been exposed as a probable fake by pigment analysis.[19] Great progress has been made since the days when one could only go by style, or simply deduce from the presence of calcite (in caves) or lichen (in the open air) on images that they had not been done extremely recently — though since calcite and lichen can form fairly quickly, their presence was not necessarily a proof of great antiquity.

In 1991, press photographs of the newly discovered 'Ice Age' art on the walls of the cave of Zubialde (Fig. 6.5) in the Spanish Basque Country were sufficient to arouse profound suspicions among most specialists that they were fakes — the doubts were based on the style

Fig. 6.6 The rhinoceros panel at Rouffignac; the three animals are, from left to right, 95, 110 and 78 cm long.

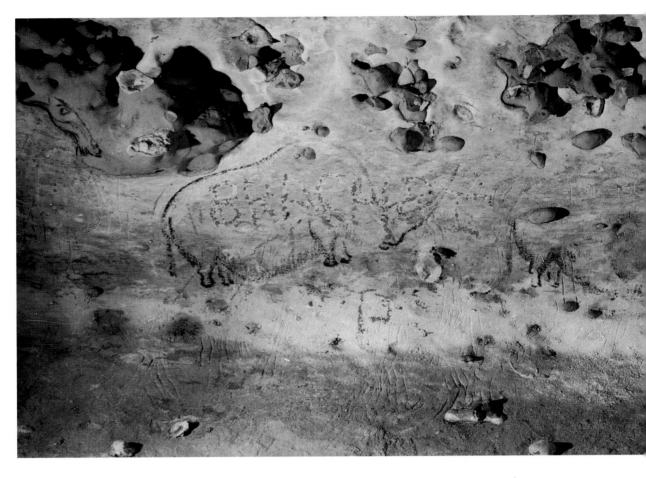

and sheer ugliness and clumsiness of many of the animal figures and hand stencils; some of the animals depicted were highly unusual for Spain (where only a couple of mammoth figures were known, and no rhinos at all); the cave seemed to contain the full range of Ice Age 'signs' which, as will be seen (Chapter 10), tend to be highly specific to a region and a period; the images had a very fresh appearance, and none of them was covered by calcite. These features in themselves, however, were not sufficient grounds for dismissing the new site, and the narrowness of the cave had clearly distorted some of the drawings in photos.

Preliminary analysis of the pigments found red ochre and manganese, which seemed compatible with the art being genuinely ancient. One can speculate that the forger used manganese rather than charcoal for black in order to avoid the possibility of exposure through direct dating; but it is equally likely that the faker still believed the traditional textbook view that manganese was always used by

Palaeolithic artists, and was simply unaware that over the past few years it has been discovered that charcoal was fairly ubiquitous as a Palaeolithic pigment.

Be that as it may, later in-depth analysis of the Zubialde paint revealed not only highly perishable materials such as insect legs, but also green synthetic fibres from a known type of modern kitchen sponge. Finally, and most curiously, it was noticed that between the time when the 'discoverer', a young amateur speleologist called Serafin Ruiz, took the photographs which he presented with his report, and the time when specialist Basque researchers began a serious study of the site, new lines and motifs had appeared on the walls, made with exactly the same pigments.[20] Equally it was scientific analysis which brought firm and final proof of the authenticity of Cosquer Cave, over which there were initial doubts, thanks to poor press pictures and to the recent Zubialde fiasco.

There are, however, examples of cave art

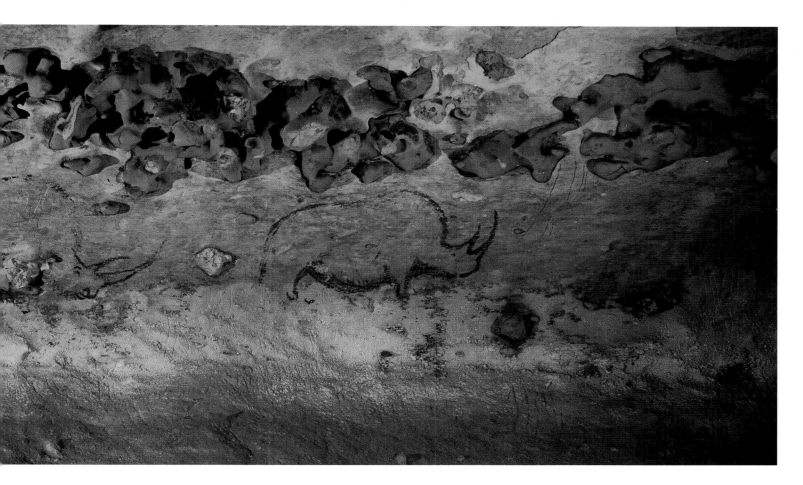

where specific doubts linger – for example, over a handful of figures in the cave of Rouffignac (Dordogne). Rouffignac is undoubtedly a genuine Palaeolithic decorated cave, and most of the art in its miles of galleries is unquestionably authentic. However, it is as illogical to claim that all the cave's figures are Palaeolithic as it is to claim that they are all fakes, and legitimate doubts remain about certain images: in particular the frieze of three rhino figures. One local landowner claimed he saw them in 1938, but we also have categorical statements from several speleologists who frequented the cave in the late 1940s that two rhinos appeared between September and December 1948, and the third by Easter 1949. In addition, there is a decorated ceiling featuring eleven ibex figures (the only ones in the cave), drawn as stiff, squarish animals, together with a (to some eyes) 'naive' mammoth figure.[21]

Breuil, with characteristic modesty, claimed that Rouffignac must be authentic because he was probably the only person capable of such successful pastiches of Ice Age art! However, since conflicting eyewitness testimony and stylistic assessments lead one nowhere in cases of this kind, the problem can only be settled through analyses and/or dating of pigment. This was called for at Rouffignac forty years ago by both its defenders and its detractors, but it has never happened. We now have the tools to examine properly these few specific images in that cave over which, rightly or wrongly, a question mark still hangs, and hence to sweep away or confirm these doubts once and for all, so that Rouffignac can at last be accepted fully by non-French specialists. All figures, whether portable or parietal, that are of dubious date and/or authenticity are so much dead wood in a subject that requires a solid base of good, reliable, chronologically sound data. Thanks to new methods of analysis, much of that dead wood can now be rooted out.

7 PORTABLE ART

Before embarking on a survey of those forms and techniques of Palaeolithic 'art' which have survived, it is worth remembering that they probably represent only the tip of the iceberg: although it cannot be proved (unless a waterlogged, desiccated or frozen site of the last Ice Age is discovered one day, in which everything is preserved), it is virtually certain that a great deal of artistic activity involved perishable materials which are gone for ever: work in wood, bark, fibres, feathers, or hides (as mentioned earlier, some decorated wooden rods are known from late Middle Stone Age levels of Border Cave, South Africa, dating to c. 37,000 to 50,000 years ago;[1]) figures made in mud, sand or snow; and, of course, body-painting which, through finds of red ochre in living sites and in graves, is thought to have very remote origins.

Music

In addition, dance and song leave no traces at all, and such things as reed-pipes, wooden instruments, and stretched-skin drums will have disintegrated; however, a few musical instruments have survived from the Upper Palaeolithic – there are over thirty 'flutes', spanning the Aurignacian and Gravettian (twenty), the Solutrean (three) and the Magdalenian; a handful come from Hungary, Yugoslavia, Spain, Austria and the former USSR, but most are from France, with twenty fragments from different layers in the supersite of Isturitz alone.[2] Indeed, the majority of flutes are broken. The French ones are made of hollow bird bones, while the eastern specimens are of reindeer or bear bone; they have from three to seven finger-holes along their length, and are played like penny whistles rather than true flutes. Experiments with a replica by a modern musicologist have revealed that, once a whistle-head is attached to direct the air-flow, one can produce strong, clear notes of piccolo-type, on a five-tone scale.[3] A 12cm fragment of the hollowed-out femur of a young bear, with a line of at least two round holes in it, was found in July 1995 at the cave of Divje Babe in north-west Slovenia, and is thought to date to at least 43,000 years ago, and perhaps as much as 82,000, according to Electron Spin Resonance dates on the enamel from bear teeth in the same layer. If the holes were made by Neanderthals rather than by carnivore teeth – a crucial point that remains to be proven – this could be an early flute.[4]

A few shaped, polished and engraved bird-bone tubes have been found which have no holes, and have been interpreted as trumpet-like 'lures' for imitating the call of a hind in the rutting season – one fine example from the Magdalenian site of Saint-Marcel even has a series of what look like cervid ears engraved on it![5] Many perforated reindeer phalanges have been interpreted as whistles in the past, though often the hole was made – or at least started – by carnivore teeth or other natural breakage; those which were intentionally made do produce a shrill, powerful note. For example, Lartet found one at Aurignac in 1860, and managed to get a strident sound out of it.[6] A few definite whistles in bird bone are also known, such as the Magdalenian specimens from Le Roc de Marcamps, Gironde,[7] while perforated shells would have made good ocarinas.

A number of oval objects of bone or ivory, with a hole at one end, have been interpreted as 'bull-roarers' (rhombes, bramaderas), a type of instrument which makes a loud humming noise when whirled round on a string – experiments have shown them to be particularly sonorous in caves. A very fine example made of reindeer antler is that from the cave of La Roche de Birol (Dordogne, Fig.7.1). The well-known parietal engraving from Trois Frères of a 'sorcerer' with a bison head (p.178) has often been interpreted as playing a musical bow, but this seems an extremely tenuous idea: since the lines go to its nose rather than its mouth, then, if it were a musical instrument, it would have to be a nose flute! — and in any case these enigmatic marks could be all manner of things (see later, p. 173).

As for percussion, a number of mammoth bones, painted with red ochre, from Mezin, near Kiev, dating to c. 20,000 BC, have been claimed to be musical instruments — a hip-bone xylophone (osteophone?), skull and shoulder-blade drums, and jawbone rattles — and have been played by Soviet archaeologists, who even cut a record of their jam-session.[8] However, doubt has recently been cast on whether the supposed marks of surface damage, polish and wear on these objects really exist.

Another category of instrument is the scraper or rasp (râcler), and claims have been made that a worked mammoth bone from a Mousterian site at Schulen, Belgium is such an instrument. Certainly the parallel grooves on it — if they are definitely man-made rather than carnivore traces — give it a close resemblance to bovine-horn rasps from Mexico and the Dutch Antilles. On the same basis, it has been speculated that the horn held by the 'Venus' of Laussel (p. 113), with its parallel lines, may likewise be a musical scraper (see p. 161–2 for a very different interpretation!).[9]

Finally, there are possible lithophones in a number of caves: 'draperies' of folded calcite formations often resound when struck with a hard object (wooden sticks seem to produce the clearest and most resonant notes), and this seems to have been noticed by Palaeolithic people, since some of the lithophones are somewhat battered, and are decorated with painted lines and dots.[10] Apart from Nerja in Spain and Escoural in Portugal, and a couple of cases in the Pyrenees (Le Portel has draperies and columns bearing traces of ancient blows, while the Réseau Clastres has numerous broken concretions),[11] all known examples are in the Lot region of France (Pech Merle, Les Fieux, Roucadour); moreover, most of them are in or near large chambers which could have held a large audience.

Red skins or tanned hides?

Virtually all peoples around the world paint their bodies on certain occasions, and we have no reason to doubt that the same was true during the Palaeolithic — indeed, this was probably one of the very first forms of aesthetic expression. Unfortunately, owing to the decomposition of bodies, we have to infer it from other evidence. As explained in Chapter 2, lumps of natural pigments are known from many archaeological sites of very remote periods. It was in the Upper Palaeolithic period that pigments became really abundant, being transported in tens of kilograms; in some French occupation sites, it is not rare to find habitation floors impregnated with red to a depth of 20 cm. Well over 100 sites with pigment are known, as well as at least twenty-five burials. As we shall see, some of these pigments can be linked to the decoration of cave walls; but what about open-air sites or unpainted caves? Can we assume that the presence of pigments here necessarily indicates body-painting?

Unfortunately, things are not so straightforward, for the simple reason that mineral pigments of this type have a number of properties. Ethnographic studies around the world show that ochre is often used in the treatment of animal skins. According to taxidermists this is not, as used to be thought, because it preserves organic tissues by protecting them from putrefaction and from vermin such as maggots — iron oxide has no antibacterial powers. Instead, experiments show that it was probably used as an additive at the end of the hide-tanning process, acting as a fine abrasive and colouring material, and giving dry tanned skins a soft, velvety finish.[12] This kind of function probably explains the impregnated soil in some habitation sites, and the traces of red mineral on many stone tools such as scrapers. Similarly, red pigment may have been applied to corpses not so much out of pious beliefs about life-blood, as is commonly assumed, or in order to restore an illusion of health and life to dead cheeks, but rather to neutralize odours and help to preserve the body.[13]

Even if, as most prehistorians believe, the people of the Old Stone Age did indeed paint their bodies, the practice may have been purely functional in some cases, rather than aesthetic:

Fig. 7.1 Engraved 'bull-roarer' with geometric/linear motifs and covered with red ochre, from La Roche at Lalinde (Dordogne). Magdalenian. Length: 18 cm, width: 4 cm.

ochre is very effective in cauterizing and cleaning injuries, and is still used in parts of Africa to dry bleeding wounds – in fact, until the end of the last century, it was still used by country doctors in parts of Europe as an antiseptic in the treatment of purulent wounds. Another function which may have been important during the last Ice Age is that of protection against the elements and insects. Peoples such as Polynesians, Melanesians and Hottentots used red pigments to maintain bodily warmth and ward off the effects of cold and rain; among some North American Indian tribes, a mixture of red ochre and fat was often applied to the cheeks of women and children as a hygienic measure to protect their skins against the sun and dry winds, while in other parts of the world red paint has been used as a protection against mosquitoes, flies and other vermin.

The more aesthetic uses of pigment on the body may well have arisen from practices such as these: for instance, the medicinal properties of ochre may have led to the painting of the dead or dying, as was commonly done by certain tribes in the New World. In Palaeolithic cases, it is often difficult to tell whether a body had its flesh painted, or merely its bones. If the whole body was painted at death, or just before, the lumps of pigment placed with the corpse may represent supplies of body-paint for the afterworld.

But what of the living? As already mentioned, most authorities agree that Upper Palaeolithic people must have painted their bodies, but the evidence is very limited. The use-wear on lumps of pigment is often shiny, indicating that they were applied to soft surfaces, which could be human skin, but also animal hide; the traces of use on a rough surface may simply indicate removal of powder to be used in a liquid paste. Some French caves have yielded bone tubes or hollow bird-bones, often engraved, containing powdered pigment, and not all of this material can be linked to cave art; at Le Mas d'Azil, excavators found a flat cake of red ochre, pitted with holes, and associated with sharp, bone needles, which they interpreted as evidence for tattooing in the Upper Palaeolithic.[14] Finally, as will be seen below, some human figurines of the period were originally painted red – but the same is true of bas-

reliefs of horses and fish!

Recently, a theory has been put forward that hide preparation was merely a marginal or secondary use for ochre in Palaeolithic times, and that its principal function was in a bizarre practice known as 'sham menstruation' – i.e. basically, women would use ochre, a blood-coloured pigment, to mimic menstrual blood and hence give off a signal of imminent fertility to attract men to bond with them.[15] The many flaws in this strange theory have already been exposed and ironically, as has been pointed out, red ochre is actually more likely to have been used as a cosmetic by males desperate to attract a mate, as in many societies worldwide.[16]

Techniques in portable art

These will be reviewed in order of apparent complexity, but this should not be taken as a chronological progression: we have seen in the preceding chapter that the phenomenon of Palaeolithic art is a complex web of forms and styles rather than a simple linear development.

Portable art is usually divided into what seem to be utilitarian and non-utilitarian objects (i.e. decorated tools versus art objects or ornaments), and their decoration is classed as figurative or non-figurative. A very wide variety of materials and of forms was employed.

The study of the techniques used in the period rests on two main types of evidence: first, the traces left on the objects or images by the tools, together with precise observation of their technological characteristics; in a very few cases, production debris or what may be the original tools have survived in close proximity to the images. Second, experiments with similar materials and tools have been carried out, followed by comparison between the modern results and the originals.

Slightly modified natural objects

Much of what is called 'parure' (jewellery) belongs in this category — i.e. fossils, teeth, shells or bones which have been incised, sawn or perforated. Such techniques are by no means restricted to the Upper Palaeolithic: a growing number of specimens are known from the preceding Mousterian period, and, as was shown in Chapter 2, can therefore be attributed to

Neanderthals, for example at Arcy-sur-Cure.

The next layer at Arcy, representing the Aurignacian (*c.* 30,000 BC), has material which features the same techniques as the earlier objects, clearly drawing on what had been developing for millennia: perforated fossils, a bone pendant with a wide carved hole, and so forth. Even older Aurignacian sites, such as the cave of Bacho Kiro in Bulgaria (over 41,000 BC), contain perforated animal tooth pendants.[17]

The Aurignacian therefore has no sudden appearance of this kind of material, but on present evidence there seems to be a marked increase in its abundance, perhaps linked to advances in lithic technology which improved or facilitated the working of decorative objects. This has led to an idiosyncratic and highly Eurocentric view, that the appearance of the quantities of beads which have survived in Aurignacian sites, particularly in the Dordogne, somehow represents not only the origins of art but also of human self-awareness, of personal decoration and of social complexity, perhaps even of modern human language, with these little perforated objects being seen as a powerful and pervasive means for constructing and representing beliefs, values and social identity.[18] Unfortunately, this theory not only ignores the earlier material of this kind (see Chapter 2) and the ready availability of natural beads in all periods, but also neglects to take into account the factor of taphonomy – i.e. the huge range of decorative forms which have vanished for ever, not just body-paint but also hairstyles, body-perforation, and anything made of wood, feathers, cordage, and so forth. There is simply no justification for seeing the Aurignacian beads as the start of anything, let alone as signifying the sudden creation of human self-awareness.[19]

The three main classes of *parure* are beads (of ivory, bone, stone, fossil wood, etc.), animal teeth, and shells.[20]

The animal teeth, perforated through the root, are mostly bovine incisors and the canines of fox, stag, wolf, bear or lion – fox teeth are often the most abundant in the Aurignacian and Gravettian, especially in Central and Eastern Europe.[21] For example, the old man buried at Sungir, near Moscow, about 24,000 years ago, had two dozen perforated fox canines sewn on

the back of his cap,[22] and more were found with the two youngsters buried at the site. Over fifty perforated fox canines were found in Kostenki XVII (Russia) and are dated to 32,000 years ago,[23] while 150 covered the head of the child buried beside Kostenki XV. In Western Europe, on the other hand, and particularly in the Magdalenian, stag canines were the favoured decoration – the best-known are the seventy perforated specimens found round the neck or chest of the woman buried at St Germain-la-Rivière (Gironde), twenty of which are engraved with crosses or parallel lines.[24]

The popularity of canine teeth continued to the end of the Upper Palaeolithic, but in the Magdalenian there was an increase in the practice of sawing reindeer incisors which, as mentioned above (see p. 25), is already present in the Châtelperronian period at Arcy: over fifty have been found at La Madeleine and Gönnersdorf, and over 200 at Petersfels (Germany). Their occasional discovery in rows shows that, as among some northern peoples in historical times, their roots were sawn through, and they were then cut from the mouth as a group, still held inside a strip of gum which was handy for hanging them as a string of eight 'pearls'.[25] Occasionally, one encounters the teeth of other species used as pendants: eight perforated human teeth are known from sites in France (one from the Aurignacian site of La

Fig. 7.2 Perforated and engraved deer canines from the Magdalenian burial at St Germain-la-Rivière (Gironde). (After Laurent)

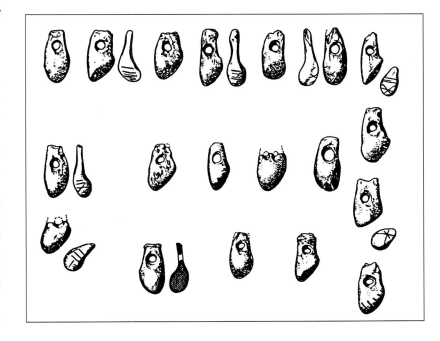

Combe, the rest from Magdalenian sites such as Bédeilhac), and one from the Gravettian of Dolní Vestonice (Czech Republic);[26] there are pierced seal teeth from the Magdalenian of Isturitz, and many sites have imitations of canines (especially stag canines) made out of ivory, stone or bone.[27]

As for shells, only a few species were selected: primarily small, globular gastropods (such as *Littorina* and *Cypraea*) which could easily be sewn to clothing; long forms (such as *Dentalia* or *Turritella* which could easily be strung; and a few scallops (*Cardium*, *Pectunculus*).[28] Many of these species are inedible, and their function was clearly decorative rather than nutritional (the same is true of several sea-shells found in Mousterian layers in Amalda Cave in the Spanish Basque Country, as well as in Cantabrian sites; they too are assumed to have an ornamental purpose).[29] Most Upper Palaeolithic specimens were perforated with a pointed tool. They are often found in considerable quantity, even in early sites – there were 300 in the Aurignacian Cro-Magnon burial alone (and hundreds more in other burials), while living-sites such as Isturitz (Pyrénées-Atlantiques) or the Abri Blanchard (Dordogne) contained hundreds of periwinkles. Fossil shells were also utilized, and sometimes came from great distances – for example, those at Mezin (Ukraine), a site dating to 21,000 years ago, came from a distance of 600 to 800 km.[30] The shells of land molluscs were rarely used, no doubt because they are thinner and more fragile.

Research shows that Dordogne sites (such as Laugerie-Basse, Abri Pataud, Abri Castanet) generally yield a high proportion of species from the Atlantic, particularly those which are common along the coast of Charente, but shells from the Mediterranean are also clearly represented. One finds the same ratio at the Atlantic end of the Pyrenees; but sites in the Central Pyrenees, such as Lespugue, at a distance of 200 km from either coast, contain a more even ratio, while further to the east, in Ariège, the proximity of the Mediterranean is reflected quantitatively in the shell collections,[31] although even here Atlantic shells dominate slightly – this is no doubt because all rivers in the French Pyrenees (apart from those at the eastern extremity), like those of Dordogne, flow out to the Atlantic, and this must have determined to a considerable extent the movement of people and materials.

It is theoretically possible that all the shells came inland in an exchange network involving 'maritime peoples' for whom we have no evidence whatsoever thanks to the drowning of the coastlines of the last Ice Age through the rise in sea-level since that time. Certainly, a great deal of exchange went on: for example, Mediterranean shells have been found in the German site of Gönnersdorf, 1000 km away, and it is unlikely (though possible) that the site's occupants travelled to that coast for them. It has been argued[32] that the great number of shells in the sites of Blanchard, Castanet and La Souquette means that this clutch of Aurignacian sites represents a market centre for exotic materials.

Equally, however, it could simply mean that these sites were centres of production, or the habitations of those who specialized in working these materials: there is ample evidence throughout the Upper Palaeolithic for craft specialization, and for repeated contact with the coasts which involved not merely vague 'exchange networks' but also probably the seasonal movements of people, following herds, and dispersing or coming together in certain places at different times of year.[33] In the past it was often assumed that the shells served as pendants and ear-rings or, in groups, as necklaces or bracelets. However, as with beads it is evident from finds such as the burials at Grimaldi on the Mediterranean coast, or those of Sungir, that many were attached to caps and clothing. At Sungir, for example, the three burials had only a handful of perforated shells, but about 3500 beads of mammoth ivory each, arranged in rows across the forehead and temples, across the body, down the arms and legs, and around the ankles. Rather than being sewn on to garments one at a time, it is far more likely that the beads were strung on lengths of sinew, which were then attached to the clothing.

It has been estimated that a Sungir bead would have required about forty-five minutes for its manufacture (cutting the tusk, drilling the hole, etc.), which means that each body had 2625 hours of 'bead-work' buried with it;[34] and the very standardized and uniform appearance of these objects suggests that they were produced by only a few people. In Western Europe, far more ivory beads are known from living sites than from burials; only the context provided by a burial can indicate an ornament's true function, but it is likely that most of the finds from occupation sites were also attached to clothing. Some idea of the production sequence involved in beads can be gained from the Aurignacian material at Blanchard, which includes pieces at different stages of manufacture: small prepared rods of ivory were divided into sections, separated into pairs, and then worked into a dumb-bell form and perforated, before the final shaping (most are round or basket-shaped).[35]

Other types of bead include fish vertebrae, which were sometimes strung together as necklaces, as in an example from a burial at Barma Grande.[36] It is worth noting that ornamentation of this kind is by no means restricted to Europe: for example, it will be recalled (see above, p.36) that perforated bone beads and shell beads are also known from the late Pleistocene of Australia (Devil's Lair, Mandu Mandu Creek), while the Upper Cave at Zhoukoudian, China, dating to 18,000 years ago, yielded over 120 decorative items such as pierced vertebrae of carp and other fish, perforated sea-shells, animal teeth, small pebbles, and engraved/polished bone tubes cut from the leg-bones of birds. Many of the perforations are coloured red, suggesting that whatever thread they were strung on was dyed.[37]

As with any other category of portable art, there is a marked differentiation in the distribution of ornaments: many sites in Europe (including some burials) have none or a few, and others have hundreds. As mentioned above, this probably reflects the presence of specialized craftsmen, as well as the varying functions of different sites (including clothing manufacture?), and perhaps even, where rich burials are concerned, some form of incipient hierarchy.

Engraved or painted stones

It will be recalled that the engraving of a bear on a pebble from the cave of Massat was one of the first pieces of Palaeolithic portable art to be authenticated; and that a few sites have hun-dreds of incised slabs — mostly of sandstone, limestone, slate or stalagmite. Few examples of portable engraving are known from the Early Upper Palaeolithic or from Eastern Europe (one of the earliest is the geometric motif on a plaquette from the Châtelperronian Grotte du Loup, Corrèze[38]; a Late Mousterian limestone slab from the cave of Tsonskaïa in the Caucasus has a cross engraved on it),[39] and in fact this particular type of art characterizes the Magdalenian of Western Europe. The incisions on stone are sometimes deep and clear, but in many cases they are so fine that they are almost invisible — this is why so many engraved pieces are found discarded in the spoilheaps of earlier, less alert excavators, as at Enlène. Only under a strong light coming in from the side can one see the lines at all: indeed, this kind of fine engraving can almost be classed as a drawing rather than an incision!

Alexander Marshack pioneered the 'technological reading' of Palaeolithic images; his studies of engraved stone, bone and antler under the microscope enabled him to follow the mechanics, micromorphology and 'ballistic trace' of each incision — its point of impact and subsequent path — and to identify marks made in different ways: from left to right, or right to left, as arcs, jabs, etc. He also claimed that he could detect changes of tool and of hand: for example, he believed that at least seven different points were responsible for a fish with 'arrows' on a stone from Labastide (Hautes-Pyrénées), and that four or five different points were responsible for renewing the horns on the rhinoceros of La Colombière.[40]

However, other scholars have argued that, at least where small stone plaques are concerned, a single burin can produce a wide variety of traces on them, depending on the part of the tool used, its position and angle, and the strength of the hand — for example, the section of the incision changes when a straight line is continued as a curve.[41] In any case, a tool may have been resharpened in the course of use. Experiments with slates of the type found at Gönnersdorf and Saut-du-Perron suggest that any changes in the incised marks are due to differences in tool-pressure, not to different tools.[42]

The new technique of placing varnish

Fig. 7.3 Engraved plaquette from Le Puy de Lacan (Corrèze), showing a duck-like bird, a fine bison head above it, and, to the right, the hindquarters of another two bison. Total width: 20 cm. Magdalenian.

replicas of engraved surfaces under the Scanning Electron Microscope (see above, p. 52) is now being used in an attempt to elucidate this problem further; criteria have been sought to identify the use of the same tool in different kinds of incisions – for example, it has been found that every tool leaves distinctive secondary striations alongside the main incision whenever parts other than the point have momentary contact with the stone.[43]

A different question concerning 'plaquettes' (defined as slabs of stone with parallel faces, under 20 cm across and 4 cm thick) is that of their usage. Scholars such as Henri Bégouën, who believed in hunting magic (see p. 171), were inclined to see evidence of ritual in everything. Since many of the engraved plaquettes from Enlène were broken, and fragments of the same specimen might be found metres apart, he concluded that the breakage and dispersal were purposeful; since most of them had been burned, and appeared to lie with the engraved face downward, this too formed part of the ritual.[44]

It is not always the case that most lie with the engraving face-down (though this seems to be true for sites such as La Ferrassie and the Abri Durif at Enval);[45] the fineness of many incisions makes it impossible to see them within the cave, and if plaquettes are collected from a cave floor, as at Labastide, it is hard to remember which way up they were – in any case, some

have engravings on both sides: at Enlène, about 16 per cent are engraved on both sides; and, of fifty which are not but whose position in situ was noted, twenty-seven had the engraving face-up and twenty-three face-down![46]

Moreover, in some caves, the upper face of a plaquette lying on the surface becomes coated with a deposit of clay or calcite, and needs to be washed or treated with acid before the engravings underneath become visible.[47] As for dispersal, it is true that pieces have often been well scattered (e.g. fragments of the same plaquette up to 30 m apart at Gönnersdorf), and in some cases missing fragments have never been recovered; this may sometimes denote purposeful breakage, but people and animals trampling around on cave or habitation floors can also have drastic effects on the material lying there. The 138 plaquettes from the cave of Gourdan (Haute-Garonne) are all highly incomplete fragments of what must originally have been quite big slabs.[48]

The question of burning is more interesting. There is a definite link between plaquettes and fire in some caves: many have marks of burning and charcoal, some (as at Labastide) have been found in hearths, while others have even been used in their construction, as at Le Mas d'Azil.[49] Breuil and Bégouën adopted the view that they had served as crude lamps, and this may be true in some cases. However, a theory has also been put forward that they represent a kind of heating device – sandstone has thermal qualities and a resistance to tension which make it suitable for a function of this kind. In France, old peasants in some areas still use heated sandstone plaquettes, wrapped in cloth, as bedwarmers![50] It may therefore be thermal tension which shattered many stones, not some arcane ritual, although a few (e.g. at Labastide and Enlène) do bear marks of blows, suggesting deliberate breakage.

It is difficult to assess how important these engraved stones were. In sites such as Parpalló, Enlène, Gönnersdorf and Labastide there are many of them (over 5000, 1150, 500 and fifty are known respectively), but there are hundreds more without engravings in some of these sites. Enlène has tens of thousands of plaquettes, brought in from a source some 200 m from the cave; excavation has shown that an area of over 5 square metres of cave floor was paved with them, no doubt a measure against humidity, and this paving includes engraved specimens;[51] a similar area of cave floor at Tito Bustillo (Asturias) had eighty-three plaquettes, twenty-five of which bore engravings ranging from animal figures to simple incisions.[52] At the Late Magdalenian open-air site of Roc-La-Tour I (northernmost France), thousands of schist plaquettes were brought in as paving, but only about 600 fragments (10 per cent) have engravings.[53] Similarly, at Gönnersdorf, several tons of schist plaquettes were brought to the open-air site as elements of construction and as foundations for structures; only 5–10 per cent of them were engraved, and these seem to be distributed at random among the others.[54] This suggests that the engravings lost all value once the ritual had been performed and they had been broken and dispersed; or simply that they never had any ritual significance, and were done simply to pass the time, for practice, for storytelling, or perhaps even to personalize one's private bed-warmer! Future finds may help to clarify the situation further.

Occasionally, engraved specimens are stuck vertically in the floor – for instance, around hearths at La Marche; and a few painted stones are known from Enlène, Labastide, Parpalló, etc. Some are clear figures, while on others simple staining by ochre in the soil may be involved: at Tito Bustillo, it was thought at first that some plaquettes had paint on, but careful study revealed that this was all contamination by pigments in the soil;[55] the same phenomenon occurred on the engraved plaquettes of the Abri Durif at Enval, where red sand had coloured the side facing downward, but at La Marche it is possible that some red traces on slabs are paint, while others seem natural.[56]

The best-known painted stones are those which characterize the very end of the Palaeolithic in Western Europe – the small 'Azilian pebbles', first identified in the 1880s at Le Mas d'Azil by Piette (see p. 20). They have been found in twenty-eight sites in France, five in Spain, three in Italy and one in Switzerland; but, of the almost 2200 known, over 1600 are

Fig. 7.4 Tracing of an aurochs and deer on a plaquette from Trou de Chaleux, Belgium. Magdalenian. Total width: 80 cm. (After Lejeune)

from Le Mas d'Azil. Their motifs, usually in reddish ochre, are simple (mostly dots and lines), and seem to have been applied with the finger, less often with a fine brush; but, as will be seen later (p. 204), a recent study has produced fascinating results from an analysis of the numbers and combinations of these marks.[57]

Three stones painted with designs in ochre have been recovered from the Riparo Villabruna-A in the Dolomites (Italy), dating to around 12,000 years ago. Two formed part of the covering of a burial, while the third, about a foot long, has three tree-shaped designs painted on it.[58] It will be recalled (see above, p. 30–31) that painted animal figures are also known on stones from the late Pleistocene Apollo 11 Cave in Namibia, while engraved pebbles have been found at Kamikuroiwa, Japan (see p. 33), and non-figurative engraved specimens are quite common in the Azilian of Western Europe.[59]

Where engraved stones are concerned, it should be noted that, although it is the fine figurative examples which tend to get published, there are far more which are undecipherable, either because they are tiny fragments, or because they are non-figurative (at Roc-La-Tour I, about 600 small engraved fragments have been recovered, but there are only twenty-five 'readable' figures so far).[60] Some have a confused mass of superimposed lines (as on the slabs of La Marche, or the pebbles of La Colombière), but experiments show that a fresh engraving is very visible due to the presence of white powder in the incisions. When this is washed off, the effect is like wiping chalk off a slate, and a new engraving can be made (the incisions can be made quickly and easily, without effort)[61] – this suggests that some of these engravings only had significance for a very brief time, and the 'associations' of superimposed animals on a given surface are not necessarily meaningful; on the other hand, there were plenty of stones available, and each figure could easily have had one to itself if desired (as is generally the case at Enlène, for example), so the superimpositions may indeed have had some significance.

Painted bone, and engraving on bone and antler

Many of the above comments also apply to incisions on flat pieces of bone; but experiments show that, unlike stone, fresh bone is hard to engrave: the tool tends to skid when it cuts bone fibres, and extremely sharp tools are required.[62] Moreover, it is necessary to pass the tool backwards and forwards, to widen the incision. One of the chief difficulties is that the initial marks can barely be seen, though it has been found that covering the bone with ochre beforehand makes incisions readily visible[63] – this may explain why some Palaeolithic specimens, such as the bone fragment from Enlène with a grasshopper engraved on it, were covered in ochre when found[64] (although, as with plaquettes, some of the colouring may come from the surrounding sediment); decoration of bones with pigment also survives occasionally – it will be recalled that the mammoth-bone 'musical instruments' from Mezin were painted with geometric motifs, chevrons, and undulations. Similarly, a mammoth skull from dwelling No. 1 at Mezhirich, Ukraine (c. 12,000 BC) is decorated with zigzags and dots of red ochre; engraved lines on bone and ivory in Eastern Europe are often highlighted with a filling of black manganese paste, whereas red ochre tended to be used for this purpose in Western Europe.[65] In Cantabria, however, only two pieces of portable art are known which seem to have colour intentionally applied to them.[66]

Shoulder-blades are the bone equivalent of plaquettes, having a smooth, large surface, and

Fig. 7.5 Selection of Azilian pebbles from Le Mas d'Azil (Ariège), each a few cm long.

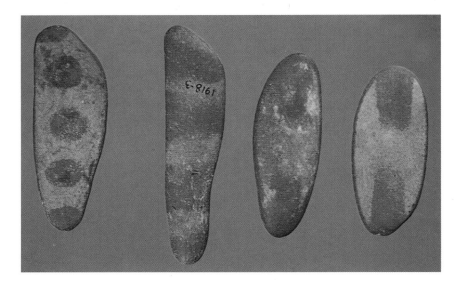

it is therefore surprising that they were not engraved in large numbers; nevertheless, decorated specimens are known from some sites, and Le Mas d'Azil and Castillo have over thirty each together with undecorated ones; their engravings are both figurative and 'abstract'. Those from Le Mas d'Azil, at least, were clustered together in a small area very poor in other finds.[67] At the Abri Morin (Dordogne), seventeen fragments with figurative engravings have been found, and twenty-nine with non-figurative marks.[68]

Although fibrous and relatively soft, such bones are by no means easy to decorate: Jean Bouyssonie, the French prehistorian, found that it took considerable muscular strength to engrave a fresh horse shoulder-blade, and the flint point often broke.[69] In some cases, the tool does not incise the bone, but compresses its surface into a furrow.

Experiments also indicate that a burin is not the only tool which can engrave bone, although it is the one which is always mentioned in this connection. In fact, a wide range of stone tools are equally effective — awls, pointed backed bladelets, and even the edge of a broken blade are just as good; it is the sharpness which counts, not the precise form. A copy of a small bison engraving on bone from La Madeleine, using different kinds of incisions and tools (which displayed no traces of wear afterwards), took four hours; but a second attempt halved that time, showing that a practised Palaeolithic craftsman would doubtless have taken little time to produce this kind of image — similar results were obtained by Louis Leguay, the pioneer of this type of work, who, in the 1860s and 1870s, was the first person to carry out experiments in engraving on bone in the prehistoric way using original stone tools from Palaeolithic sites (following Lartet who had affirmed for the first time in 1861 that prehistoric engravings were done with flint rather than metal). Leguay found that it could be done quickly with a little practice, and observed the use-wear traces that resulted.[70] However, we still have much to learn about the engraving of bone and antler, such as whether specimens of different kinds and ages vary in their 'incisability', or how special processes such as soaking may have affected the work — it has been found, for example, that

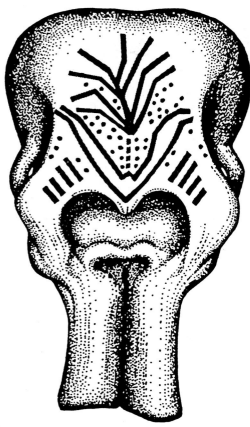

Fig. 7.6 Reconstruction of the mammoth skull painted with red ochre, from Mezhirich (Ukraine). Maximum width: 60 cm. (After Pidoplichko)

dampened fresh antler is quite easy to engrave in any direction.[71]

Although engraving on bone has a very long history (Chapter 2), only the Upper Palaeolithic has so far produced definite figurative designs; and although there is no real evolution in the technology of portable art from the Aurignacian to the Magdalenian,[72] figurative designs become particularly abundant and impressive in the latter period. The technique of champlevé was invented, where bone around the figure is scraped away, making the design stand out as in a cameo (see Fig. 7.7); and the skill was developed of engraving on bone shafts and on batons of antler, not only lengthwise (there are numerous compositions involving lines of animals, or heads) but also around the cylinder. Here, amazingly, perfect proportions were maintained, even though the whole figure could not be seen at once: the finest examples include that of Lortet, with its deer and salmon; that of Montgaudier, showing seals and other figures; and the gannet bone from Torre, Spain, with its fine collection of human and animal heads (Fig. 7.10)[73] — all decorated bird bones belong to the Magdalenian, as do most

Fig. 7.7 Detail of a
perforated antler baton
from Duruthy (Landes)
which bears a depiction
in *champlevé* of two ibex,
perhaps fleeing a preda-
tor at left. Magdalenian.
The photo shows one of
the ibex (*c*. 4 cm long).

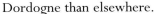

Fig. 7.7 Detail of a
perforated antler baton
from Duruthy (Landes)
which bears a depiction
in *champlevé* of two ibex,
perhaps fleeing a preda-
tor at left. Magdalenian.
The photo shows one of
the ibex (*c*. 4 cm long).

Fig. 7.8 Reindeer
shoulder-blade from the
rock shelter of Duruthy
(Landes) with a reindeer
engraved on it.
Magdalenian. The animal
measures 9.5 cm from
muzzle to tail.

depictions of birds.

A certain number of regional differences in these engravings have been apparent since Piette's time[74] (although inevitably there are exceptions to the rule): Dordogne specimens often have very deep incisions, almost bas-reliefs, whereas in the Pyrenees and Cantabria engravings are generally finer, with a mass of detail provided of the animals' coats. Scenes or 'tableaux' are also more common in the Dordogne than elsewhere.

It should also be noted that the art of engraving on teeth, already seen in the simple motifs incised on pendants, was further developed in the Magdalenian, as in the series of bear canines from Duruthy (Landes), with their engravings of a seal, a fish, 'harpoons', etc.[75]

Carved bone and antler

As mentioned earlier (see p. 27), a carved and engraved bone is known from the Pleistocene of Mexico; but once again it is Europe, and particularly the Magdalenian of France and Spain, which can boast the finest and most numerous specimens of the art. Apart from the perforated and sectioned bones mentioned earlier, probably the earliest examples are a bear-like animal head carved on the vertebra of a woolly rhino from Tolbaga, in southern Siberia, dating to about 35,000 years ago; and the 'phallus' from Abri Blanchard, carved in a horn-core. In the Gravettian, tools and weapons began to be decorated with both figurative and geometric motifs,[76] although this practice really came into its own in the Magdalenian – for example, almost all portable decoration in Cantabrian Spain comes from this phase, and increases through the period. However, it should be noted that much of the simplest 'decoration' of

tools and weapons – particularly transverse incisions near the base – was probably intended to strengthen the adherence to the haft, no doubt aided by gum or resin, and to improve the grip of the user: it has been called 'technical aesthetics'.[77]

Presumably, decorated objects were for long-term use, since there is little point in investing time and effort in engraving an implement which can be easily lost or broken – but this is not an absolute rule, since some sites, such as Abri Morin, have 'harpoons' and spearpoints with carefully made, figurative engravings.[78] It is worth noting that Garrigou interpreted the small, perforated Azilian harpoons of

La Vache (Ariège) as ear-pendants! Semi-cylindrical rods of antler were carved with a variety of motifs, such as the well-known 'spiral' decoration found in a cluster of Pyrenean sites; perforated antler batons were also decorated, especially in the Magdalenian.[79] As mentioned above, antler, unlike bone, is relatively easy to engrave.

In the Middle Magdalenian, one encounters figures of animals and fish (*contours découpés* or *perfiles recortados*)[80] and circular discs (*rondelles* or *rodetes*)[81] cut out of thin bone. The discs were often cut from shoulder-blades, and several examples of the latter are known with circles removed from them (see Fig. 7.13); many are engraved, either with animal or human figures or abstract designs like sun-rays, and some have tiny perforations round the circumference. Those with a central perforation have occasionally been interpreted as buttons, which seems an unlikely function in view of their fragility. Similar discs are also known in other materials: for example, the grave of Brno II (Czech Republic), dating to c. 23,000 BC, had specimens not only in bone but also in stone, ivory, and cut/polished mammoth molar,[82] while Gönnersdorf has some in slate.

Animal figures are occasionally large, such as the 22-centimetre bison from Isturitz, found in two pieces 100 m apart, and probably cut from a pelvic bone;[83] but the majority of the (approximately) 150 contours découpés known are animal heads (about two-thirds of them horses) cut from a horse hyoid (bone of the tongue), the natural shape of which already bears some resemblance to a herbivore head.[84] Many are perforated – some through the nostril

Fig. 7.9 Engraved and carved depiction of a bison calf, only a few cm in length, from Le Grand-Pastou (Landes). Magdalenian.

Fig. 7.10 Tracing of figures (deer, horse, isard, ibex, aurochs and human) engraved around the gannet bone from Torre (Guipúzcoa). Magdalenian. Length: c. 16 cm. (After Barandiarán)

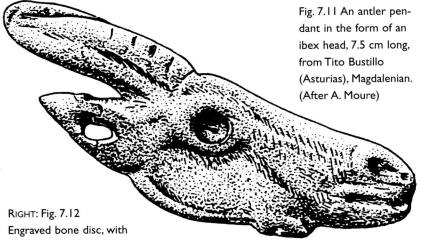

Fig. 7.11 An antler pendant in the form of an ibex head, 7.5 cm long, from Tito Bustillo (Asturias), Magdalenian. (After A. Moure)

RIGHT: Fig. 7.12 Engraved bone disc, with a central perforation and sixteen others around the edge, from Le Mas d'Azil (Ariège). The engraved motif may be non-figurative. Magdalenian. Maximum diameter: 5 cm.

BELOW: Fig. 7.13 A shoulder-blade from which a bone disc has been cut. Le Mas d'Azil (Ariège). Magdalenian. Total length: 24 cm.

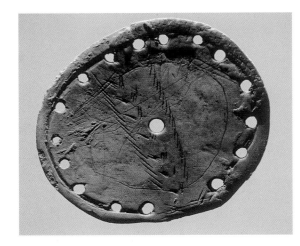

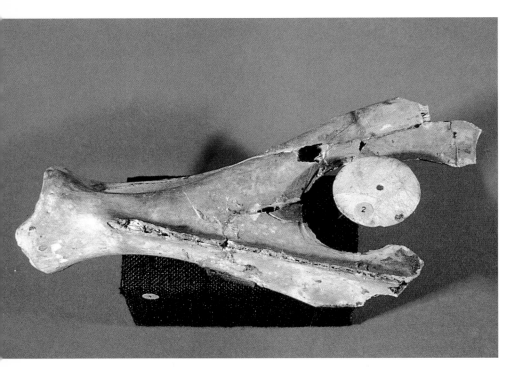

or eye, presumably for figurative effect, and others probably to serve as pendants – and they have differing degrees of detail engraved on them: eyes, muzzle, coat, and so forth.

Over 80 per cent of them have been found in the French Pyrenees, although sites in Asturias and Cantabria have recently produced a few very fine specimens of exactly the same type.[85] No doubt the most outstanding find, hidden in a corner of the cave of Labastide (Hautes-Pyrénées), is what seems to be a necklace of nineteen identical perforated heads, apparently of the isard (Pyrenean chamois) with its cold-season markings, together with one perforated bison head, all cut from horse hyoids;[86] this remarkable ensemble was clearly made by a single artist.[87]

It is worth noting that, like portable art as a whole, the distribution of contours découpés is extremely uneven: of those in France, about two-thirds come from three Pyrenean sites (the 'supersites' Le Mas d'Azil and Isturitz, plus Labastide because of its necklace); if those from Arudy are added, it means that over 75 per cent come from only four sites.[88] Bone discs are rather more widespread, but over half of them come from the Pyrenees, with Isturitz and Le Mas d'Azil again the richest sites. Enlène has yielded four intact specimens, but more than eighty fragments!

The antler spearthrowers of the Magdalenian tend to have two kinds of decoration: animal heads or forequarters carved in relief along the shaft (a type found both in France and Switzerland); or figures carved in the round at the hook-end of the object, where the roughly triangular area of available antler dictates the posture and size of the carving.[89] However, within these constraints the artists produced a wide variety of images – fighting fawns, a pheasant, mammoths, a leaping horse, and so forth.[90]

Many of the finest of these carvings have been found in the Pyrenees, and none finer than the intact spearthrower from Le Mas d'Azil with its image of a young ibex which stands, turning its head to the right and looking back to where two birds are perched on what seems to be an enormous turd emerging from its rear end (see Fig. 7.16); this composition is all the more startling because of the almost identical specimen found a few years later at Bédeilhac, a few miles away – this one had lost its shaft, and the ibex is kneeling, and turns its head to the left, but otherwise is identical in all respects. Broken specimens have also been tentatively identified from other sites in and near the Pyrenees, with the result that up to ten examples are known; if one allows for preservation, recovery, recognition and publication it becomes obvious that these must represent a tiny fraction of the scores – perhaps hundreds – originally produced. One can therefore argue for a high output by an individual artist or a small group of artisans with a favourite theme,

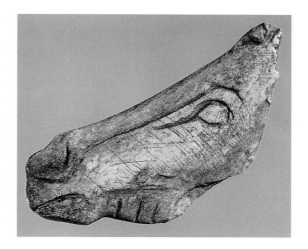

Fig. 7.14 Bone *contour découpé* of a horse head, from Enlène (Ariège). Magdalenian. Length: 5 cm.

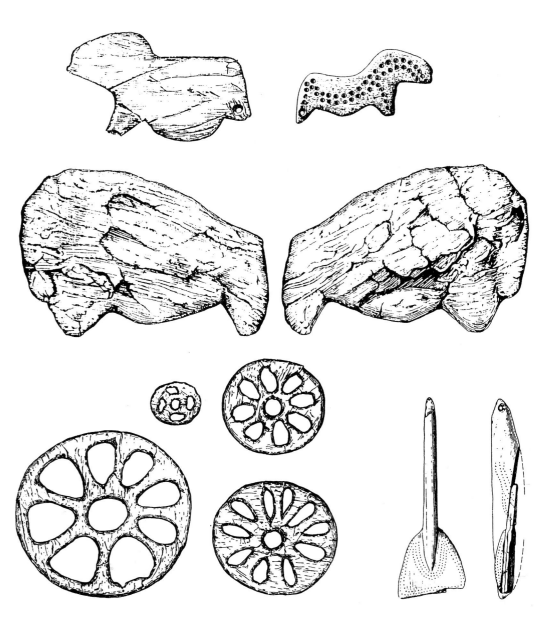

Fig. 7.15 Some of the small carved ivory objects found in the Sungir burials. (After Bader)

OVERLEAF: Fig. 7.16 Antler spearthrower showing a young ibex with emerging turd, on which two birds are perched. Le Mas d'Azil (Ariège). Magdalenian. Total length: 29.6 cm.
Fig. 7.17 The detail shows the other side of the ibex. Length: *c.* 7.5 cm. The figure resembles a pose in Walt Disney's *Bambi*, featuring the fawn with two birds: this cartoon feature was released to movie

theatres in 1942. The Mas d'Azil spearthrower was discovered between 1939 and 1941, but the first illustration of it was published only in 1942 (by M. and S-J. Péquart in *La Revue Scientifique* vol. 80, Feb. 1942); the authors had not yet seen *Bambi* but made a comparison (pp. 94–5) with the fawns in Disney's *Snow White* of 1937. In a subsequent publication (1960/3, p. 299) they also refer to *Bambi* and reveal that illustrations of the carving and of Bambi had been placed side by side in the archives of the Louvre as a comparison!

since all the examples are attributed to the Middle Magdalenian, a period which spans a few centuries in the Pyrenees.[91] Even more remarkable is the virtually identical pose struck by Bambi and two birds (though without the turd!) in the Disney cartoon, made before the two intact spearthrowers were discovered...[92]

Statuettes and ivory carvings

The simplest free-standing figurines known from the Upper Palaeolithic are terracotta models; their existence originally came as a surprise, since one of the dogmas of archaeology was that fired clay was invented later, in the Neolithic period, and that the people of the Ice Age were incapable of making it. This was clearly nonsense, as shown by the mastery of clay in that period (see below, p. 109) and by the fact that any fire lit in a cave floor will have hardened the clay around it: indeed, lumps of fired clay around hearths bear a marked resemblance to crude potsherds. The recent discovery

that pottery in Japan dates back at least 16,000 years merely underlines the fact that, if Ice Age people did not make pottery, it was through lack of need rather than ignorance.

Nevertheless, a number of terracotta figurines have survived and been recovered in different areas: a few examples are known from the Pyrenees, North Africa (see p. 31), and Siberia (fig. 7.18);[93] but considerable quantities – seventy-seven fairly intact, together with over 10,000 fragments (and others poorly fired, which have disintegrated) – have been found in the Czech Republic at the open-air sites of Dolní Vestonice (more than 5700 fragments), Pavlov (more than 3500) and Predmostí and Petrkovice (a handful each), where they are securely dated to the Gravettian, c 22,400 BC. They comprise small figurines of animals and a few humans, and display some spatial differentiation (i.e. herbivores in one hut, but human figurines in the centre of another, together with carnivores); a hearth or 'oven' for their manufacture has also been found.[94] The best-known of these figurines is the 'Venus' of Dolní Vestonice, made, like the rest, of wetted local loess soil. Recently, fragments of three terracotta animal figures have been reported from the Gravettian of Wachtberg, Austria, and others from Kostenki I (Russia).[95]

Tests on the Czech figurines indicate that they were fired at temperatures from 500 to 800°C, and the shape of their fractures implies that they were broken by thermal shock – in other words they were placed, while still wet, in the hottest part of the fire, and thus deliberately caused to explode. Rather than carefully made art objects, therefore, their lack of finish and the manner of their breakage suggest that they may have been used in some special ritual.[96]

The Upper Palaeolithic also produced carvings in other materials such as soft stone, as at Isturitz and Bédeilhac (figurines which seem to have been deliberately and systematically broken),[97] and occasionally in steatite, coal, jet or even amber;[98] there are also great numbers of pierres figures, entirely natural objects which bear a fortuitous resemblance to something figurative – it is an open question as to whether this resemblance was noticed or considered significant by Palaeolithic people.[99]

Sculpture is almost impossible in bone, which is hard and fibrous (although a mammoth from Avdeevo, Russia, was carved from the patella or vertebra of a mammoth).[100] Consequently, the great majority of figurines are in limestone, sandstone or ivory. The rock shelter of Duruthy, Landes, has yielded fine horse carvings in each of these materials (Fig. 7.20 and 7.22),[101] while the cave of Lourdes produced the best-known ivory horse. Human figures were carved out of horse teeth at Le Mas d'Azil and Bédeilhac.

The famous 'Venus' of Willendorf is made of a particular type of oolitic limestone not found in Austria and was thus brought in from elsewhere.[102] It is worth noting that this figurine, like a few others, still bears traces of red ochre. Stone figures were presumably carved with powerful flint tools, and some traces of the process can occasionally be seen, as on the 'Venus' of Tursac, made from a pebble. Limestone was sometimes used for female figurines in Russia, such as the huge 'Venus' from Kostenki I; Avdeevo has two small mammoths in sandstone, and small carvings, including many mammoths, were made in marl or limestone at Kostenki.[103]

One particularly important find of recent years is the small 'Venus of Galgenberg' (Austria), carved in green serpentine, and nicknamed 'Fanny' after a nineteenth-century Austrian dancer, which has been dated by charcoal around it to c. 32,000–31,000 years ago[104] – its lively pose, so different from those of the later, more symmetrical and static female figurines, is quite remarkable for this period. A few researchers have claimed this to be a male figure, despite the presence of a breast and an engraved vulva!

Mammoth ivory was used quite extensively in the Ice Age, not only for statuettes but also for beads, as we have seen, and for bracelets and armlets such as those found on the Sungir skeletons, or the various objects of Mezin (Ukraine) with their rich decoration of chevrons, zigzags and other geometric motifs.[105]

Ivory is easy to engrave along its fibres, but not across them, with any kind of sharp flint (not necessarily a burin). It is so extremely hard, however, that for any kind of carving it

Fig. 7.18 Terracotta human figure from Maininskaya (Siberia), 96 mm high, and dating to c. 14,500 BC.

Стоянка Майнинская
Женская статуэтка
Глина

Fig. 7.19 Tiny terracotta lion head from Dolní Vestonice, Czech Republic.

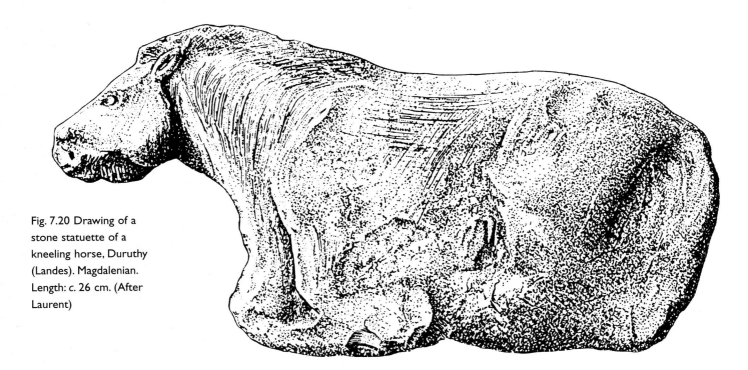

Fig. 7.20 Drawing of a stone statuette of a kneeling horse, Duruthy (Landes). Magdalenian. Length: c. 26 cm. (After Laurent)

Fig. 7.21 OPPOSITE: Ivory statuette of a human with a feline head, from Hohlenstein-Stadel (Germany). Opinions differ as to whether it is male or female. Aurignacian. Height: 29.6 cm.

needs to be softened by soaking or long-term storage, to accelerate its ageing and facilitate working.[106] A kind of champlevé could be achieved, as in the Aurignacian human figure with raised arms from Geissenklösterle, Germany. However, three-dimensional figurines were mastered in this material at a very early stage: indeed, some of the earliest known pieces of Ice Age art are ivory statuettes, including the human head and female figures from Brassempouy (Landes), the animal and human figurines from Vogelherd (Germany), those of Geissenklösterle (fragments of two mammoths and a feline), and the remarkable statuette from Hohlenstein-Stadel, opposite.[107]

Scholars have usually assigned the Brassempouy statuettes to the Gravettian, simply because one or two 'Venuses' such as that of Tursac, and perhaps that of Lespugue, came from a layer of that period;[108] among the very few 'Venuses' in Western Europe with a stratigraphic context, there are also specimens from the Magdalenian (e.g. Angles-sur-l'Anglin),[109] and consequently all 'Venus' figurines have been unjustifiably lumped into either the Gravettian (for the most part) or the Magdalenian. It is fiendishly difficult to discover from Piette's excavation reports precisely where the Brassempouy statuettes came from (assuming that those he found came from the same layer as those dug out like potatoes by earlier workers); but Breuil, who visited Piette's dig in 1897, saw the relevant layer, and kept samples of the associated industry, stated quite clearly on several occasions that the figurines came from the earliest Aurignacian, and perhaps even the Châtelperronian.[110] In short, Palaeolithic figurative art produced some of its greatest masterpieces – the Brassempouy head (if authentic – see Chapter 6), the Galgenberg statuette, the Vogelherd animals,[111] and the statuette from Hohlenstein-Stadel, Germany[112] which dates to c 30,000 BC – in its initial phase, which suggests strongly that they must have been preceded by a long tradition in carving materials which have not survived. As was seen in Chapter 5, this situation fits very well with the analogous early sophistication of the parietal art in Chauvet Cave.

Experiments have been carried out in carving 'Venuses' in ivory as well as in stone and other materials; to make an ivory figure, a piece of tusk was first prised out by cutting two deep grooves into the material; this was then rubbed into a rough shape with sandstone, and a burin and broken blade were used to carve the legs and other areas. It is thought that long handling of the figures will have smoothed away any striations left by the work.[113] Since carving ivory is so very hard, it takes many hours and even days

Fig. 7.22 Horse head carved in limestone, from Duruthy (Landes). Magdalenian. Length: 7.1 cm, height: 6.2 cm.

of effort.

Ivory statuettes were also made in composite form, as in the case of the Czech male figurine from Brno II, dating to c 23,000 BC, whose head, trunk and arm were fitted together to make an articulated 'doll'.[114]

Some idea of how the human figurines of the former Soviet Union were made is provided by the finds at Kostenki and other sites which are so numerous that there are examples of every stage of manufacture, from rough-outs to final polishing;[115] one piece of ivory recovered from Gagarino, a site which has yielded eight ivory 'Venuses', has been carved to form two human figures of different lengths, attached at the head like Siamese twins.[116] Although it is possible that this may have some special significance (especially since the two children at Sungir were buried head to head), it seems more likely that the piece is simply two separate figurines, made together, which have not yet been separated (Fig. 7.23).

It is worth noting that there are regional differences in the material used for the 'Venuses' (which has doubtless affected their survival and chances of discovery, and hence their apparent distribution pattern) and also in their locations: in the Pyrenees, Eastern Europe and Siberia they are predominantly of ivory, while those of Dordogne are in stone, those of Grimaldi are mostly steatite, and those of Central Europe employ ivory, stone and terracotta.[117] Most Western specimens have been found in caves or rock shelters, while almost all those from Central and Eastern Europe come from open-air settlements, and sometimes seem to have had a special role in the home (see below, p. 163).

As mentioned above, many of those from Western Europe, including most Italian specimens, have no context, but the dated examples are scattered through the entire Upper

Fig. 7.23 Two ivory figurines still joined at the head, from Gagarino (Russia). Probably Gravettian. (After Tarassov)

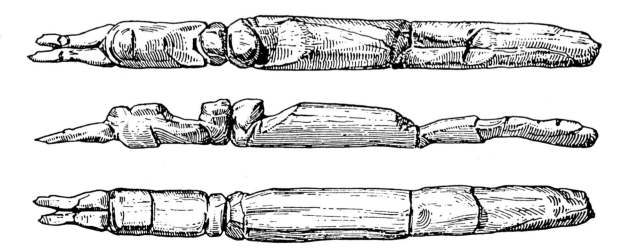

Palaeolithic, spanning more than least ten millennia from 25,000 to 12,000 years ago. It is therefore a somewhat pointless exercise to treat them as a homogeneous or contemporaneous group, as has often been done, since they have no stylistic or temporal unity, and certainly represent cultural traditions and, doubtless, different purposes.[118] However, treated simply as a very artificial category of object, like spearthrowers or jewellery, they have a strange distribution: a virtually uninterrupted spread over 3000 km from Brassempouy (Landes) to the numerous specimens from Kostenki and other Russian sites; and then a 5000 km gap between Kostenki and the Siberian sites (such as Mal'ta) around Lake Baïkal.

There are many such enigmas in the regional variations and the distribution of different types of portable art. In some cases, it is the availability of materials which is the cause, as well as their ability to survive in the archaeological record: Central and Eastern Europe clearly had great quantities of mammoth bones, which were often used for the construction of dwellings,[119] and this helps explain the dominant use of ivory for carving; the raw material's properties, shape and volume will, in turn, have determined the nature and extent of the decoration or images. Some materials (ivory, antler, certain stones) were suitable for sculpture, others (bone, stone, antler) for engraving. In some cases, the shape placed clear limits on the composition – figures on spearthrowers have already been mentioned, and the same is true of antler batons, long bones or ribs which tend to have linear arrangements of figures, usually by a single artist. Similarly, hyoid bones were merely made to look more like animal heads than they already do, which explains why some contours découpés are slightly too long or short to be truly naturalistic.

The basic question, of course, is whether the composition was chosen to fit the support, or vice versa: this is particularly pertinent to plaquettes, which, unlike bones, come in a wide variety of sizes. Was a composition planned, and then a stone of suitable size selected? Or was the stone chosen first, and the figures made to fit? No doubt both scenarios occurred fre-quently, and it is sometimes impossible for us to tell. In a few cases, however, reconstruction has cast light on the problem – at Gönnersdorf, for example, some big slates, more than a metre in length, have been reconstituted from over fifty fragments; on one of them, a large horse figure could only be seen after the reconstruction, whereas some figures were engraved on fragments, their form and size respecting the borders of the plaquettes. Clearly, these slabs were being engraved at every stage of the breakage process, and the figures corresponded to the size and shape of the slates available.[120]

Almost all of the Gönnersdorf figures are perfect, with only a few badly proportioned 'failures'. Occasionally, however, we can see apparent mistakes in the engraving process (as distinct from multiple versions of some features – see below, p. 206): for example, a pebble from Le Mas d'Azil bears an engraving of a horse whose body turned out to be too big for the stone, so that the head's position had to be altered accordingly.[121] In other cases, when a drawing proved too big, it was continued on the sides and even on the other side of the stone. If nothing else, slip-ups of this kind show that our ancestors were all too human, and that Palaeolithic art does not consist exclusively of masterpieces!

To sum up, Palaeolithic people were skilled at working with every material naturally available to them, and could even employ pyrotechnology when it suited them. It seems that much of the portable art was made rapidly and discarded, or deliberately broken as part of some ritual, while other pieces seem to have been treasured and carried around for a long time. It is possible that, in some cases, the fact that carvings or engravings were done with material that was once alive may have played a role in the beliefs or meanings attached to them – especially antler, which is such a mystifying and impressive substance, growing bigger each year and giving the animal an aura of great power, yet also falling off every year. One can speculate endlessly, and indeed pointlessly, about the role such phenomena may have played in Palaeolithic art production!

BLOCKS, ROCK SHELTERS AND CAVES

8

As mentioned earlier, the division between portable and parietal art is simply one of convenience, since there is an area of overlap which principally comprises blocks of stone which, although movable, could not be carried around, and which may in some cases be fallen fragments of decorated wall. All are from well-lit rock shelters.

Intermediate forms – 'pseudoparietal art'

Most of the best-known blocks are thought to date to the Aurignacian although, if they fell from the walls on to Aurignacian layers, they could be considerably older.[1] Certainly, as we have seen (Chapter 2), the decoration of blocks is known at the Mousterian site of La Ferrassie

(Dordogne). Some Aurignacian blocks have deep engravings (almost bas-relief), others have paint on them. The best-known painted specimen is that from the Abri Blanchard, where the rock was painted red, and then an animal figure (the bottom part of which survives) was painted on top, with a black outline and a reddish-brown infill.

Occasionally – as in the case of an Aurignacian block from Ferrassie, where the 'broken side' has been painted, not the smooth side which formed part of the shelter vault[2] – it is clear that the artwork was done after the fragment fell from the wall; but by and large one cannot be certain. However, there are so many painted chunks of rock in these early collapsed shelters that it seems highly probable that some of the back walls and vaults must have been painted (and perhaps engraved) over wide areas – the fact that nothing remains on the walls themselves is simply an indication of how remote this art is: around 30,000 years old, more than twice as ancient as most surviving Palaeolithic parietal art is estimated to be.

Observation and experimentation show that some of the blocks were prepared by grinding and rubbing to produce flatter and better surfaces for engraving.[3] Fine engraving is rare – most of the incisions are deep and wide, some of them clearly made by crudely joining together rows of cupmarks or hammer-blows, and all very weathered.

Brigitte and Gilles Delluc have devised a system of representing the different techniques used on these blocks – as well as in parietal deep engravings and bas-reliefs – in cartographic form, with different symbols and conventions to indicate the depth of the incisions, their shape in section (curved, angular, etc.)

Fig. 8.1 Deeply engraved motifs on limestone block (60 cm high) from the Abri Cellier (Dordogne). The principal oval is c. 10 cm long. Aurignacian. These motifs have often been interpreted as vulvas.

Fig. 8.2 Section of the Altamira ceiling, showing standing figures on the flat surface, and curled-up bison on the natural bosses in the rock. Magdalenian.

and so forth.[4] Such 'maps' avoid many of the problems mentioned in Chapter 4 concerning copies of the art, and at the same time convey a great deal of information about the figure's three dimensions.

Parietal art

The principal advantage of parietal art over portable is that, whereas the latter may have been made far away from its final resting-place, and indeed in a period earlier than the layer in which it is found, wall art is still just where the artist placed it. As will be seen in Chapter 11, this simple fact provides us with a great deal of information. Like portable objects, parietal art encompasses an astonishing variety and mastery of techniques.

Slightly modified natural formations

In a few cases, it is hard to tell whether the use of natural features was intentional or not, but this is usually pretty clear: for example, in the bosses of the Altamira ceiling used as bulging bison bodies (Fig. 8.2); the shape of cave walls and stalagmitic formations incorporated into animal backs and legs at Cougnac (p. 148), a bison back and tail at Ekain, or a bird at Altxerri; a bison at Pech Merle, almost 4 m

long, which uses the natural rock for its neck, back and belly, with only its stylized head being drawn in; the little stalagmites turned into ithyphallic men at Le Portel (Fig. 8.3); the antlers added to a hollow at Niaux which resembles a deer head seen from the front (Fig. 8.4), or, in the same cave, the engraving of a bison on the clay floor (p. 110) which began with, and was

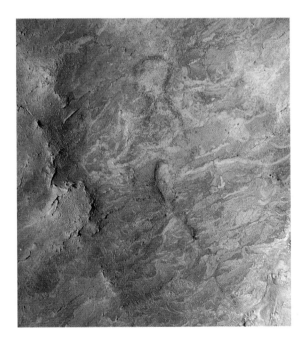

Fig. 8.3 Figure of a human drawn around a natural phallus-like stalagmite protruding from the wall in Le Portel (Ariège). Probably Magdalenian. Height: 38 cm.

Fig. 8.4 This figure in the Salon Noir at Niaux (Ariège) looks something like a deer head seen from the front, and the artist added two antlers. Probably Magdalenian. Total length: c. 40 cm.

Fig. 8.5 ABOVE RIGHT: Vertical bison painted at Castillo (Santander), utilizing the shape of a stalagmite for its back, tail and hind leg. Probably Magdalenian. Length: 80 cm.

composed around, an eye formed by a cupmark left by water droplets. There are countless other examples.[5]

Finger-markings

The simplest technique of marking cave walls was to run one or more fingers over them, leaving traces in the soft layer of clay or 'mondmilch' (a white, clayey precipitate of calcium carbonate): the technique is probably extremely ancient, and indeed may have been the first used – it is certainly a method which requires neither great effort nor any kind of tool. Finger-markings in the European caves may well span the entire Upper Palaeolithic and even part of the Mousterian; most scholars followed Breuil in attributing them to a very early phase, but those of caves such as the Tuc d'Audoubert and the Réseau Clastres are

almost certainly Magdalenian.

It is possible that in some cases the mondmilch was actually removed in this fashion – either for body-decoration, or for medical purposes, since in some Alpine regions in historic times it was used as an ophthalmic analgesic[6] – but more often the finger-marks are splayed, and the mondmilch compressed rather than removed, suggesting that marking was the aim.

In Australia, the lines made by fingers appear purely non-figurative; but in some European caves such as Gargas and Pech Merle – the latter has about 120 square metres of them – they also include definite animal and anthropomorphous figures.[7]

Engraving

One possible source of inspiration for the finger technique is the abundance of clawmarks of cave-bear and other animals on the cave walls; in a few cases these seem to have been incorporated into designs, such as a circle at Aldène (Hérault), or a hand at Bara-Bahau (Dordogne) (a very similar, well-carved hand is known in Koongine Cave, South Australia);[8] and it is thought that some marks on cave walls, at La Croze à Gontran (Dordogne) and elsewhere, are engraved imitations of these clawmarks.[9] In fact, it is important to make sure at the start of any study that incised lines are definitely man-made rather than natural cracks or animal claw-marks – this may seem self-evident, but even experienced researchers can be fooled by apparent V-shapes or curved 'backs' which were actually 'drawn' by animals.[10]

Engraving itself, which, as in portable art, is by far the most abundant technique on cave walls, encompasses a wide variety of forms, the choice of which was largely dictated by the nature of the rock; incisions range from the fine and barely visible to the broad, deep lines already seen on the Aurignacian blocks; scratching and scraping were also used at times.

The deep engravings on the walls of Pair-non-Pair seem to have been done with stone picks, whereas the finer incisions of later periods can be attributed to sharp instruments producing a fine, V-shaped section. Jean Vertut was developing new techniques to visualize the profile of fine engraved lines, using micro-photographic analysis and even computer enhancement of contrast and equidensities.[11] As mentioned above (p. 49), his innovative macro-photography reveals the very slight damage done by direct tracing methods, which affects our ability to assess the tools used in the incisions.

A robust tool must have been used for the deep engravings and bas-reliefs in the caves of Domme (Dordogne), and therefore the flint burin found in a narrow fissure in the Grotte du Pigeonnier here, in the immediate proximity of the parietal figures, may simply have been used for initial sketches, or for final details.[12] In any case, such finds are not necessarily always of relevance to the art: a burin-scraper found in a fissure between the legs of the engraved feline of Trois Frères was traditionally assumed to be the artist's tool, left there when the job was done, but in fact nine other objects (flints, bones, a tooth and a shell) have also been found in fissures in this little 'sanctuary'.[13]

As in portable art, the burin was not the only possible tool for the engraver: almost any sharp flint could have been used, from carefully made retouched pieces to simple waste flakes. This is seen clearly at Lascaux, where use-wear was found on the sharp angles of twenty-seven stone tools, including burins, backed bladelets,

and simple blades and flakes.[14] Experiments showed that a similar wear and polish could be produced by passing a tool up to fifty times across the rock, thus making an incision 1 or 2 mm in depth. These tools were found only in those zones of Lascaux which have parietal engravings; they were absent in parts which do not. A few engraving tools were also found at Gouy, though their use-wear was less pro-

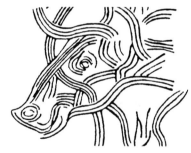

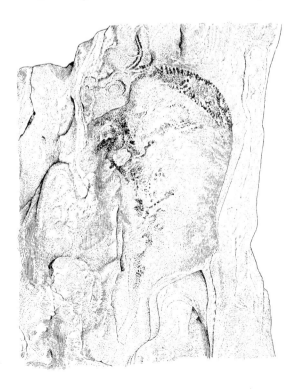

Fig. 8.6 & 8.7 Digital tracings at Altamira (Santander) with depiction of bovine head, and tracing by Breuil. The head is just over 1 m in length.

Fig. 8.5 & 8.8 OPPOSITE PAGE AND LEFT: Vertical bison painted at Castillo (Santander) utilizing the shape of a stalagmite for its back, tail and hind leg. Probably Magdalenian. Length: 80 cm. (Explanatory drawing after E. Ripoll and A. Bregante.)

Fig. 8.9 Detail of one of the engraved owls at Trois Frères (Ariège) (see Fig.10.24), showing the striations in the lines. The head is about 10 cm wide.

Fig. 8.10 'Kneeling' reindeer engraved and scraped on a wall in Trois Frères (Ariège). Probably Magdalenian. Length: c. 20 cm.

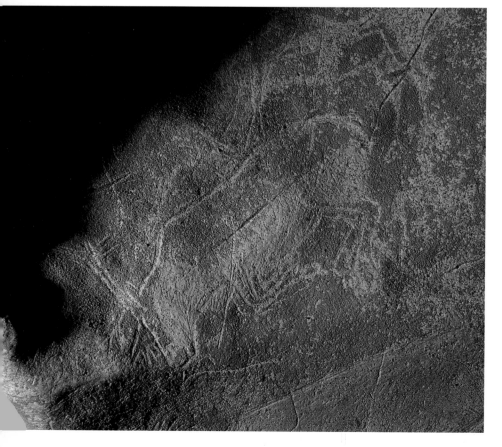

nounced than on the Lascaux tools, no doubt because the limestone here is softer: experiments on rock from Gouy have shown that an initial fine incision is needed to serve as a guide to the engraving tool, but that the whole process takes very little time.[15]

It will be recalled that fresh incisions on bone are almost invisible unless the surface is covered with ochre; a similar suggestion has been made for parietal engravings: namely, that the wall may have been coated with a substance such as clay or blood so that new lines stood out from those beneath.[16] It is the great numbers of engraved figures on top of each other at Trois Frères or Lascaux which have led to hypotheses of this type, but in fact they are exceptions to the rule: by and large, parietal figures, whether engraved or painted, carefully avoid superimposition: even at Lascaux juxtaposition is far more common than overlaps, while at Niaux there seems to be a positive concern to avoid pre-existing figures.

In most cases, fine engravings are almost invisible when lit from the front, but 'leap out' when lit from the side. This fact is of some importance, for it provides an indication of whether the artist was right- or left-handed. Right-handed artists tend to have their light-source on the left, to prevent the shadow of their hand falling on the burin (or brush), and accordingly the majority of Palaeolithic parietal engravings are best lit from the left (in portable engravings, too, the proportion of right- and left-handers is similar to that of today).[17] Occasionally, however, one comes across the work of a southpaw – for example, the Pyrenean cave of Gargas has many engravings, including a fine, detailed pair of front legs of a horse; these had been known and admired for decades; but one day, when visiting scholars lit the figure from the right instead of the left, they were suddenly confronted with the rest of the horse, which nobody had seen before![18] It is possible, of course, that Palaeolithic artists used the lighting of engravings to their advantage, making them appear or disappear to great effect – alas, we shall never know.

In places where walls were too rough for fine incisions, engraving became almost a form of painting: it was either done more crudely, so that it is best lit from the front, or it was done by scraping. In both cases, the image came not from the relief and light/shadow of an incision, but from the difference in colour between the white engraving and the darker surrounding panel area. This was the technique used on the Trois Frères feline, mentioned above, and on many other figures.[19]

There are also different regional/chronological styles of engraving. The best example, one easily recognized and recently dated (see p. 63–4), is 'striated engraving', using multiple traces, found in a wide area of northern Spain, both in portable art (Rascaño, El Cierro, Castillo, Altamira, etc.) and in parietal (Altxerri, Castillo, Altamira, Llonin, etc.).[20] However, it is very dangerous to assume that every use of a technique such as this belongs to a single chronological phase.

Work in clay

A different example of regional specialization is work in clay, which (apart from the terracotta figurines discussed in Chapters 3 and 7) is restricted to the Pyrenees on present evidence – presumably because the limestone of most of the Pyrenean caves other than Isturitz was unsuitable for sculpture or bas-relief.

The simplest forms of clay decoration were finger-holes, finger-tracings, and engraving with tools in the cave floor or in artificially set-up banks of clay. Patterns of dots made by finger-tip have been found in the floor at Fontanet (Ariège), and apparently random holes were punched by fingers into the frieze on clay at Montespan (Haute-Garonne), which itself comprises horse figures and vertical lines drawn by finger. One of the finest engravings known is the vertical horse head which begins this frieze (Fig. 8.11).[21]

Figures and signs were drawn in the floor at Bédeilhac and elsewhere, but by far the best known cave-floor engravings are those of Niaux,[22] mostly located around the Salon Noir; apart from the bison mentioned above, they include figures such as two fish, horses, an ibex head, an aurochs, and something vaguely resembling a human fist. Drawings in clay at Massat include three heads of isard (the Pyrenean chamois).

Such drawings, for the most part, have survived only under rock overhangs or in recesses where the feet of unwary visitors could not reach them. It is therefore almost certain that originally there must have been many more of them, and in far more sites, although at present they do seem to be limited to the Pyrenees, like other work in clay. Untold numbers of fine figures have probably been obliterated by casual visitors to caves in the days before Palaeolithic art became known; only in caves such as Fontanet, the Tuc d'Audoubert and Erberua, where the utmost care has been taken since their discovery, and visits severely restricted, can one be certain that all existing ephemeral traces of this kind have survived.

Since, as we have seen, Palaeolithic people could model clay into figurines, it is hardly surprising that they were equally capable of parietal modelling. Bas-relief exists in cave floors, as in the series of little horses at Montespan (some almost in haut-relief, since they rise to 9 and 10 cm in height), and also on banks of clay: the best-known example is in a side chamber at Bédeilhac, where there were four bison (two of them now destroyed), with other marks and a carefully modelled and accurate vulva nearby, with a small fragment of stalactite inserted at the position of the clitoris.[23]

Fig. 8.11 Engraving of a vertical horse head in clay, Montespan (Haute-Garonne). Probably Magdalenian. Length: 15 cm.

Fig. 8.12 Clay bank in the cave of Bédeilhac (Ariège), showing modelled vulva on the left, and a bas-relief bison about 30 cm in length. Probably Magdalenian.

including heel-prints, a crude engraving of a third bison in the cave floor, and a small clay statuette of a fourth.

The clay was brought from a neighbouring chamber; some 'sausages' on the ground have traditionally been interpreted as phalluses, or occasionally as horns, but a study by a sculptor[24] has suggested, far more plausibly, that they are simply the result of testing the clay's plasticity, and the position of palm- and finger-prints on them supports this. Marks on the haut-relief bison show that they too were modelled with fingers, and further shaped with some sort of spatula; a pointed object was used to insert the eyes, nostrils, mouths, mane, etc. It is likely that the cracks in the figures occurred within a few days of their execution, as they dried out. Most of the artist's detritus around the figures seems to have been carefully cleared away.

Finally, one fully three-dimensional large clay figure is known: the famous bear at Montespan. However, although a fully rounded form, in a Sphinx-like posture, it cannot definitely be classed as a statue — at present it is simply a bear-shaped headless figure, 1.1 m long and 60 cm high. Opinions differ over

Montespan has some broken haut-relief figures in clay, including at least one possible feline (thought to be a horse by some scholars), but they are so damaged that they pale into insignificance beside the classic examples of the technique: namely, the two bison of the Tuc d'Audoubert (Fig. 8.14), which are about one-sixth normal size (63 and 61 cm long), and placed at the centre of a distant chamber, against some rocks. Around them are marks

Fig. 8.13 Bison drawn in the clay floor at Niaux (Ariège), 57 by 20 cm. Probably Magdalenian. It was drawn around a natural cupule for the eye; other cupules were turned into 'wounds', but some postdate the drawing (e.g. on a hindleg).

Fig. 8.14 The two clay bison of the Tuc d'Audoubert (Ariège) in their context, as the focal point of a small, low-ceilinged chamber. Probably Magdalenian. The bison are 63 and 61 cm long respectively.

whether it was originally a full statue, whose head disappeared when people collected clay from the cave about 5000 years ago, or simply a headless dummy over which the hide (perhaps with head still attached) of a real bear was probably draped.[25] This would account for the polish on the clay. Nevertheless, a great deal of work went into shaping this considerable mass of clay (c. 700 kg if it is all clay, and not built up around a rock), and its haunchesand stomach area are particularly fine (Fig. 8.17); its claws were also incised into the clay paws.

Bas- and haut-relief in stone

Just as work in clay is restricted to the Pyrenees, so parietal sculpture is limited to other parts of France such as the Périgord, where the limestone could be shaped in this way. One difference between the two techniques is that clay figures are found (or have only survived) inside the dark depths of caves; sculptures are always in rock shelters or the front, illuminated parts of caves. The reason for this discrepancy remains obscure, since the artists were clearly capable of working for long periods far inside caves, and, as we shall see, had light-sources adequate to the task.

Among the earliest examples of parietal stone-carving are the bas-reliefs of Laussel (Dordogne), including the well-known woman holding a 'horn' in her right hand (p. 113); this specimen was originally carved on a large block of 4 cubic metres in front of the rock shelter, which was not, therefore, a 'movable wall';[26] the figure includes a number of techniques —

her left side is merely deeply engraved, like the Aurignacian blocks mentioned earlier, while her right is in 'champlevé', since material has been removed next to it which brings her body into relief. The figure seems to have been regularized and perhaps even polished; finally, fine engraving was used for her fingers and for the marks on the horn.[27]

The carvings of Fourneau du Diable (Dordogne), on the other hand, are on a block

Fig. 8.15 Detail of one of the Tuc d'Audoubert bison, showing how the clay was worked with fingers and implements.

Fig. 8.16 & 8.17 The front and right flank of the clay bear of Montespan (Haute Garonne), about 1.1 m in length. Probably Magdalenian. The view of the 'dark side' (near the cave-wall) of the bear shows the left forepaw and shoulder, and the finely modelled belly and haunch. Note the traces of red pigment nearby.

Fig. 8.19 OPPOSITE: 'Venus with horn' from Laussel (Dordogne). Probably Gravettian. Height: 44 cm. Although now detached, it should be classed as parietal art since originally it was carved on a block of 4 cubic metres.

Fig. 8.18 The sculpted bas-relief horse head of Comarque (Dordogne), 70 cm long. Probably Magdalenian.

of only half a cubic metre, weighing less than a ton, and had therefore been placed precisely where the artist wanted them,[28] while the blocks of Roc de Sers (Charente) had originally been arranged in a semi-circle at the back of the rock shelter, but subsequently some had rolled down the talus slope.

One of the best examples of haut-relief is the frieze at Cap Blanc (Dordogne) (Fig. 8.20), where figures reach a depth of 30 cm in places. By and large, we do not know precisely what implements were used for work of this kind, although it is a safe bet that they were pretty robust percussion tools (hammers, chisels, saws and abrasives), judging from the traces of impact still visible on and around many sculpted figures. Few of the sites have yielded plausible tools in their fill, but at Laussel some Aurignacian 'picks' have battered ends; the

picks of Angles-sur-l'Anglin were mentioned earlier (see above, p. 63), and worn flint tools including burins and scrapers were recovered from the Magdalenian of Cap Blanc.[29]

The engraving on the Laussel female has echoes in incised lines on the sculpted blocks at Roc de Sers; far more common, however, are traces of pigment on these carved figures, just like on portable carvings of various types. The Laussel woman has red ochre on her whole body, especially visible on the breasts and abdomen, as well as in the groove around her, and some of the other Laussel figures have traces too. In the magnificent sculpted frieze of Angles-sur-l'Anglin (Vienne, p. 2) many figures (including the male portrait) have traces of paint, and vestiges of red ochre have also been found on the fish in the Abri du Poisson, and on the friezes of La Chaire à Calvin (Charente) and Cap Blanc — indeed, at the latter the ochre seems to cover large areas, both on the figures and around them, but can now only be seen under very intense light or when the wall is humid.[30]

Pigment was also used in combination with engraving, as at Pair-non-Pair, where there are many traces of ochre inside the incisions as well as on the areas in relief; but in many sites it is difficult to decide whether the engravings were 'redone' with colour, or whether they were originally done on painted surfaces, thus trapping some pigment in the grooves.

PIGMENTS
Identification and preparation

As soon as Palaeolithic parietal art began to be accepted, at the end of the last century, analyses were undertaken to identify the pigments used: Rivière, the pioneer of La Mouthe, was the first

to take samples, in 1898, and had them analysed by Adolphe Carnot, who found that they were iron oxide, with no trace of lead or mercury, and had been applied in the form of a watercolour which had penetrated the rock; later, samples from Gargas and Marsoulas, submitted by Regnault to C. Fabre, were also identified as an iron oxide paint.[31] H. Moissan did analyses for Font de Gaume in 1902, with similar results.

Since that time, the red pigment on cave walls has consistently proved to be iron oxide (haematite, or red ochre), not only in Western Europe but also in remote caves like Kapova, Ignatiev and Cuciulat.[32] Black, on the other hand, was until recently usually assumed to be manganese dioxide, based on the earliest analyses. Nevertheless, a few pioneers did mention the possibility of charcoal having been used: in 1908, Cartailhac and Breuil suggested that it had been used in Niaux, mixed with manganese, while in 1929 the Abbé Amédée Lemozi described a mammoth in Pech Merle as having been drawn with charcoal, since he had found a fragment of charcoal at the end of its trunk.[33] However, it was relatively recently that the use of charcoal was finally proved: first at Altamira, where in 1977 J. Martí took samples of paint from the ceiling, and found that two types of

manganese, as well as pine or juniper charcoal, had been used for black. The following year, J. M. Cabrera Garrido analysed paint samples from the cave's excavations and ceiling with X-ray diffraction, and found the same pigments in both, with some of the black being animal charcoal (i.e. burnt powder of horn, bone or teeth).[34]

Shortly afterwards, analysis of pigment in the Salon Noir, Niaux, revealed that wood charcoal was used for figures there: the plant cells are visible under the microscope, while the Scanning Electron Microscope has shown that it was a resinous wood similar to juniper; subsequently, charcoal was also detected in paint at Las Monedas,[35] and indeed has been found to be almost ubiquitous at sites where pigment analysis has been carried out. As we saw in Chapter 5, this has proved a godsend for obtaining direct dates from figures in many caves. Wood charcoal has even been detected in one of the great bulls of Lascaux and may therefore help solve the problem of that cave's chronology (see Chapter 5).[36]

White was used far more rarely; kaolinite was brought into the Abri Pataud and may have been used as a white pigment, or as an 'extending pigment' mixed with other colours;[37] at Altamira, some white paste found in a shell

Fig. 8.20 Photomontage of the sculpted frieze at Cap Blanc (Dordogne), showing the series of horses, some facing left and others right. Magdalenian. Total length: c. 8 m. The central horse is 2.15 m long.

proved to be a mix of mica and illite,[38] and one nodule from Lascaux is also thought to be kaolin.

It has often been assumed that the Palaeolithic artists also used blues and greens extracted from plants, but that these have not survived on the walls – however, it requires complex chemical treatments developed in modern times to extract these colours, and the plants concerned grow in warm or tropical areas; so, apart from the charcoal mentioned above and the possibility of a little woad, plants were probably not used for pigment.[39] Similarly, copper and lapis-lazuli were unknown as sources. Consequently, Palaeolithic artists had only five basic colours to work with: red, yellow, brown, black and white, but the latter is so rare that the choice was really four.

The main mineral colouring materials were usually readily available, either casually collected as nodules or exploited from known sources (or even mined occasionally, as we have seen in Africa and Hungary, p. 26); lumps of them have been found in abundance in some sites, and tens of kilograms of them, bearing traces of how they were used, lie unstudied in museums.

At Tito Bustillo, in front of the great Magdalenian painted frieze, there were colouring materials, some of them still in the barnacle shells in which they were mixed. In the far earlier cave site of Arcy-sur-Cure (Yonne), dating to the very beginning of the Upper Palaeolithic, there was a limestone plate which had clearly been used for carrying ochre.

However, the greatest assemblage of evidence of this kind, and the best studied so far, is that from Lascaux, where 158 mineral fragments were found in various parts of the cave, together with crude 'mortars' and 'pestles', stained with pigment, and naturally hollowed stones still containing small amounts of powdered pigment.[40] There are scratches and traces of use-wear on thirty-one of the mineral lumps. Black dominates (105), followed by yellows (twenty-six), reds (twenty-four) and white (three). It was found that there were sources of ochre and of manganese dioxide within 500 m and 5 km of the cave respectively.

The shades vary considerably: the colour of ochre is modified by heat, and Palaeolithic people clearly knew this, since even in the Châtelperronian of Arcy there were fragments at different stages of oxidation still in the hearths. Yellow ochre, when heated beyond 250°C, passes through different shades of red as it oxidizes into haematite.[41]

A further stage in pigment preparation, in Lascaux at least, involved the mixing of different powdered minerals, since unmixed pigments are rare here. Chemical analysis of ten samples produced some surprising results: for instance, one pigment contained calcium phosphate, a substance obtained by heating animal bone to 400°C. It was then mixed with calcite, and heated again to 1000°C, thus transforming the mix into tetracalcite phosphate.[42] A white pigment was found to comprise porcelain clay (10 per cent), powdered quartz (20 per cent) and powdered calcite (70 per cent); while a black pigment found in a mortar was charcoal (65 per cent) mixed with iron-rich clay (25 per cent) and a few other minerals including powdered quartz. Clearly, therefore, the Palaeolithic artists were experimenting and combining their raw materials in different ways. Analysis is now proceeding further, and the Scanning Electron Microscope is being used on tiny chips of some Lascaux paintings to study the microstructure of the pigments, and compare them with natural samples in an effort to assess the nature and extent of the prehistoric processing.[43]

Whereas analysis of pigments at the turn of the century could only be done in relatively crude ways which depended on chemical reaction, today, only a tiny amount of pigment is required, and the new techniques that can be applied to it – such as Scanning Electron Microscopy, X-ray diffraction and proton-induced X-ray emission – produce a highly detailed analysis of the paint's content. A new phase of analyses in France was begun by Michel Lorblanchet in the Quercy[44] where it was found, for example, that the red figures of Cougnac were done with the same paint which matched a reserve of ochre on the cave floor (Fig. 8.21), and that it had been obtained locally. Brown motifs were added later, and some figures were retouched with this colour, with subsequent repainting or retouching in black – some in pine charcoal, some probably in bone charcoal. At Pech Merle, on the other hand, while some black is charcoal, two hand stencils are done with manganese dioxide, as was a horse in Marcenac. Since 1993, optical analysis has also been tried in the Quercy caves, using spectrometry with infra-red and ultra-violet, in order to differentiate pigments without even removing samples.

A similar programme of analyses was also undertaken by Jean Clottes and his team in the the French Pyrenees,[45] and subsequently elsewhere; it has also been applied to residues of paint on portable art objects.[46] For example, in Niaux, in the Ariège, a number of different mixtures or 'recipes' of pigments and minerals have been detected – recipe 'B', in which biotite was used as an extender; 'F' in which feldspar was used, and 'T' where talc was employed. Extenders would have been added to pigments not only to make paint go further,

but also for other advantages: for example, adding biotite makes red paint spread easily when wet, produces a darker colour than pure red ochre, improves adhesion to the wall, and stops the paint cracking as it dries. There is still some debate as to whether the Niaux recipes are indeed fully intentional mixes, natural mixes (where the extender may be naturally present in clays with the pigments), or caused either by retouching of existing drawings, or, at least in part, by minerals becoming detached from the grinding instruments used.[47] The Quercy team believes the composition of its pigments to be entirely natural, because the mineral components discovered exist in the same proportions in local sediments. The Pyrenean team, on the other hand, insist that their recipes contain associations of components which it is quite impossible to find in nature. The claims are not necessarily contra-

Fig. 8.21 Red pigment among broken stalagmites in the cave of Cougnac (Lot).

dictory – artists in the Quercy may have been content to use natural pigments, whereas those in the Pyrenees may have concocted new ones.

Be that as it may, in Niaux's famous Salon Noir, most of the animal figures were first sketched in charcoal, with manganese paint added on top using recipe B. This was clearly a special place where the figures were carefully planned, whereas the other figures in the cave and in the Réseau Clastres (the latter also done with B) were done without preliminary sketches. It has proved possible to date the recipes, not only by the direct dating of figures at Niaux (see Chapter 5 and Table, p. 75), but also because the occupation site of La Vache, directly across the valley from the cave, has that same extender used with red and black paint on nine pieces of portable art on bones from layers dating to 12,850–11,650 years ago, the Upper Magdalenian. Recipe F, on the other hand, has been detected on Middle Magdalenian portable art objects from Le Mas d'Azil and Enlène, as well as in the famous 'sorcerer' of Trois Frères. In other words, the recipes seem to have a certain chronological value, confirmed to some extent by the differing radiocarbon results from figures in Niaux – the Salon Noir is mostly Late Magdalenian. However, claviform signs in Niaux seem to contain both B and F, so either the recipes are not restricted to one period, or claviforms span a considerable length of time!

Analysis of the red and black pigments in Russia's Ignatiev Cave by X-ray diffraction and infra-red spectrometry has revealed that they sometimes contain gypsum, feldspar and clay minerals which are not naturally present in the cave, and which may thus have been added as extenders or to produce different shades. The black pigment is not manganese but a mixture of pine charcoal, gypsum and calcite.[48]

A different problem is the binding medium used. In the past it was often assumed that some form of fatty animal product was used for the purpose; however, a series of 205 experiments in two caves was carried out by Claude Couraud, involving a variety of pigments and binding substances (including fish glue, arabic gum, gelatin, egg white, bovine blood, and urine), and a range of wall-types and degrees of humidity. Observation of the results and deteriorations over three years led him to the con-clusion that fatty and organic substances were totally unsuitable binding agents, and fail to adhere well to humid walls. In fact, the only substance which seemed to be good at fixing and preserving the pigments on the rock-face was water – especially cave water, which is rich in calcium carbonate, and which was probably used at Lascaux.[49] It was also found that pigments adhered better if they had been finely ground.

In 1978, Cabrera Garrido's analyses at Altamira led him to suggest the possible use here of powdered fossil amber as a binder.[50] However, in some Ariège caves, recent analyses of paints, using gas chromatography and mass spectrometry, have detected traces of what are thought to be oils of animal or plant origin, presumably used as a binder.[51] On portable art from Enlène and in parietal figures at Trois Frères it seems to be a plant oil, whereas parietal figures at Fontanet seem to contain an animal oil. Other sites such as the Réseau Clastres have no trace of any binder, as at Lascaux.

Application

The simplest way to apply pigment to walls was with fingers, and this was certainly done in some caves: for example, the animal figures of La Baume-Latrone (Gard) were clearly painted with fingers, as were the *serpentiformes* of La Pileta (Málaga), versions in red clay or black pigment of the simple finger-markings or 'macaronis' which also abound in these caves.[52] La Pileta and other caves such as Cougnac also have series of double finger-marks in pigment or charcoal on or around animal figures.[53]

Experiments suggest that painting by finger tends to produce poor results. Nevertheless, when Lascaux II was being made, it became apparent to the modern artists that some painted lines had been done with the finger in the original figures.

Normally, however, paint was applied with some sort of tool, and again, since none has survived, it is experimentation which suggests what was used, although one must always remember that the original artists had years of experience and familiarity with their materials, and acquired 'knacks' which cannot be revealed by modern short-term exercises. Some lumps of pigment are in the form of 'crayons' (there are nineteen with use-wear at Lascaux alone), and these may have been used to sketch outlines; however, they do not mark the rock well (and may even damage it), and only really work on humid walls. Some were definitely used as crayons, as is shown by the presence of calcite in their striations.[54] However, they wear down very fast – in tests, 1 centimetre of length produced a mark of only 15 cm![55] Consequently, lumps of pigment must have been used mostly as sources of powder, which explains the traces of scraping on many of them – and flints with ochre on their edge have been found at Lascaux, Tito Bustillo and other sites, close to the minerals.

As mentioned earlier, mortars and other 'vessels' have been found at Lascaux containing crushed pigment, and similar objects with traces of colour inside are known from Altamira (including some vertebrae), Villars (concave fragments of calcite), Tito Bustillo (barnacle shells) and elsewhere. What we do not always know is whether the powder was made into a paste or a liquid – doubtless it depended on the circumstances and the desired effect, whether a dot, an outline or a flat-wash. In some cases, the paint was probably in liquid form, since it is diffuse, or it has run – for example on some figures of the Frise Noire at Pech Merle – although trickling may also have been caused by wet walls at times.

Whether in paste or liquid form, how was the paint applied? Experiments with a variety of materials produced their best results – i.e. solid, precise and regular marks – with animal-hair brushes (especially badger); brushes of crushed or chewed vegetable fibre were next best. On the other hand, brushes of human hair

Fig. 8.22 Three hinds drawn with dots at Covalanas (Santander). Total length: c. 2.5 m. The central hind is 60 cm long.

were too supple and fragile, while pads of bison fur transferred colour to the rock efficiently, but quickly became flaccid and unusable.[56]

Nevertheless, there are cases where some sort of pad must have been used because other methods were unsuitable. Some surfaces at Lascaux have a cauliflower-like covering (the crystalline calcite coating of the Hall of the Bulls made it unsuitable for engravings), and once figures had been outlined on them – presumably by crayon or brush – the infill was done with hundreds of circular, diffuse spots that join up and give the impression of an even wash. They appear to have been done with a pad and dampened powder: for example, a red cow in the 'Diverticule Axial' has been filled by about 200 such patches. Yet these figures have sharp edges, which suggest that a hide or some such object was placed along the desired line.[57] This pad-and-paste method would certainly have been easier to use on ceilings and inclined walls than brushes and liquid paint. In some lines of big dots in various parts of Lascaux, one can see a change after two or three similar ones, which suggests that the pad or brush held enough pigment for only a few at a time.

In some cases, such as the Covalanas and Arenaza deer (outline figures made of dots) or the famous bison of Marsoulas (p. 21), composed almost entirely of of red dots, each 2 or 3 cm in diameter,[58] it is hard to decide whether fingers, thumbs, pads or brushes were responsible. In any case, this technique predates the French Impressionists by 15,000 years! As mentioned earlier (see p. 53), use of infra-red film has revealed that some apparently unbroken lines, such as in 'tectiforms' at Bernifal, were originally made with dots, whose pigment ran together because the earthy red pigment was soluble in humid conditions and spread (unlike manganese dioxide, which is insoluble and thus does not run and fade). At Combel (Pech Merle), infra-red analysis showed that a group of red dots had been made with a pad dipped in ochre, whereas an adjacent set had been applied by spraying.[59]

Hand stencils

A technique of spraying paint was also clearly employed for most of the hand stencils which are so numerous in certain caves (occasionally with forearm included, as at Maltravieso and Fuente del Salín,(Fig. 8.23);[60] the comparatively rare positive hand-prints (for example at Chauvet, Altamira, Santián, La Pasiega and Fuente del Salín) were simply made by applying a paint-covered palm to the wall. Very occasionally, hand stencils seem to have been made with a broad brush or fur pad, as in the case of two black specimens at Gargas which have a regular area of paint around them; a white specimen in the same cave apparently had white material crushed around the fingers; but most have a 'diffused halo' which results from spraying.[61] There are two possible methods: through a tube, or directly from the mouth; and was the pigment in dry or liquid form? Once again, observation and experiments have helped to clarify the situation.

When dry, the powder can only have been applied through a tube; the site of Les Cottés (Vienne) contained a bone tube with powdered red pigment inside it, which seems to indicate that this technique was indeed used at times. If dry powder was projected, then a humid wall was required, or there would have been no adhesion; and in any case, dry pigment leaves 'fallout' beneath the hand, which does not exist on the Palaeolithic stencils.

On the other hand, if liquid paint is used, blowing it through a tube concentrates it too much; experiments show that spraying liquid paint from the mouth, about 7 to 10 cm from the wall, is not only the easiest method, but also the one which produces results which best resemble the original stencils;[62] pursing the lips slightly projects a spray of fine droplets which forms the required halo with diffuse edges. On average it takes about 3 grams of pigment and between thirty and forty-five minutes to do each hand in this way, whereas the 'dry method' can sometimes be quicker, but uses 9 or 10 grams. On the other hand, filling one's mouth with paint is hardly pleasant, and it is possible that a combination of the two techniques was used: i.e. placing a hollow tube upright in a paint-holder, and then blowing across its top through a second tube forces the paint up and out in a fine mist; or perhaps blowing into a bent reed so that a spray emerges from a split at its base.[63]

We have already seen that most engravings

Fig. 8.23 Hand and forearm stencils from Fuente del Salín (Santander). (After Bohigas et al.)

were probably done by right-handed artists; and it has traditionally been assumed that because most stencils are of left hands (at Gargas, 136 left have been identified, and only twenty-two right), this too denotes predominantly right-handed people. Although this is likely to be correct, the hands are poor evidence, partly because the painting may have been done by mouth, not by the dominant hand, and partly because some hands may have been stencilled palm-upward! However, one should also note that in the very rare Ice Age depictions of people holding objects – particularly those of Laussel, such as the 'Venus with horn' (p. 113) – it is usually the right hand which is doing the holding.

Unfortunately, the stencil experiments have not settled the old debate about the apparently

mutilated or deformed hands, most abundant at Gargas but also known in other caves such as Tibiran, Cosquer, Fuente del Trucho (interestingly, located just across the Pyrenees from Gargas!), possibly Erberua and Fuente del Salín, etc. There are three basic explanations for the missing phalanges on these hands: either the fingers were bent (probably palm-upward) when the stencils were done, or the phalanges were actually missing, through ritual mutilation of some kind or through pathological conditions.

Many prehistorians have thought the fingers were deliberately bent, for one reason or another, perhaps as a kind of language of gestures or signals,[64] and some experimenters claim to have duplicated all the 'mutilations' at Gargas by bending their fingers, and point out

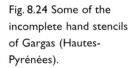

Fig. 8.24 Some of the incomplete hand stencils of Gargas (Hautes-Pyrénées).

that there are stencils with clearly bent thumbs at Gargas and Pech Merle.[65] Other experimenters, however, consistently found that placing the hand palm-upward and bending the fingers caused pigment to infiltrate behind it,[66] which, as we have seen, was not the case in the originals. They therefore support the view that the phalanges were actually missing.

If they were, then ritual mutilation may have been responsible; but at Gargas the cause is more likely to be pathological. Investigation by Ali Sahly showed that the stencils here represented people of all ages – adults, women and/or youths, and children including infants; of the 124 best preserved, only ten are free of abnormality; the various combinations of missing phalanges are only on the fingers, not the thumb. Almost 30 per cent had the last two phalanges of all four fingers gone. He claimed that repetitions of the same hands were detectable, so that fewer than twenty people produced 231 prints between them.

His conclusion was that conditions such as frostbite, gangrene and Reynaud's disease (which only attacks the fingers and very rarely the thumb) were responsible.[67] However, a more recent medical analysis of the case has concluded that – if the phalanges were indeed missing, and as we have seen, many researchers contest this point – the phenomenon cannot be accounted for by any known pathological cause, and so must be due to deliberate mutilation.[68] So the debate rolls on, as it has continued for decades, without achieving any consensus of opinion whatsoever.

Some painting techniques

Strictly speaking, figures made by finger or in simple outline should simply be seen as coloured drawings; the term painting might better be limited to those cases with infill of various kinds, and, as we have seen, painted infill dates back to at least the Aurignacian block from Abri Blanchard, and perhaps beyond, as does the practice of colouring wide areas of rock wall. The Chauvet Cave, also of the Aurignacian, likewise displays a great deal of infill and very subtle shading, where the paint was spread by hand or with a tool to depict relief or the nuances of an animal's coat. The well-known bichrome and polychrome (i.e.

three-coloured) figures of the Magdalenian arrive relatively late, and are rare in Palaeolithic iconography in comparison with engravings and outline drawings, being known only at Altamira, Tito Bustillo, Ekain, Font de Gaume and Labastide.

Engravings and outline drawings were undoubtedly quick and easy things to do for an artist of experience and talent – they required far less time and preparation than painting. In an already famous experiment, Michel Lorblanchet memorized every mark in the 'Frise Noire' of Pech Merle (there are twenty-five animals on this panel measuring 7 m by 2.5 m; see Fig. 8.25), and then reproduced the whole thing on an equally smooth panel of similar dimensions in another cave, lit by a lamp in his left hand. Each figure took an average of one to four minutes (compare with the figures for hand stencils, above), so that the whole frieze required about an hour, including initial sketching with a stick.[69]

More recently, Lorblanchet has drawn on knowledge of paint-spitting techniques of the Aborigines, acquired during a period in Australia, to produce accurate reproductions of various Palaeolithic figures – not only hand stencils but also the spotted horse panel of Pech Merle (in a total of thirty-two hours) and a Lascaux-type bichrome horse, which were clearly originally made by this method. He uses his hands to direct and guide the paint, create sharp or blurred edges, etc. He has found that blowing through a bent tube may work for hand stencils, but for large panels only spitting from the mouth is efficient enough. For red, he uses iron oxide, but for black he only spits charcoal, since manganese is highly toxic.[70]

Similarly, Claude Couraud has reproduced a Lascaux horse and a Niaux bison both full-size on paper, and at reduced scale in a cave: the horse required thirty minutes of preparation and another thirty to do on paper, but only took thirty minutes in the cave; the bison took twenty minutes on paper, and only ten in the cave.[71] Therefore, even quite detailed figures seem to have needed little time for their execution.

Experiments have also been done to investigate problems of superimposition of different paints. At Lascaux, it was long uncertain whether the red cows were on top of the black

Fig. 8.25 Photomontage of the 'black frieze' in the cave of Pech Merle (Lot), showing mammoths, bison, aurochs, horse and red dots. Probably Solutrean. Total size: c. 7 m by 2.5 m.

bulls, or vice versa, because the two pigments had mixed; some scholars thought the cows were done first, but infra-red pictures suggested the opposite. Using samples of similar pigment, it was found that red on top of black did not mix, but black on top of red did so – and so the bulls were clearly painted after the cows.[72] As mentioned earlier (see p. 53), infra-red with special cut-off filters has also been used by Marshack, Vertut, Lorblanchet and others to determine the use of different ochres and mixes of ochre within a single composition or panel, where all looks the same to the naked eye. Similarly, ultra-violet light helps one to separate out over-painting and over-engraving more clearly. However, it must be stressed that these

techniques cannot be applied universally or at random, but need to be developed for each particular problem or cave.[73]

Combinations of painting and engraving are quite common and varied in parietal art; at Trois Frères, clearly visible traces show that a panel was scraped extensively before a large red 'claviform' was painted on it (p. 54);[74] simi-larly, several panels at Altxerri were scraped extensively to produce a light background for painted figures,[75] while some of the black draw-ings in Chauvet Cave are surrounded by lighter areas of wall which seem to have been rubbed to produce increased contrast.[76] In some cases, an engraved figure is filled in with ochre; else-where, as in the spotted bison of Marsoulas, a

Fig. 8.26 Michel Lorblanchet spitting dots (through a hole in a piece of leather) for his re-creation of the spotted horse panel of Pech Merle.

Fig. 8.27 Michel Lorblanchet with his re-creation of a Lascaux bichrome horse, produced by spitting charcoal and red ochre.

few incisions delimit certain parts of a painted figure such as the head and leg; while at Tito Bustillo, some of the bichrome horses on the main panel are silhouetted by wide areas of engraving which make them stand out from the darker background and from earlier figures.[77] There are even painted animals which are redone or 'highlighted' with engraving, as at Trois Frères, Lascaux, Le Portel, or Santimamiñe, for example.

As mentioned earlier, just as perfectly proportioned figures could be engraved around a cylinder of antler, so the 'falling horse' of Lascaux is painted around a rock (p. 176): the artist could never see the whole animal, yet its proportions remain sound. At Pergouset (Lot), a very detailed horse-head was even engraved in a narrow fissure at arm's length by an artist whose head could not possibly have been inserted there, so that it was, in effect, drawn blind.[78] It is possible that an even more sophisticated technique was occasionally used in the caves — namely, anamorphosis, or deliberate distortion. Claims have been made that four red cows at Lascaux and a black horse at Tito Bustillo were purposely deformed in this way so as to look normal from ground level or from the side;[79] given the other accomplishments of Palaeolithic artists, there is no reason to suppose that touches of this sort were beyond their capabilities. On the other hand, in the Réseau Clastres, the horse only looks normal from the sitting position in which it was drawn, whereas from a distance it looks extended and deformed.[80]

Just as with portable art, there are clear regional differences in artistic technique (quite apart from style and content): for instance, in the Pyrenees, there are roughly equal numbers of engraved and painted figures in Le Mas d'Azil, but engraving dominates in the decorated caves to the west, and painting in those to the east, although there are exceptions (Tibiran has more painting than engraving, while Fontanet and Massat are the reverse).[81] The regional difference in distribution between clay and sculpture has already been mentioned; perhaps linked to this is the fact that art on cave floors is largely restricted to the Pyrenees, but unlike Cantabria, the Périgord and Cosquer, this region has no decorated ceilings – there are occasional engraved or painted figures on ceilings (engravings at Montespan, for example), but there is nothing remotely like the Lascaux passage, or the great painted ceiling of Altamira which Joseph Déchelette baptised the 'Sistine Chapel of Rock Art'.[82]

The Altamira ceiling was within easy reach – indeed, it was so low that the artists could not have seen its whole surface at once; but how did they manage to paint on high ceilings? Occasionally, as with a sign in Lascaux, it might have been accomplished with a brush on the end of a long pole. At La Griega (Segovia), it is thought that an engraving on a ceiling 3 m up could have been done by a person straddling the passage with a foot on ledges on either side.[83] Elsewhere, as at Roucadour (Lot) where the engravings are 6 m above ground level, or at Baume-Latrone where there are broad sweeps of painting on a ceiling over 3 m high, it is probable that the floor has sunk down since Palaeolithic times. This is also the case in Chauvet Cave, where some engravings were probably made by a standing adult, but after subsidence they are now 5 m above the ground.[84] But for truly monumental work such as the great Labastide horse on a rock 4 m high, the two black mammoths painted 5 m up on the vault of Bernifal, or much of Lascaux's decoration, it is clear that ladders (perhaps tree-trunks with branch-stumps as rungs) or scaffolding must have been used. Abundant wood residues found in Lascaux may come from these constructions (some are from big oaks) as well as from torches and so forth, but the clearest evidence is the series of about twenty sockets cut into the rock on both sides of the 'Diverticule Axial', about 2 m above the floor; these were packed with clay. Holes about 10 cm deep in this clay suggest that branches long enough to span the passage were fitted into the sockets, and cemented into place with the clay. This series of solid joists could then support a platform providing easy access to the upper walls and ceiling.[85]

Shedding light on the subject

Portable art may all have been done in daylight, like the art on blocks and the decoration of rock shelters. But the work inside caves required a reliable source of light. In a few cases, a hearth at the foot of the decorated panel may have sufficed: at La Tête-du-Lion (Ardèche) a concentration of pine charcoal only 1.2 m from the decorated wall may be a *foyer d'éclairage* of this type, though it could simply be the remains of torches.[86] Certainly, portable light was necessary in most caves. What did they have at their disposal?

Despite Rivière's discovery at La Mouthe, the very existence of Palaeolithic lamps was not

Fig. 8.28 The red sandstone lamp found in a Magdalenian hearth 17 m inside the cave of La Mouthe (Dordogne) in 1899 - the first lamp found, or at least recognized, from the Ice Age, it measures 171 by 120 mm. Residues inside it were analysed in 1900 and found to be an animal-fat fuel. Note the fine engraving of an ibex on the exterior.

fully accepted until 1902, the same year as cave art was finally validated. This was no coincidence: once the pictures deep inside caves, often hundreds of metres from daylight, were seen to be authentic, it was obvious that the artists must have had some means of illumination.

As often happens, the establishment of a previously rejected notion led scholars to the opposite extreme, so that all hollow objects, and lots of flat ones, were identified as lamps indiscriminately! A critical study of the hundreds of objects claimed to be lamps has resulted in only 302 being considered as possibilities, of which a mere eighty-five are definite and thirty-one others probable. This is a very poor total when seen against the 25,000 years of Upper Palaeolithic life and the hundreds of decorated caves, especially when it is noted that 70 per cent of them come from open-air sites, rock shelters or shallow caves.[87] It is certain that a different method was usually employed in deep caves – probably burning torches, which have left little or no trace other than a few fragments of charcoal or black marks on walls.

In one experiment undertaken in 1978, French researcher Claude Andrieux went barefoot to the farthest depths of the cave of Niaux (2 km into the mountain), carrying six pine torches, each 80 cm long and impregnated with

beeswax, and a reserve of 200 g of soft wax. One torch was lit before entering. Walking slowly, he did the 4-km round trip in three hours, before he ran out of illumination. He found that each torch gave off an orangey-yellow flame which provided quite enough light for walking.[88]

In the Grande Grotte d'Arcy-sur-Cure, an interesting item found recently is a vertical stone lamp which was either held in the hand or stuck into the ground.[89] This specimen is rather crude, but there are a few beautifully carved lamps, and some such as those from La Mouthe or Lascaux even have engravings on them. Many of them are of a red sandstone from the Corrèze region of France, which thus seems to have been a centre of production.

Combustion residues in some specimens have been subjected to analysis, which indicated that they were fatty acids of animal origin; while remains of resinous wood or of non-woody material clearly come from the wicks (residues in the carved Lascaux specimen proved to be juniper). Experiments have been carried out with replica lamps of different types, different fuels (cow lard, horse grease, deer marrow, seal fat), and a variety of wicks (lichen, moss, birch bark, juniper wood, pine needles, dried mushrooms, and kindling). The results led to a number of interesting insights, which were confirmed by study of the lamps used by Eskimos.[90]

First, a good fuel needs to be fluid, and easy to light. The initial melting of the fat needs to absorb the wick material which, by burning, continues to melt the fuel. This cycle can continue for hours providing both wick and fuel are replenished from time to time: one estimate is that 500 grams of fat will keep a lamp going for twenty-four hours.[91] As mentioned earlier (see p. 19), animal-fat fuel produces no soot.

There are two basic types of lamp: the open-circuit model, in which the fuel is evacuated as it melts; and the closed-circuit, where the fuel is kept in a cavity. The open type seems very rare in the Ice Age, although many simple slabs of stone may have been used like this (about 130 limestone slabs at Lascaux were interpreted in this way, although many are now lost, and only about thirty-six seem likely

Fig. 8.29 An experiment with a replica of an Ice-Age stone lamp, burning animal fat.

lamps); most recognizable Palaeolithic lamps are of closed-circuit type. In Eskimo communities, the open types are for occasional use, while the closed types – in which far more work has been invested – are used daily.

But how bright are these lamps? The answer, surprisingly, is that they are pretty dim, even in comparison with a modern candle. The power of the light given off depends on the quality and quantity of fuel; the flame is usually unstable and trembling. Experiments with a stone lamp using horse fat produced a flame of one-sixth the power of a candle, according to measurements with a photometer.[92]

With such limited radiance, it would have been necessary for Ice Age people to use several at once, or to resort to burning torches. In some deep caves, such as Labastide, there are also the remains of hearths of various kinds, which would certainly have provided strong light at times.

We, of course, are spoiled by artificial light, and are no longer accustomed to dimmer sources. But in fact it is surprising how much can be seen by the light of a single candle, and a large cave chamber could be lit adequately with two or three. It has been found that with only one lamp one can move around a cave, read, and even sew if one is close enough to the light – the eye cannot really tell that the flame is weaker than a candle.

One important factor is that flickering flames of this type have the effect – no doubt observed and perhaps exploited in the Palaeolithic – of making the depicted animals seem to move, an eerie experience. For the really big figures, over 2 metres in length, more than one lamp would have been needed to see the whole thing at once. The lamps would also have affected colour perception – they give off a warm yellowish light which makes yellow look orangey, and makes red pale or brown. The eye adapts well to red in this light, and this may explain the frequent use of red dots and signs at various points in the caves – by lamp-light they were readily visible signals.

We still have much to learn about how Ice Age people – including many children (see p. 10) – managed to make their occasional forays into the remote depths of some caves. Even armed with bundles of torches, or lamps with a plentiful supply of fat and wicks, it seems a very risky thing to do. However, if archaeology has taught us anything, it is that we should never underestimate the capabilities of our ancestors. We must not lose sight of the fact that caves and darkness formed part of their environment, and they clearly knew how to cope with them very successfully, even though they spent most of their lives out of doors (Chapter 9).

Conclusion: portable versus parietal

In the past, portable art was sometimes seen merely as a 'sketch-book', a repetition or rehearsal in miniature of parietal work. Today, however, we see them as parallel and equally important artforms, clearly part of the same world, but perhaps quite different in function.

As we have seen, one basic difference between the two is that portable objects have been moved around, whereas parietal figures are precisely where the artist wanted them, whether on the floor, the walls or the ceiling of the cave. Another difference is that portable art, by definition, is composed of small, light objects, and the figures and motifs drawn on them are of a corresponding scale; as we have seen, the size and shape of the stones, bones and antlers either determined the form of the design (lines of figures are common) or were chosen according to requirements.

The same applies to parietal art (especially where rock-shape was incorporated into a figure), but the height and width of the available surfaces meant freedom from many of these constraints; hence, parietal figures could be arranged in a variety of ways (individuals, clusters, panels), and simple lines of animals are rare. Moreover, there was no restriction on size: figures range from the tiny (such as those of Fornols) to the enormous (such as the great Labastide horse, almost 2 metres long, for which scaffolding was required).[93] The biggest known engraving in Palaeolithic art is a horse in Pair-non-Pair, which is 2.16 m long and 1.58 m high.[94] The Hall of the Bulls at Lascaux has relatively small deer and horses, together with the huge bulls, over 5 m in length; but of course, not all of the figures on a panel of this type are necessarily of the same date, meaning or composition, and this is the fundamental problem addressed in Chapter 10.

9 ART IN THE OPEN AIR

We should not let the terms 'cave art' or 'cavemen' blind us to the fact that Ice Age people very rarely lived far inside caves but, rather, occupied rock shelters, cave mouths, and a variety of open-air habitations, from the tents of Western Europe to the mammoth-bone huts of Central and Eastern Europe.

Since they spent almost their whole lives in the great outdoors, it has always been assumed that they must also have produced art outside the caves, but that it has not survived the millennia of weathering and erosion. As mentioned earlier (p. 45), there have been occasional claims that open-air figures were of Palaeolithic age – for example, at Chichkino, Siberia, where hundreds of animal depictions over a distance of about 3 kilometres include a horse and a wild bovid considered by some local scholars to be characteristic of the end of the Ice Age – but few scholars have been prepared to take them

seriously. In the past fifteen years, however, a series of important finds in Western Europe has finally brought the proof that Palaeolithic people did produce art in the open air, and that it can survive in exceptional circumstances.[1]

Since one of the original arguments against the authenticity of both the Altamira ceiling and the painted pebbles of the Azilian was that parietal art could not possibly survive from such a remote age even inside a cave (see Chapter 1), it goes without saying that virtually nobody was prepared to entertain the possibility that Ice Age depictions in the open air could have survived; and certainly no Ice Age painting has survived outside – all the sites in question consist of figures engraved and pecked into rocks. Painting was almost certainly done in the open air too, and indeed some of these engravings may originally have been coloured, like Palaeolithic bas-reliefs and much portable art (see Chapters 7 and 8).

It is hard to say, at this early stage of the investigations, how open-air art relates to cave art – especially since there is still no consensus about the meaning of cave art after a century of study, with new caves like Cosquer and Chauvet constantly bringing surprises and modifying our knowledge. There are some similarities – the recognizable figures are primarily adult animals drawn in profile, with stylistic traits that correspond to those of known portable and cave art; they are dominated by horses and bovids; there are a few 'signs' and apparently abstract motifs; there are no scenes, virtually no humans, and no ground-lines; and the art seems to cluster in 'panels' (i.e. separate rocks). By virtue of its location, the open-air art appears inherently less mysterious or magical than the art in deep caves, but of course this is no guide to its meaning, since we know from ethnography that open-air art can be enormously powerful, religious or taboo, just as it

Fig. 9.1 Open-air engraved horse at Mazouco, Portugal. Length: 62 cm, height: 37.5 cm. Unfortunately, the figure has suffered much from chalking and painting since being discovered.

Fig. 9.2 & 9.3 The open-air engraved horse of Domingo García (Segovia), almost a metre in length. (Tracing after Moure Romanillo)

can also be simply decorative or narrative.

The first such finds occurred at a number of rock shelters and caves (including Murciélagos) in the Nalón Valley, Asturias, which have some eroded lines and animal figures deeply engraved in their exterior areas;[2] nevertheless, these engravings were still in or near shelters and cave mouths, rather like the sculpted friezes and carved blocks of France. The next discoveries, however, were at truly open-air sites.

In 1981, three animal figures, including a fine horse, 62 cm long and 37.5 cm high, were found on a rock-face on the right bank of the Albaguera, a little tributary of the river Douro, at Mazouco in north-east Portugal, at an altitude of 210 m above sea-level (Fig. 9.1);[3] they had survived thanks to their position which protected them from the elements. They were hammered out, though the marks were subsequently scored into continuous lines. Since its discovery, the horse has been badly damaged by chalking, scoring and painting. Its style has been attributed to the Early Magdalenian.

Around the same time, at Domingo García, Segovia (Spain), the figure of a horse, almost a metre in length, was found hammered into a rocky outcrop, 960 m above sea-level (Fig.

9.2).[4] In style, it resembles the engraved horses in the cave of La Griega, Segovia, in the same region. A crude engraving of a different style and period is superimposed on its outline. Since then, a closer examination of this rock and other rocks in the region (some of them up to 15 km away) as well as a removal of lichens has revealed at least 150 figures; most of them are fine engravings, and there are many

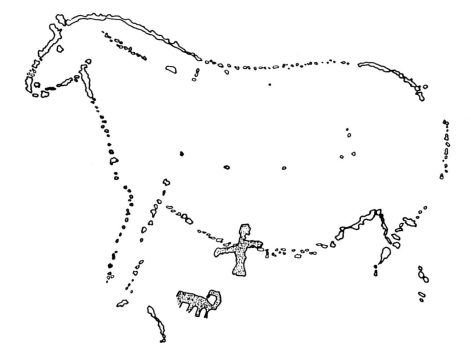

Fig. 9.4 Engraved deer head, facing right, at Domingo García; it is crossed by the horns of an ibex head, facing left. The latter measures 16 by 8 cm.

Fig. 9.5 The engraved rock on the mountain-side at Fornols-Haut (Pyrénées-Orientales), altitude 750 m, with Mont Canigou in the background. The rock is 2.3 m high and 3.9 m wide at the base.

hundreds of unidentified lines or fragments of figures.[5] Stylistically, most of the figures have been ascribed to the end of the Solutrean and to the Early Magdalenian. They are dominated by horses, but cervids and caprids are also well represented (Fig. 9.4). Bovids are rare.

A series of fine incisions was found at Fornols-Haut, Campôme, in the eastern French Pyrenees, on a huge block of schist located at an altitude of 750 m on a mountainside (Fig. 9.5).[6]

The rock has been greatly weathered by wind but, because the eastern face is sheltered, its engravings, although eroded, are clearly visible. The face is covered in engravings, drawn in all directions, and comprising about ten small animals – none complete – as well as signs and zigzags. The finest figures include the head of an isard, 7.5 cm high, and that of an ibex (Fig. 9.6). Stylistically they seem to belong to the Magdalenian.

More recently, two further sites of the same nature were found in Spain. In Almería, an area previously bereft of Palaeolithic art, a horse figure was discovered at Piedras Blancas.[7] Situated on an inclined block, at an altitude of 1400 m near the town of Escúllar, the horse is made with multiple deeply incised lines (Fig. 9.7). Stylistically it has been ascribed to the Final Gravettian or the Solutrean, through comparisons with engraved plaquettes from the cave of Parpalló. In view of developments at Domingo García, it is likely that more figures will eventually be found in the vicinity.

This is certainly what has occurred at Siega Verde, near Ciudad Rodrigo, Spain, about 60 km south of Mazouco. Here, in rocks along the left bank of the River Agueda (another tributary of the Douro), what were at first thought to be a few hammered-out figures, discovered in

1989, have turned out to be a minimum of 540 pecked and incised images, most of them located within a 1300-m stretch, though 75 per cent are concentrated within 400 m;[8] no fewer than half of the identifiable figures are horses, with bovids and cervids in second place. Like the other sites of the region, the style has been attributed to the Final Solutrean/Early Magdalenian.

The most recent discovery – first made public in November 1994 – occurred in the Côa Valley in north-east Portugal, yet another tributary of the Douro. Here, the pecked figures and engravings – there are at least 150 of them, on schist blocks spread over *c.* 17 km – are of horses, ibex and, especially, of aurochs; they measure from 15 cm to over 2 m in size. From their style, it has been estimated that they are probably Solutrean,[9] though in fact it is extremely likely that several phases of Palaeolithic and perhaps even Epipalaeolithic artistic activity are represented. They include some very fine images, particularly two large horses with overlapping heads (Fig. 9.9). The vast majority of known figures are the visible, pecked examples, but the area also contains great quantities of exceedingly fine, almost invisible engravings similar to those of Domingo García and Fornols.

Some of the Côa figures have been under 3 m of water for the past twelve years, since the Pocinho dam was built 2 km away; but fluctuations in water-level have exposed them several times during that period, so that they are still more or less accessible. A new hydro-electric

dam, that of Vila Nova de Foz Côa, was planned for the valley, and work began on it in September 1994. Had it been completed, the engravings would have been irrevocably lost under 100 m of water. An international campaign which lasted a year finally led to victory for the art; thanks to a change of government in Portugal in October 1995, work on the dam was stopped, and an archaeological park was created in the Côa Valley.

We only know of six open-air sites of this

Fig. 9.6 Ibex head engraved on the rock at Fornols-Haut. Length: *c.* 7.5 cm.

Fig. 9.7 The horse pecked into a rock at Piedras Blancas (Almería); it is about 70 cm long.

Fig. 9.8 Panel 29 at Siega Verde (Salamanca); the central aurochs figure is about 1 m long.

kind, and since three of them (Mazouco, Fornols and Piedras Blancas) comprise single figures or rock-faces, our extremely limited knowledge of this newly discovered phenomenon really rests only on the other three, which are currently being studied so that we can learn more about 'normal' Palaeolithic art.

But is this really Palaeolithic art? The six known sites have all been assigned to this period simply on the basis of the style of their pecked or engraved figures; on the other hand, the same is true of the vast majority of Palaeolithic cave art, and it is safe to say that if most of these figures had been found inside caves they would have been classed as Palaeolithic without hesitation. All parietal art is notoriously difficult to date, open-air examples especially so since they have no context whatsoever. However, there is nothing remotely similar to these figures in the Middle Stone Age and later art of eastern Spain, or in the schematic art of more recent periods. Every single European Ice Age art specialist who has

visited these sites or even seen photographs of them has unhesitatingly attributed some of the figures to the Upper Palaeolithic, giving age estimates varying from 20,000 to 10,000 years.[10]

In 1995, an attempt was made by the electricity company building the Côa dam to have a few of the valley's figures dated directly by four dating specialists using different methods, all of them highly experimental, such as AMS dating of organic deposits and the micro-erosion technique. Unfortunately, some of the panels chosen had already been much affected by latex moulding, chalking and other damage. The results were mixed, to say the least, with two specialists concluding that the art is indeed Palaeolithic in age, the third claiming they are probably only a few thousand years old, and the fourth insisting they date back only a couple of centuries![11] Although some see this as the last nail in the coffin of stylistic dating, others have pointed out the many flaws and uncertainties in the particular methods used, and dismissed the results as useless.[12]

Stylistic comparisons are always somewhat subjective, unless the two kinds of art are from the same site (see Chapter 5), but the same applies to any typological scheme and, as in anything, including direct dating, one ultimately has to rely on the experience and expertise of the practitioner. The similarities of some Côa figures — and figures from the other open-air sites — to solidly dated drawings from Parpalló,[13] La Tête-du-Lion (see pp. 62–3) and elsewhere are so striking that, for the moment at least, they must join the accepted corpus of Palaeolithic art unless and until some convincing, concrete evidence of a more recent age can be mustered.

In short, a series of major discoveries over the past fifteen years have transformed our conception of the parietal art of the last Ice Age in Europe, confirming what had long been suspected by some researchers — that the well-known art surviving in roughly 300 caves in Western Europe is unrepresentative and uncharacteristic of the period, owing its apparent predominance in the archaeological record to a taphonomic fluke. In reality we have no idea how important or frequent the decoration of caves was in Ice Age Europe, but it is extremely probable that the vast majority of that period's rock art was produced in the open air. This in turn has very profound implications for our interpretation of the possible functions of Palaeolithic art as a whole. It was by no means limited to dark, mysterious caves — they are merely the places where it has been best preserved.

Fig. 9.9 Two horses with overlapping heads at Ribeira de Piscos, Côa Valley (Portugal). The full horse figure is about 60 cm long.

10 | WHAT WAS DEPICTED?

Fig. 10.1 Three bison drawn in the Réseau Clastres (Ariège) in different degrees of completeness (see also Fig.11.23). The most complete, on the left, is 1.14 m long. Probably Magdalenian.

For convenience, Palaeolithic images are normally grouped into three categories, although, as usual, there is a degree of overlap and uncertainty in this division: the categories are animals, humans and non-figurative or abstract (including 'signs'). Each presents its own problems, but the first two share fundamental questions of zoology (identification of the species) and ethology (what are they doing?). Neither question is straightforward.

Animal figures

The vast majority of animal figures seem to be adults drawn in profile; there are a number of possible reasons for this – it is easier than drawing them full-face, particularly if one wants to convey the important features of the animals' anatomy, which are also far clearer on adults than on the young; and it is certain that this was how the animals were observed and identified by Palaeolithic people. Nevertheless, animals were sometimes drawn full-face – primarily in the Final Magdalenian (mostly ibex and horse), but also in the Aurignacian of Chauvet Cave, where a number of bison heads are drawn in this way.[1]

Allowing that we can never know what any prehistoric image was meant to depict – only the artist could tell us – it is obvious that most of the animals drawn seem easily 'recognizable'

at genus level; but there are great numbers of figures which are badly drawn (in our eyes), incomplete, or (perhaps purposely) ambiguous. Numerous examples can be found of figures which have received markedly different attributions – beaver or feline, fish or porcupine, bear or reindeer, big cat or young woolly rhino, and crocodile, reindeer or horse![2] The same is true in Australian rock art, where there can be difficulties in deciding between possums and dingos, or tortoises and echidnas, since figures have few diagnostic features, and Aboriginal identifications can differ from those reached by zoological reasoning.[3] However, in the absence of informants about Ice Age images, zoological reasoning is all we have to go on – comparison with present-day animals, or, in the case of mammoths or woolly rhinos, with their closest living relatives or with frozen carcasses from Siberia – but such comparisons have an inevitable degree of subjective assessment.

It is therefore no surprise that, just as scholars differ on the numbers of figures in a cave, they also tend to produce very different totals and percentages for each species: at Marsoulas, for example, the four most recent studies have claimed 21/32/30/35 bison respectively, 11/32/26/24 horses, and 16/18/12/21 humanoids; at Le Portel, three recent studies claim 33/26/31 horses, 27/23/30 bison, and 6/5/4 humans![4] As with different tracings of a single figure (see above, p. 54), which version should one believe? The most recent is not necessarily the most reliable.

Another problem is that many authors, including some of the foremost authorities on Palaeolithic art, admit in the text their doubts about particular interpretations; but in captions, statistics and tables these uncertainties often become definite attributions, and thus distort the results.[5]

A hierarchy of identification should therefore be adopted – definite, probable and uncertain – and must be maintained consistently throughout a piece of work. Even so, differences of opinion will remain.[6] Identification of figures in art can never fully become an exact science, although expert knowledge of animal anatomy, as exemplified by the remarkable studies of Léon Pales, can help to clarify many issues.[7] As shown in Chapter 4, it is also wisest

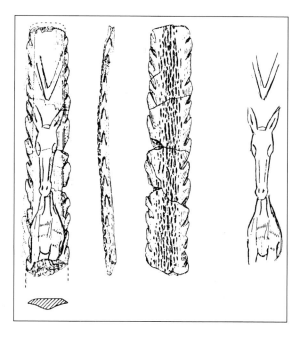

Fig. 10.2 Frontal engraving of a horse on an antler fragment from Las Caldas (Asturias). Late Magdalenian. 63 mm long. (After Fortea et al.)

to work from originals or from photographs, rather than from someone else's tracings.

Another type of evidence to bear in mind is the range of species present in the faunal remains of the site or region in question during the Upper Palaeolithic. Inevitably, however, there are cases where bones are of no help – for example, nobody denies that the ibex is depicted at Pair-non-Pair – indeed, it is dominated by ten ibex figures[8] – although no bones of the species have ever been found in Gironde; similarly, pictures of seals do not coincide with the few remains of the creature found in Palaeolithic sites; and despite the existence of a few depictions of mammoth in northern Spain, remains of the animal have been found only at one or two sites there.

Where complete or detailed pictures of animals are concerned, there is generally a broad measure of agreement, providing they are well preserved: no one could doubt the horses of Ekain or the bison of Niaux. Problems arise where there is a choice of more than one type of animal such as species of deer, as will be seen below; or where there is an extreme degree of stylization in the figure.[9] A figure's size is of no help, since Palaeolithic art has no scale or ground-line, and, as we have seen, tiny figures are found next to large ones, even of the same species, and sometimes on the same panel.

Occasionally, complete animals may defy

identification; they resemble no known species, are presumably imaginary, and have been left undetermined, though some have been given bizarre names: the spotted 'unicorn' of Lascaux with its two straight horns (which has been interpreted as a Tibetan antelope, a lynx, and even two men in a skin – see below, p. 181); the 'antelopes' of Pech Merle, with their swollen bodies, long necks and tiny goat-like heads, which seem to be an amalgamation of horse, ibex and megaloceros;[10] or the two strange beasts of the Tuc d'Audoubert, with their shapeless muzzles and short horns.[11]

There are also figures of beasts with extremely long necks at sites such as Pergouset and Altxerri; and a few images seem to be composites, such as a deer at Trois Frères which, according to Breuil's tracing, has a bison's head; a fish at Pindal which has the fins and tail of a tuna on a salmonid body; a bear or feline at Chauvet with the hooves of a bison; an animal at La Marche which seems to have the body of a horse and the head of a bovid, while another has the body of a horse or bovid, together with an antler; and a few horses have the cloven hooves of cattle, such as one in the Lascaux Nave, and an engraved example in Pergouset.[12]

The many incomplete figures offer a different set of problems. Lack of detail in apparently unfinished animals can lead to different views, as in Niaux's Salon Noir, where a sketchy outline of a quadruped, 1.5 m long, above a horse high on one panel has been interpreted as a feline by some scholars, but as a definite horse by others.[13] Fragmentary animals are not usually a problem when the head is present; where the head is missing, as in a bison on the Altamira ceiling, a bear engraved on bone from Abri

Fig. 10.3 The so-called 'antelopes' of Pech Merle (Lot); this panel is c. 1 m long.

Morin, or the bears of Ekain (p. 151) which were purposely left headless, the characteristic shape of the rest of the body leaves no room for doubt; but if only part of the body survives, it can be difficult to decide: one such figure at Bédeilhac which lacks both head and forelimbs is generally thought to be a horse. The back legs on the Laugerie Basse engraving called the 'Femme au Renne' are thought by most specialists to belong to a reindeer, though Leroi-Gourhan preferred it to be a bison in order to fit with his theories (see p. 192).[14]

However, it is all too easy to make mistakes with incomplete figures, particularly on fragmentary plaquettes, where much remains uncertain: a piece from Labastide, for example, had the back half of a quadruped whose shape was thought characteristic of a horse, but when later a further piece was found, the figure proved to be a bison![15] Interpretation can also depend on which way up a plaquette is held – for example, an engraving on a specimen from Lourdes was seen only as a horse head by some scholars, but a fine ibex also appeared when it was turned upside down.[16]

Schematization, which involves reducing a figure to its essential traits and leaving out the rest, is well known in Palaeolithic art, and is perhaps clearest in the isolated neck and back-lines of horses, bison or mammoths (or, for horses, simply the mane), for example at Marsoulas and Pech Merle – indeed such 'abbreviation' is particularly common in the Quercy, where horns, manes or back-lines comprise no less than 40 per cent of the animal figures of Cougnac.[17] Nevertheless, identification of the animal is still possible: in the Réseau Clastres at Niaux, there are three painted bison (Fig. 10.1); the first is complete and 'naturalistic'; the other two are very incomplete, but even without the proximity of the first they could confidently be identified as bison through comparison with complete examples elsewhere: the shape of the back and position of the mane are quite distinctive. Such figures perhaps constitute a kind of Palaeolithic 'shorthand' in which a part stands for the whole; however, this implies that we know the artist's intention – and it is possible that the figures simply depict a part of the animal, since the shape of the back and head would be fundamental in recognition

at a distance.[18]

As will be seen below, there have been many attempts to derive detailed information about different species or 'races' of animals – especially horses – from Palaeolithic figures; all were doomed to failure, precisely because this is art. We cannot assume that the Palaeolithic view of 'realism' was the same as our own,[19] or that they were concerned in any way with presenting accurately the external features of the fauna. There is no justification for creating new species or subspecies on the basis of these images, as has sometimes been proposed: the international rules of zoological nomenclature require that species be defined on the basis of anatomical remains; it is unscientific to attempt it from pictures. For bison or horses, each art site could present at least one subspecies of its own, but the diversity probably owes more to the artists than to nature.

Similarly, little reliable information can be obtained about the animals' coats – which in any case vary with sex, age, season and climate – because of omissions, distortions and conventions in the figures.[20] The magnificent antlers and horns, and the optimal body growth suggested by the prodigious amounts of fat on some animals, together with the lack of emaciated animals in the art, have been interpreted as proof of the rich environment and long growth season in these parts of Europe.[21]

The age of the animals can almost never be estimated, except in the case of the few juveniles such as the calf with two adult bison on a bone fragment from Abri Morin, the calf on a bone disc from Le Mas d'Azil (with an adult bovid on the other side), or the fawns apparently suckling at does drawn on the wall at La Bigourdane (Lot) and a plaquette from Parpalló.[22] In other cases, one can only make a subjective assessment of the size of the antlers or horns, which may be an artistic convention rather than an accurate record.

The animals' sex is sometimes displayed directly but almost always discreetly – usually on male bison or deer,[23] but rarely on horses for some reason (e.g. at La Sotarriza). Only a few animals are known which have female genitalia depicted;[24] this may be partly due to the predominance of profile views, in which the vulva is not visible, though this could easily be overcome through twisted perspective if it was important to convey gender. Even udders are very rarely shown, even where they could be, as in the 'jumping cow' of Lascaux (p. 67).

One therefore has to look for secondary sexual characteristics, such as antlers in red deer, differing horn shapes in ibex, or sheer size and proportions in other species – though there is always a potential overlap between females and juvenile males.[25] These secondary features were sometimes exaggerated – particularly deer antlers and bison 'humps' – and thus comparisons of size cannot be undertaken between sites, since the massive mammoth-like humps of some Font de Gaume bison, when placed

beside some of the more streamlined, 'boar-like' bison of Niaux, make the latter appear to be the females to Font de Gaume's males.[26]

Recently, the expertise of an animal ethologist has been brought to bear on the bison depictions in Ariège caves and at Covaciella, which has led to a number of interesting suggestions as to the animals' age, sex and posture: for example, although the vulva is almost never indicated and only nine figures have a possible phallus, it has been claimed that most bison depicted are males, and more adults than young, often in active poses. At Covaciella (Fig. 10.5), the main figures are thought to be an adult female followed by an excited adult male,

Fig. 10.4 Three bison drawn in Le Portel (Ariège). From left to right, they measure 71 by 43, 51 by 25 and 73 by 36 cm. Probably Magdalenian.

while the third bison, below, is an old male. At Fontanet, the heavy fur denotes a winter coat, and the animals are reckoned to be males around four to seven years old.[27] Such studies have a tendency to assume that the Palaeolithic artists were striving for total realism and accuracy, and that these artistic depictions can therefore be read like photographs. On the other hand, it is always interesting to have input from experts in different fields, and it is certainly true that simply to label a figure as 'bison' is somewhat reductionist, when the Sa'ami of the Arctic, for example, have not one but hundreds of words for reindeer, depending on the animal's age, sex, colour (85 words), coat (34), antlers (102), etc.[28]

So much for identification; but what are the animals doing? Many of them are 'motionless', to the point where it has been claimed that they are dead, or at least copied directly from carcasses[29] – primarily because they seem to be standing on tiptoe, and show no sign of weight-bearing in the limbs; these features are not to be found in living animals. The hypothesis also accounts for some of the twisted perspective in hooves (since the underside of the foot is frequently turned up by a fallen animal), for the fact that nearside feet are sometimes higher than the others, for the prominence of the belly, and for projecting tongues and raised tails. It is, of course, possible to attribute all of these features to muddled memories of living and dead animals,[30] or simply to artistic conventions, but it must be admitted that close observation of carcasses was certainly the artists' principal source of anatomical detail.

One cannot, however, imagine them dragging such dead-weights far into caves to be copied directly! Perhaps some portable depictions are outdoor sketches of carcasses, and were later carried into the caves and copied on to the walls, but this is pure speculation. The fluent, effortlessly drawn and very well-proportioned animal figures in parietal art suggest that the artists carried everything in their mind's eye – as mentioned above (p. 124), they could sometimes even 'draw blind'. As Breuil, who used the same technique, said of the Palaeolithic artists: they never took a measurement – they projected on to the rock an inner vision of the animal.[31] Even if the Palaeolithic

figures were ultimately based on dead specimens, it is noteworthy that their motionless figures are somehow imbued with an impression of life and power. It has been pointed out that there are very few clear depictions of blood, pain and death, and even the supposedly 'wounded' animals with possible missiles in them do not seem to be suffering.[32]

Nevertheless, a recent study by a veterinarian has returned to the notion that many of the figures represent animals which are dead or dying, based on the notion that the depictions are 'so realistic that a precise diagnosis is possible'. Although it is admitted that there is a huge predominance of animals which are well, and the author is aware of the pitfalls of treating caves as veterinary clinics, it is claimed that the fish of the Abri du Poisson is dead; that animals (such as a few reindeer in the sanctuary of Trois Frères) with heads down and bent legs are in

Fig. 10.5 Part of the main panel at Covaciella (Asturias). The upper bison measure 62 cm (left) and 76 cm (right) in length; the lower bison measures 69 cm. Magdalenian.

distress; and that bovids with stiff legs and extended tongue are sick or dead.[33]

In their turn, some hunting specialists have chosen to interpret many of the features of these images in behavioural terms: for example, the raised tail is seen as a signal of alarm, anger or threat, of submissiveness, or of a copulatory invitation by females[34] – and thus as an illustration of autumn, when such displays are common due to the interactions of rutting behaviour, although if the threat were directed at a hunter it could occur in any season.

There have been two particularly famous examples of zoological deliberation about posture in Palaeolithic images. The engraved reindeer from Kesslerloch (Switzerland) was traditionally thought to be grazing, until it was pointed out that its stance was exactly that of a male in rut – either sniffing a female's track or, more likely, in a position for attacking a rival.[35] The curled-up bison on the bosses of the Altamira ceiling have been described as sleeping, wounded, dead or dying (according to veterinarians, the ceiling reeks of death,[36] see below, p. 175), or as clear pictures of females giving birth![37] Another prominent view is that they are males, rolling in dust impregnated with their urine, in order to rub their scent on territorial markers – even though one of them has udders according to Breuil's copy![38] In fact, they may simply be bison figures drawn so as to fit the bosses – they have the same volume, form and dorsal line as those standing around them, but their legs are bent and their heads are down.[39] In any case, it has been pointed out that their configuration, with females and juveniles in the centre, and adult and adolescent males around the edges, corresponds to August/ September, the rutting season – the only time when male and female bison are found together.[40] However, one wonders what a hind – the biggest animal in the group – is doing next to a rutting bison herd!

'Animated' figures are fairly rare, and although they have occasionally been placed in the centre of a decorated 'panel', they are more usually to be found at the edges.[41] Such movement already exists in the Aurignacian, as exemplified by the aurochs with splayed legs and the fighting rhinos of Chauvet Cave (Fig. 5.8); it grows more common in Solutrean and Early Magdalenian times (as at Roc de Sers), and is most abundant in the Middle and Later Magdalenian, although this kind of 'realism' never predominates: in the Pyrenees, for example, it is found in just over a quarter of figures in both parietal and portable art.[42] It is Lascaux which has the most abundant and varied animated figures; the cave of Gabillou, in the same region and thought to be of roughly the same period, also has lots of movement in its figures.

'Scenes' are very hard to identify in Palaeolithic art, since without an informant it is often impossible to prove 'association' of figures rather than simple juxtaposition. In rare cases, however, this can be done: for example, the Laugerie-Basse engraving mentioned above clearly shows a supine woman beneath and beyond a deer. It has often been claimed that a scene depicting a supposed 'wolf' menacing a deer has been found in portable engravings from the Pyrenees (Lortet, Le Mas d'Azil), Cantabria (Pendo) and Dordogne (Les Eyzies), but it has now been shown that the 'wolf' is non-existent in three cases, and any links between the objects are illusory.[43] A few other portable engravings such as the 'bear hunt' of Péchialet (Dordogne), or the 'aurochs hunt' of La Vache have also been interpreted as scenes, as has the parietal group of figures in the shaft at Lascaux (see below, p. 178–9), though one cannot be sure that a narrative is involved rather than a symbolic mythology.[44]

In short, therefore, one has to be very careful in making zoological and ethological observations from Palaeolithic images. If one has photographs of animals, one can accurately identify them and assess what they are doing; but here we are dealing with what has been called 'stylized naturalism', drawings by artists with a message to convey, and using stylistic conventions: nobody would assume, from the spotted specimens at Pech Merle, that the horses of the period had big bodies, small legs and tiny heads; so how reliable are the other features on display? These are not 'Palaeolithic photographs' – we need to allow for convention, technique, lack of skill, faulty memory, distortion, and whatever symbolism and message were involved. Only the most complete and 'naturalistic' figures are really informative, but even the finest Palaeolithic pictures may prove a disappointment to the zoologist. Ice Age images are exciting because they allow us to 'see' the fauna of the time, to put flesh on the bones which we dig up;[45] but we are always seeing the animals through the artists' eyes.

Despite the different identifications and percentages of species produced by each specialist, one fact remains clear: the overwhelming overall dominance of the horse and bison among Palaeolithic depictions. In Leroi-Gourhan's sample of 2260 parietal figures, no fewer than 610 were of horse, and 510 were of bison,[46] which thus account for about half between them. No other type of animal comes close, although, as we shall see, other species do dominate at particular sites. The next most numerous include deer, ibex and mammoth, followed by a cluster of rarer species.

The horse

Study of the most important animal in Ice-Age iconography has, in the past, wasted a great deal of effort on attempts to establish exactly which 'races' were depicted; Piette had a try at this in the nineteenth century,[47] and was later followed by other scholars who distinguished up to thirty-seven varieties![48] However, all such exercises were eventually abandoned; as mentioned above, they were based on the shape and size of the figures, their manes, and the colour and pattern of the coats, all of which were subject to artistic whim – for example, Gravettian figures tend to have elongated heads, while those of the Solutrean and Lower Magdalenian tend to have tiny heads! The only firm basis for subdividing horses into different species is palaeontological differences, chiefly in the skull and backbone, and it is therefore a matter best left to those working with faunal remains.

More recent studies have avoided this topic,

Fig. 10.6 & 10.7 BELOW: Perforated antler baton with carving of horse head with ears raised and a stylized mane; the shaft is also decorated. Le Mas d'Azil (Ariège). Magdalenian. Total length: 20 cm.
LEFT: Detail of the horse head. From muzzle to ear-tip it measures less than 4 cm.

and instead approached important collections of horse figures from a more rigorous knowledge of equine anatomy.[49] At Gönnersdorf, seventy-four mostly fragmentary horse depictions are known from sixty-one plaquettes, making this animal dominant among those depicted at the site. Horses likewise dominate the animal figures at La Marche, where at least ninety-one whole or partial specimens have been found on sixty-four stones; 62.4 per cent of them face right, a percentage similar to that (57-59 per cent) at sites such as Ekain, Lascaux and Les Combarelles.[50] Not one of the La Marche examples can be sexed; and it has been found that, like artists throughout prehistory

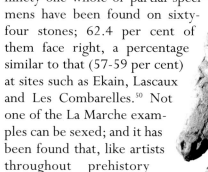

and history, those of La Marche consistently made the horses' bodies too long[51] – perhaps to give an impression of speed and lightness. Certainly, in Palaeolithic art as a whole, horses strike a number of different poses, which can be interpreted from a behavioural point of view, such as the jumping horse on a spearthrower from Bruniquel, or the neighing horse head from Le Mas d'Azil.[52]

One piece of zoological information which is probably reliable in these horse depictions is that of the length of the coat, since present-day Przewalski horses (thought to be among the closest surviving equivalents to the Palaeolithic types) do display striking seasonal differences : hence the shaggy ones recorded at sites like Niaux must be the long pelt of winter, and, interestingly, no distinct 'winter horses' are known in Cantabrian caves.[53] On the other hand, these differences might also denote different climatic phases, as has been suggested for a similar range in bison depictions (see below).

Another frequent feature of Palaeolithic horse figures is the 'M' mark on their side, which denotes a change in colour between the dark hide and the lighter belly area – a shape and change still clearly visible on modern animals such as the little semi-wild 'pottoks' of the Basque Country.[54] This mark is already present on some horses in Chauvet Cave, as are the stripes which are sometimes drawn on horses' shoulders (for example, at Ekain), a feature that Darwin noticed on living specimens, but which the Palaeolithic artists may have exaggerated.[55]

It is certain that the horse was of tremendous cultural importance to people in Western Europe during the Upper Palaeolithic – horse teeth and bones have been found carefully placed in Magdalenian hearths in a number of deep Pyrenean caves such as Labastide and Erberua,[56] as well as near hearths at the Magdalenian open-air camp of Pincevent near Paris (despite the fact that reindeer account for almost 100 per cent of this site's faunal remains).[57] At Duruthy (Landes), the carved horses (see above, p. 99) were found in a kind of 'horse sanctuary' dated to the twelfth millennium BC; the biggest, a kneeling sandstone figure, rested against two horse skulls and on fragments of horse-jaw, while three horse-head pendants were in close proximity (p. 100).[58]

Finally, it should be noted that the horse may well have been under close control during the Upper Palaeolithic, as suggested by a variety of evidence including some possible depictions of simple harnesses on certain figures, especially one from La Marche, and most recently in parietal figures at the neighbouring caves of Erberua and Oxocelhaya.[59]

The bison

Like the horse, depiction of the bison varied through the Upper Palaeolithic, in terms of shape, size and shagginess; some scholars have argued that not only artistic convention was involved, but also perhaps climatic change, so that the heavy, shaggy bison would denote cold phases and the lighter forms would be linked with milder phases[60] – it is a nice idea, but a lot of work and more accurate dating would be required to prove that climate was indeed responsible, rather than style or different species/subspecies. As mentioned above, different sites have their own characteristic bison 'construction', seen in an extreme form at Font de Gaume, and attempts to differentiate specific varieties have led no further than similar

Fig. 10.8 Apparent scene of bison (or musk-ox?) pursuing a human, carved in bas-relief around a block from Roc de Sers (Charente). Solutrean. The block is c. 35 cm thick, the human is 50 cm high, and the bison 54 cm.

studies of horses.

The best study of bison figures, albeit of a small sample (thirteen), concerns those of La Marche;[61] an earlier attempt to establish the proportions and rules of construction that characterize Palaeolithic depictions of bison proved of limited value because it was derived from the figures themselves; it is in fact necessary, in studies of this kind, to proceed from precise anatomical markers taken from living animals or at least from photographs of them in profile.[62] Thus nature, rather than artistic convention, is the reference point. Using this method, Léon Pales found that Palaeolithic bison figures, even within a single site (such as Niaux or Fontanet), showed differences in execution and style, whereas other groups (such as at Altamira) were fairly homogeneous. Of all those examined, the La Marche specimens were the 'closest to nature'.

It was also found that in this site, as at Trois Frères and Font de Gaume, there was a tendency to have bovids facing left. Depiction of gender varies from site to site: at Font de Gaume, twelve of the thirty-nine bovids have a phallus, whereas none of Niaux's forty-seven bison has one (although, as mentioned above, many look male from secondary sexual features

Fig. 10.9 Bison engraved and painted in Trois Frères (Ariège). Probably Magdalenian. Length: 1.5 m.

such as the hump, the massive head and the short, thick horns); only one at Trois Frères, standing on its hind legs, has a phallus – one of the rare erect phalli in Ice-Age art – and only one or two of the La Marche bison may have the sex indicated.[63] Anatomical details seem to be incorrect occasionally: for example, horns that point forwards, or which are different in volume from those found among faunal remains.

Most bison are 'immobile', but a few do show movement – some hypotheses about the curled-up specimens at Altamira have been mentioned above, and we shall return to them later (p. 175). One famous bison, a carving from La Madeleine, has turned to lick what is probably a wound or parasite on its flank; while that in the 'shaft scene' of Lascaux appears to have lost its entrails and perhaps to be charging

a man (pp. 178–9). Other 'scenes' involving bison attacking people may be depicted on the cave wall at Villars (Dordogne), on a block from Roc de Sers (Fig.10.8, though many see this animal as a musk ox because of its horn-shape), and a portable engraving from Laugerie-Basse.[64]

The aurochs

Wild cattle are the other large bovids frequently represented in Palaeolithic art;[65] the most lifelike are perhaps those of Teyjat, and the portable engraving from Trou de Chaleux (Belgium, p. 91) but, the biggest and best known are those of Lascaux, including the 5-metre bulls. Nobody has ever questioned the attribution of the latter to *Bos primigenius*, but for some reason, certain leading scholars thought the smaller cattle of Lascaux to be *Bos longifrons* or *brachyceros*, a domestic type with long forehead and short horns which is not dated earlier than the Neolithic period.

The simple explanation is that of the pronounced sexual dimorphism of *Bos primigenius*: the big cattle of Lascaux are bulls, while the smaller are probably cows of the same species. Studies of similar cattle in Morocco with the coloration of their wild ancestors shows that the great bulls are black or reddish-black, with a light-coloured strip along the back and some

Fig. 10.10–12 Engraved aurochs cow (left) and bull (right) from Teyjat (Dordogne), with tracing by Breuil. The animals are 55 and 50 cm long. Magdalenian. The detail opposite shows the cow's head.

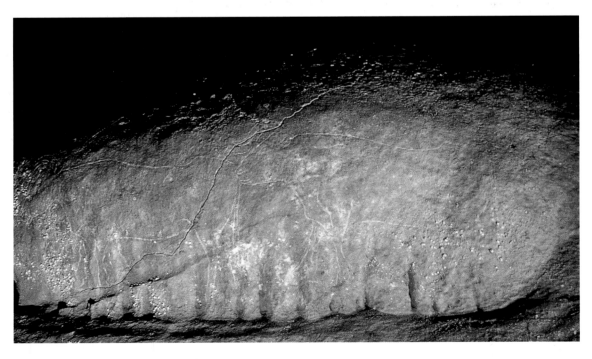

light hair between the horns; the cows are usually reddish-brown, with darker legs, a darker and slimmer head, and thinner horns. All these features are on display at Lascaux, some to an exaggerated degree, together with other specimens which are slightly abnormal in colour.[66] The sexual dimorphism can also be seen, to a lesser extent, in the bull and two cows engraved at Teyjat.

Deer

Palaeolithic art includes some notable depictions of the giant deer (*Megaloceros*), such as those of Cougnac which cleverly use natural rock-shapes for the dorsal line and hump, and the numerous specimens depicted in Roucadour (Lot) and Cosquer Cave. Apart from these, and sporadic claims for elk, and for fallow deer, as at Tursac,[67] all Palaeolithic depictions are either of red deer or of reindeer.[68]

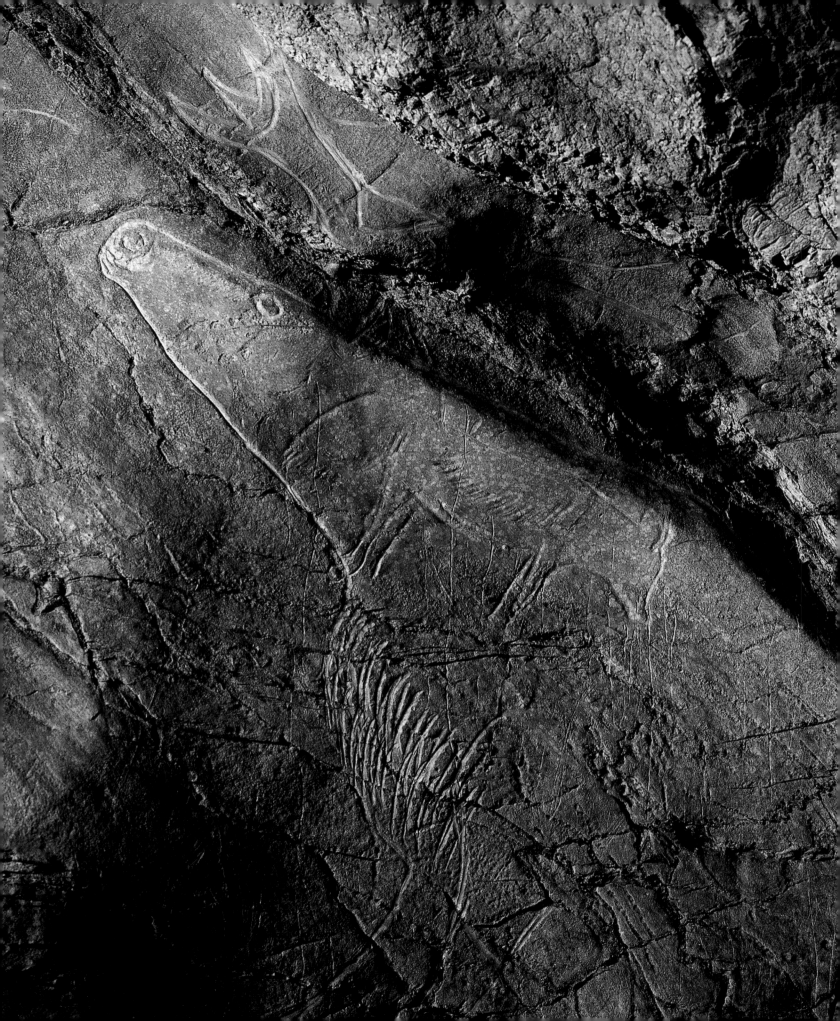

In red deer, only the males have antlers, so that these can denote sex, although some figures taken to be hinds may actually be males whose antlers have been shed. In the reindeer, however, both sexes have antlers.

Many depictions of reindeer show clearly the characteristic shape and markings of the species, as in the series of engravings at Trois Frères; ironically, some of the best figures are to be found in Cantabria (Altxerri, Tito Bustillo, Las Monedas), a region where this animal was never plentiful, according to faunal remains.[69]

Occasionally, it is difficult to differentiate the two types of deer; on the famous baton from Lortet, with a scene of deer and fish engraved around it, some scholars see stags, but others see reindeer (indeed Piette first saw them as reindeer, and later as stags!). In the portable engravings of La Colombière, most of the animals thought to be red deer have recently been reinterpreted as reindeer, while some of the hinds at Covalanas and La Haza may likewise be reindeer.[70] Similarly, all but one of the ninety-one deer on the walls of Lascaux have been assumed to be red deer by most specialists of Palaeolithic art:[71] but a number of experts on caribou instead see some of them as reindeer — particularly the series of heads, often called the 'swimming stags', since groups of swimming stags are very rare, whereas swimming caribou are a common sight in the migration season. This alternative identification may well be correct since seventy-four of the Lascaux cervids (81 per cent) have antlers, and the wide variety of shapes includes some which closely resemble those of tundra caribou, and depictions of them by Eskimo hunters during prehistoric and historic times.[72]

As usual, some authors have chosen to see these Palaeolithic figures as accurate depictions of individual animals; thus, the remarkable examples in caves such as Lascaux were detailed records of prize stags (mostly at least fifteen years old) with fine antlers — indeed Breuil claimed that the Lascaux figures were not exaggerated at all, as he had seen a stag in Germany with antlers of the same kind[73] — whereas it is far more reasonable to assume that the style and ability of individual artists played a major role, and that the antlers are merely an artistic translation of reality.[74] Attempts to differentiate tundra and forest reindeer types from antler shape in the art have also been made, but are unlikely to be reliable.[75]

In the light of a detailed anatomical study of deer depictions,[76] one can make a number of observations: many features, such as the antlers, tail, hooves, ears and eyes, are often incomplete or done badly — yet occasionally, even the gland below the eye of the rutting reindeer is depicted, as on the engraving of Kesslerloch, and on seven or eight of the Lascaux deer.[77]

A number of different positions are known for cervids, such as the male bending to lick a female's head at Font de Gaume, the rutting male of Kesslerloch, and other variations at Limeuil and elsewhere — grazing, bellowing, and so forth — and it has been pointed out that both reindeer and red deer are most often depicted in what appears to be the rutting season (autumn/early winter), characterized by features such as the white mane on the reindeer bulls and the thick mane on mature stags, and bodies rounded with fat:[78] for example, a stag from the Abri Morin has been described as having its antlers thrown back, its ears down, nostrils dilated, mouth open and neck extended.[79]

Fig. 10.13 OPPOSITE: In the cave of Altxerri (Guipúzcoa) this rare parietal engraving of a reindeer is clear evidence of the cold conditions which existed in northern Spain during the last Ice Age. A small fox, 25 cm long, has been engraved on top of the reindeer's neck. The figures probably date to the Magdalenian period.

Fig. 10.14 The so-called 'swimming' deer of Lascaux (Dordogne). Probably Magdalenian. This panel is c. 5 m wide.

Fig. 10.15 Photomontage of the decorated panels at Cougnac (Lot): note the use of natural rock formations for the neck and chest of the giant deer. The female ibex at the top (facing left) is 50 cm long; the ibex on the right (and facing right) is 81 cm in length. Various periods (see Chapter 5)

Ibex

One should really use the term 'ovicaprids' here, since it is quite possible that some Palaeolithic depictions represent the mouflon or even the Tahr *Hemitragus*, and a few have been attributed to these species;[80] but by and large, animals of this type are automatically assumed to be ibex and quite common in Palaeolithic art – the newly discovered cave of El Bosque (Asturias) has at least twenty-four on one panel!

The question then arises whether they are the Alpine variety (*Capra ibex*) or the Pyrenean (*Capra pyrenaica*); the former has curvilinear horns, while the latter's horns are more lyre-shaped, and rise at the tips. However, they are often difficult to differentiate in art, for the simple reason that in profile one cannot assess how sinuous or divergent the horns are – indeed, this may be another role for twisted perspective – and in many cases, particularly in portable art, the shape of the available space caused distortion in the shape and curve of horns: one example from Le Mas d'Azil even has them sticking straight out in front![81]

Some depictions appear ambiguous – an ibex carved on an antler shaft from Le Mas d'Azil has an Alpine horn-shape when seen from the front, but a Pyrenean shape in profile, while an engraving from Limeuil appears to be a Pyrenean type which has been transformed into an Alpine ibex.[82]

It is, of course, possible that the Limeuil example is really a young animal which was turned into an adult through extension of its horns, or that it represents the bone-stumps within the horns. However, there is evidence that different varieties of this animal coexisted in Western Europe – at Isturitz, for example, there seem to be depictions of a number of types, even if one allows for distortion, artistic licence, and so forth; and certain figures seem to be accurate portrayals of the Alpine type (e.g. the figure on the La Mouthe lamp, p. 125), and of the Pyrenean (e.g. some of those in the Salon Noir at Niaux).[83]

Horn-size serves to distinguish males from females, and where two animals with horns of different size are found together, as at Cougnac, it is reasonable to suppose that they represent one of either sex; but, as mentioned above, small horns may also denote a young male. An occasional phallus was depicted (as on a La Marche animal with big horns), and a beard is a

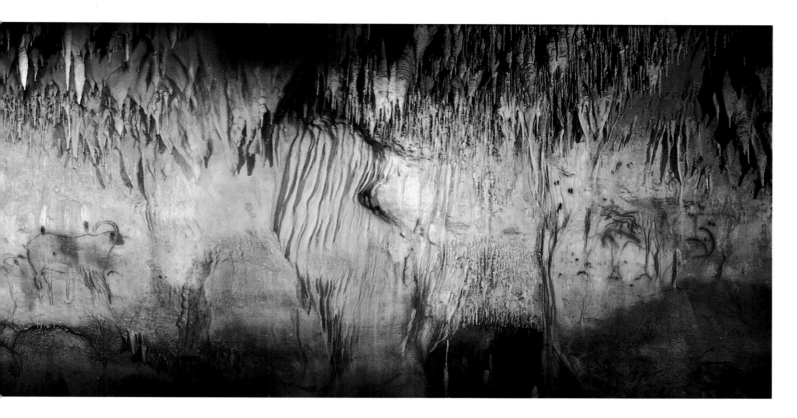

useful secondary sexual characteristic. All these features are clearly depicted on the ibex sculptured in bas-relief at Angles-sur-l'Anglin (p. 2).

Ibex figures were sometimes animated – the animals carved on spearthrowers, with head turned back to see the emerging turd and birds, are probably young ibex (p. 98), as are the two headless animals fighting or playing together on another famous spearthrower fragment from Enlène. A baton from Duruthy shows two ibex fleeing in apparent terror from some large animal (p. 94); elsewhere, ibex are often found in pairs, facing each other, head to head.[84]

The mammoth

About 400 depictions of mammoth are known from the Palaeolithic of Eurasia.[85] The most thorough studies of this species in Palaeolithic art are those concerning the magnificent adult mammoths of La Marche,[86] and the larger collection from Gönnersdorf, comprising sixty-one specimens on forty-six plaquettes.[87] About half of the latter are complete figures, the rest having become fragmented along with the plaquettes. It proved possible to divide them into adult and young animals; a few are grouped together on the same plaquettes, but most are

individual depictions.

Some details, such as the 'anal flap', coincide with observations on frozen mammoths in Siberia; but others are troubling: a few of the Gönnersdorf mammoths have small, short tusks, but most have none at all, although the depictions seem very naturalistic; similarly, there are no tusks on the far older mammoth figures from Vogelherd and Geissenklösterle,

Fig. 10.16 Mammoth sculpted in bas-relief in the Grotte du Mammouth (or de Saint-Front), Dordogne. It is 1.25 m high, 1.1 m long, and located 4 m above the present cave floor.

Fig. 10.17 Highly stylized mammoth in La Baume-Latrone (Gard), 1.2 m long, drawn with fingers dipped in red clay. The body is in profile, but the short tusks are in twisted perspective. Date uncertain.

Fig. 10.18 One of the seven mammoths painted in red ochre in Kapova Cave, Russia, about 50 cm long. Probably c. 14,000 years old.

on an engraving from Kostenki, and on a number of parietal depictions.[88] It has therefore been suggested that some mammoths had no tusks, perhaps through a depletion in natural resources; but other scholars prefer to see it as artistic licence – or perhaps even as a sexual dimorphism – since at La Marche some mammoths are depicted with long tusks, some have short tusks and some have none at all! One might compare this to the lack of hair on figures from Pindal, Cougnac, La Baume-Latrone and other sites, which led to their erroneous identification as warm-climate elephant species such as *Elephas antiquus*.[89]

The Gönnersdorf mammoths show little sign of 'movement', and mammoth depictions in general are not found in 'scenes'. They are, however, sometimes found head to head, as at Rouffignac, the Dordogne cave that contains almost half of all known mammoth depictions.[90] Other signs of animation include a raised tail, and the trunk either raised or curled up; occasionally, as in Pech Merle, one finds both detailed depictions and simple abbreviations in the form of characteristic dorsal lines.[91] Many of the thirty-seven engraved specimens in Les Combarelles are likewise limited to a cursory, partial outline.

Among recent discoveries, one must mention the important series of at least thirty-seven mammoths in Chauvet Cave – 14 per cent of the animal figures found in the site so far – including one with three tusks (a phenomenon which can occur in nature), and what seems to be a very young animal drawn with large round feet, perhaps in twisted perspective.[92] Finally it is perhaps noteworthy that almost all of the few parietal mammoth figures known in Spain all face left,[93] as do six of the seven principal mammoths in Kapova Cave.

Big carnivores

Large carnivores in the art comprise bears and big cats. In the past, some pretty uncritical inventories of supposed depictions of these animals were drawn up, including many figures which even the authors found highly dubious, and omitting others[94] — inevitably, they relied heavily on old interpretations and unreliable tracings; they have since been superseded by more objective studies based on anatomical expertise[95] — once again, primarily that of Léon Pales in his study of the La Marche figures.

Bears, according to faunal remains, com-prised two principal species during the Upper Palaeolithic of Europe — the cave-bear (*Ursus spelaeus*), and the brown bear (*Ursus arctos*). The former is characterized by the domed shape of its skull, and traditionally it was assumed that any depiction of a bear with a similarly domed forehead could safely be assigned to that species. However, it has now been shown that a whole spectrum of head-shapes are present in the art — only a few, such as those on pebbles from Massat and La Colombière, have really domed heads — and this criterion is thus not sufficiently reliable to permit identification.[96] In the Chauvet Cave, which contains at least fifteen bear figures, including a superb specimen in red, with sophisticated shading and the use of a convexity in the rock for its shoulder, there is one animal which, from its skull shape, appears to be a cave-bear, but whose spots have led

Fig. 10.19 Two headless bears painted in the cave of Ekain (Guipúzcoa). The panel is 1.2 m wide. Probably Magdalenian.

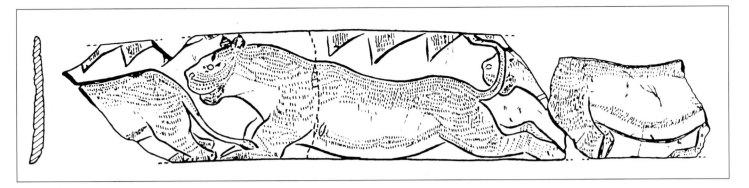

Fig. 10.20 Engraved bone
fragments from La Vache
(Ariège), showing a line
of three felines. Late
Magdalenian. (After D.
Buisson and S. Rougane)

some researchers to dub it a hyena — although, of course, we have no idea whether Palaeolithic bears had spots or not![97]

It is noteworthy that all the La Marche bears face right; a number of different postures can be found in Palaeolithic art, including bears walking, or with nose raised, or with head down — the latter seen as the threatening posture it would adopt towards humans.[98] The Péchialet 'scene', mentioned earlier, seems to show a bear on its hind legs and menacing two 'humanoids'.

Bear and lion heads can appear very similar, and thus it has been hard to differentiate them in the fragmentary terracotta figurines of Dolní Vestonice; this site, together with that of Pavlov, is important because its art is domin-

ated by carnivores: of seventy-seven miniature figurines in good condition, no fewer than twenty-one are thought to be bear, nine lions, five wolves and three foxes (= 50 per cent).[99] Only those of bear and lion are pierced and have little 'decorative' notches, but all of the figures are caricatural, with few details, and thus little can be deduced from them about the species depicted.

Indeed, in Palaeolithic art as a whole, carnivores, and especially felines, are consistently among the most inaccurate figures — perhaps because they were far less easy to observe at close quarters than herbivores: the commonest mistake is the position of the canine teeth in both felines and bears — in nature, the lower canines are in front of the upper when the

Fig. 10.21 Frontal drawing of a big cat from Trois Frères (Ariège), measuring 50 cm. Probably Magdalenian.

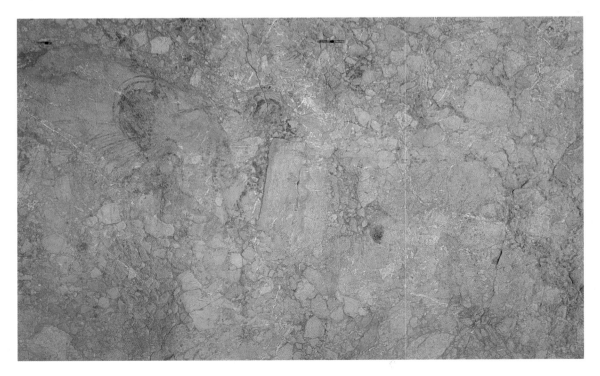

mouth is closed, whereas in these figures, if the jaws were to be closed, they would be in contact with the upper, or even stand behind them.[100] Carnivore canines clearly impressed Palaeolithic people, who, as we have seen, often used them for jewellery; they are the only teeth ever drawn in carnivore figures, and their size is often exaggerated, like deer antlers or bison humps – in one case, however, the large teeth on a figure from Isturitz have been taken at face value as a depiction of the scimitar cat.[101]

Nevertheless, a few felines are accurately portrayed, including those of La Colombière and Labouiche; the fifteen of La Marche are fairly realistic – but only 43 per cent of them face right (as opposed to 100 per cent of the bears). However, the study of Palaeolithic cats has been revolutionized by the recent discovery of Chauvet Cave, which contains at least thirty-six specimens (almost 14 per cent of the animal figures), many of astounding quality and size, and including one small depiction which appears to be a spotted panther.[102] We cannot tell for certain whether or not the male big cats had manes, because although one or two depictions in Palaeolithic art seem to have one, a group of three big cats, over 2 m long, shown in overlapping perspective in Chauvet Cave, include two which have a scrotum but no mane.[103]

The rhinoceros

Another topic which has been similarly revolutionized by the Chauvet Cave is Palaeolithic depictions of the rhinoceros. Before the cave was discovered, fewer than twenty rhinos were known in the parietal art of the Ice Age, and only a few portable specimens, including twelve at Gönnersdorf.[104] Chauvet Cave alone contains at least sixty – more than a quarter of the cave's animal figures, and the dominant theme of the site. They include examples of tremendous power and sophistication, including the fighting pair (p. 73), a rhino with a huge horn which follows the curvature of the cave wall, and a whole group of overlapping rhinos drawn in perspective (pp. 11–12).[105] Many of them feature a very puzzling broad black stripe across the middle of the body – so carefully drawn that it is sometimes engraved at the edges and filled with flat wash.[106] This has never been seen before, and even leading experts on

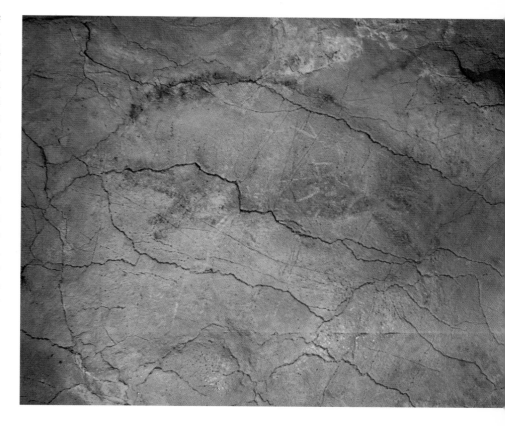

the Pleistocene woolly rhino are at a loss to explain it – it has been tentatively suggested that the stripe may represent folds in the skin, or, with tongue firmly in cheek, a saddle!

The rare mammals

Finally, there are a whole range of animals which are depicted extremely rarely, though often well.[107] Many of them belong to the Magdalenian, and their existence may thus simply be a result of the overall increase in numbers of figures in this period: one can mention the musk-ox, ass, saïga, chamois, wolf and fox, hare, glutton, otter and possibly the hyena; there is also the remarkable weasel-like animal of the Réseau Clastres – if that interpretation is correct, this is a very rare example of an animal drawn bigger than life size (like the Lascaux bulls). The 'boars' on the Altamira ceiling are now generally thought to be 'streamlined' bison – at least one of them has horns! – but one or two possible boars are known in portable art, as at Parpalló. So far, animals such as small rodents seem to be absent from Palaeolithic art.

Where sea-mammals are concerned, most are seals of various species,[108] some of the

Fig. 10.22 Painted animal on the Altamira ceiling, 1.6 m long; often thought to be a boar, it is actually a bison.

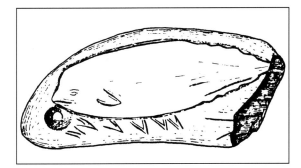

portable depictions found hundreds of km inland. For example, three engravings of seals have been found at Gönnersdorf, located 450 km from the North Sea coast during the Ice Age.[109] Among the finest examples are the male and female grey seals (*Halchoerus grypus*) on the famous baton of Montgaudier (Charente); they are characterized by their elongated muzzle, their size difference and the male's neck folds. Three enigmatic small shapes above them were recently claimed to be whales seen in the distance.[110] A definite whale has been found engraved on a whale-tooth pendant from Las Caldas (Asturias),[111] while a large cetacean is thought to be depicted on the wall of Tito Bustillo. Some parietal figures at Nerja (Málaga) have been interpreted as dolphins, but others believe them to be seals.[112] Finally, eight extremely doubtful engravings of what may be seals have been found recently on the walls of Cosquer Cave.[113]

Fish, reptiles, birds, insects and plants

Fish are quite plentiful, particularly in the form of portable depictions – unfortunately, Breuil's corpus also includes a mass of decorative motifs which he believed to have been derived from schematized fish; some of them are quite plausible, many others are not.[114] Fish species are especially difficult to identify in Palaeolithic art: most of them seem to be salmonids – sometimes males with a kipe on their lower jaw in the characteristic state of exhaustion after spawning, and thus perhaps representing the autumn.[115]

Marine fish are also represented, both on portable objects (such as the Lespugue flatfish figure, in the central Pyrenees) and on the walls of caves, particularly those near the sea, most notably at Pileta.[116] A number of painted shapes

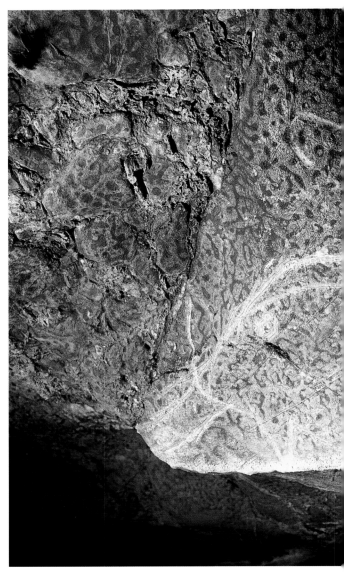

in Cosquer Cave, another coastal cave, have tentatively been interpreted as jellyfish.[117]

One or two tortoises (St Cirq, Marsoulas) and perhaps a turtle (Mas d'Azil) occur in portable art, and one possible salamander from Laugerie-Basse; there are also a number of snakes,[118] although once again Breuil's attributions are excessive, with all manner of wavy lines being interpreted as these creatures – one wonders why eels or worms were not envisaged: the specimens next to the seals on the Montgaudier baton are clearly males.

Birds occur in both types of art;[119] the most recent survey of French specimens found 121 depictions in forty-six sites, and once the doubtful cases had been weeded out there

Fig. 10.24 The owls engraved at Trois Frères (Ariège) (see also Fig. 8.9). Panel width: 87 cm.

Fig. 10.25 Three birds drawn in the Cosquer Cave and interpreted as Great Auks; they are each c. 27 cm long.

remained eighty-one birds in thirty-one sites[120] – remarkably few compared to numbers of animal figures. Portable depictions are more abundant than parietal (the latter comprised fifteen figures in 1987, or 18.5 per cent). Most are Magdalenian (86.4 per cent; 69 per cent are from the Middle and Upper Magdalenian alone), while the oldest are all parietal (such as the newly discovered owl, drawn with fingers, in Chauvet Cave).[121] Waterbirds such as swans, geese, ducks and herons seem to be the most numerous (about 37 per cent). Cosquer Cave contains four parietal depictions of birds which have been interpreted as the extinct Great Auk, although some researchers advise against excessively precise attributions and suggest calling

them simply auks or even just sea-birds.[122] It is certainly true that birds' species and even genus can be very difficult to identify, as the depictions are often mediocre: indeed, in some cases it is not sure whether faces on cave walls belong to owl-like birds or to humanoids and 'phantoms', although Leroi-Gourhan's belief that the owls of Trois Frères are 'anthropomorphs' seems to rest on his theories about cave topography (see below, p. 190) rather than on resemblance.

Another feature which birds can share with humans is a vertical, bipedal posture; some

Fig. 10.26 Plant-like form painted on a stone plaque from Parpalló, Spain. Upper Magdalenian. It measures 10.9 cm long, 6.6 cm wide and 2.6 cm thick.

portable engravings from Les Eyzies and from Raymonden, traditionally seen as groups of tiny humanoids approaching a bison (in the former) and standing around a bison carcass (in the latter), have been reinterpreted by a specialist on birds as lines of swallows:[123] certainly, their stance is identical, and microscopic study revealed tiny incisions by the heads which seem to be beaks; another piece from Raymonden seems to include two swooping birds.

Only about a dozen definite or probable depictions of insects are known, all in portable art, and mostly Magdalenian.[124] Apart from what seems to be the larva of a reindeer-botfly from Kleine Scheuer, and the famous grasshopper of Enlène, most are beetle-like bugs. However, even the grasshopper, the clearest and most detailed depiction, cannot safely be identified to the genus level.

Finally, plants are also rare in Palaeolithic art,[125] especially in parietal art, and only a very few, such as those on a baton from Veyrier and a pebble from Gourdan, can be classed as definite rather than possible. None can be identified accurately.

To sum up, therefore: the 'animal' category in Palaeolithic art is of limited information value to zoology, and vice versa. The greatest caution and rigorous objectivity must be applied to any attempt at identification of the genera or species, and assessment of other external features and their posture.

It is clear that large herbivores dominate the art, and that other types of creature are either comparatively rare (carnivores, fish, birds, etc.), extremely rare (reptiles, insects, plants) or totally absent (rodents, and many other species of mammal and birds). Since Palaeolithic people were certainly familiar with all aspects of their environment, and since birds, fish and plants were important resources for them, at least in the Magdalenian, it follows that Palaeolithic art is not a simple bestiary, not a random accumulation of artistic observations of nature. It has meaning and structure.

This becomes even more obvious when one looks at the differing frequencies of species in a variety of contexts. It has already been mentioned (see above, p. 60) that the Tuc d'Audoubert has different species in its portable and its parietal art, and this observation has been repeated elsewhere. Overall, reindeer, fish, birds and plants are primarily found in portable art rather than parietal,[126] while insects are limited to portable depictions. In general, therefore, portable art's themes may lie closer to the everyday life and activities of Ice Age people than the more conservative and less varied parietal themes.[127]

Some themes are associated with particular forms of artifact: the horse is rare on harpoons and *baguettes demi-rondes*, while bison are not found on spearpoints, are rare on perforated batons, but predominate on plaques.[128] Horses and fish are common on perforated batons, and also (in the Pyrenees and Dordogne respectively) in *contours découpés*.

Different species also predominate in different periods and regions: for example, the bison tends to dominate on cave walls in Cantabria, although it is rare in the region's portable art, where hinds are predominant on shoulderblades.[129] The aurochs is more important than the bison in the art of the early phases, but the situation is reversed in the Middle and Late Magdalenian. Spanish Basque cave art (which has only a couple of hundred animal figures) is dominated by horse and bison, with only six depictions of stags, while among more than a thousand animal figures in the caves farther west there are over 200 red deer figures, and hinds reign supreme.[130] The mammoth is only of importance in the Rhône and Quercy regions during the early phases, but in Dordogne during the Magdalenian, when the bison dominates in the Pyrenees.[131] In the Quercy, the mammoth and megaloceros seem to disappear gradually over time, while reindeer and hinds appear and multiply.[132] The horse and ibex seem to maintain roughly the same frequency in most places and periods, although the horse appears particularly important in a cluster of Dordogne caves in the Magdalenian (Cap Blanc, Comarque, St Cirq, etc.),[133] like the ibex in the late Magdalenian of Cantabria. Such regional and chronological differences may be at least partly ecological in origin, rather than purely cultural or a matter of 'preference'.

However, the same cannot be said of the massive change which has become apparent through the early date and parietal repertoire of Chauvet Cave: although horse, bison and deer are prominent in this cave, its art is dominated by rhinos, mammoths and big cats; when combined with the bear figures, these four categories account for up to two-thirds of the cave's animal figures (it is also worth noting that while most caves contain depictions of around half a dozen species – Niaux, for example, has 120 figures of six species – Chauvet Cave has around fifteen species represented, which is quite exceptional). It is apparent, when one looks at other parietal and portable depictions which can definitely or plausibly be attributed to the Aurignacian (particularly the figurines from south-west Germany), that there is a marked emphasis on large, powerful or dangerous animals.[134] It is in the Gravettian period that this phenomenon seems to alter radically in western Europe (towards the heavy emphasis on herbivores that is so well known in the later caves), though it persists somewhat longer in Central Europe, as seen in the portable art of Dolní Vestonice and the Kostenki culture.

Leroi-Gourhan's calculations concerning species frequency in different regions have been presented in the form of graphs; and cumulative percentage frequency graphs have also been produced to compare the depictions in different sites; exercises of this type obviously have to assume that all the figures have been correctly identified, but cumulative percentages are a particularly risky statistic.[135]

Finally, it should be noted that different species are not scattered at random through the sites; this is well known for parietal art (see below, p. 190), but the same phenomenon occurs with portable depictions. The differential distribution of depicted species at Dolní Vestonice has already been mentioned (see p. 98); at La Madeleine, the horse and bison dominate on the objects of bone and antler, but the reindeer was predominant on a cluster of engraved limestone slabs found in part of the shelter.[136] An excellent example of this phenomenon occurs at Gönnersdorf, where the mammoth, rhino and seal engravings were only found inside a habitation near the hearth, with another concentration a few metres away; depictions of birds, on the other hand, were concentrated precisely in those zones devoid of mammoths, whereas horses and humans were found all over the excavated area.[137] Since the mammoth engravings came from a winter habitation (according to the faunal remains) and the birds from a summer one, it is possible that there is a link between season and depicted species at this site.

Humans

The category of humans has traditionally been given the unwieldy title of 'anthropomorphs', comprising not only unmistakable human forms but also images which overlap with the other two categories, and many which seem to have been included simply because they looked figurative but did not resemble any known animal. The fact that humans lack the specialized anatomical features of other mammals (horns, antlers, etc.) meant that all kinds of vague, amorphous images were unjustifiably lumped together with clear depictions of people.[138] Instead, a distinction should be made between definite humans, 'humanoids' and composites, and the rest should be left undetermined.

Definite humans

Perhaps the clearest images in this group are the painted stencils and prints of hands, which, as we have seen (p. 119), are quite numerous and can dominate the art of whole caves (e.g. Fuente del Salín) or parts of caves (Gargas, Castillo). Apart from a stencil on a block from Abri Labattut (which may have fallen from the wall), found between two Gravettian layers, all hands are parietal; only the Labattut hand and some of the fifty-five examples in Cosquer Cave have firm dates, though others at Fuente del Salín and Gargas have been dated indirectly (see Chapter 5). However, studies of superimposition, composition and association in different caves suggest that this simple motif spans the entire Upper Palaeolithic.

As we saw in Chapter 8, a few caves have incomplete hands, which may be due to mutilation, pathologies, or some system of signals and gestures using bent fingers. Ethnography provides a wide range of possible explanations for hand stencils, both intact and incomplete: e.g. a signature, a property mark, a memorial, a wish to leave a mark in some sacred place, a record of growth, 'I was here', or simply 'just put there'. In view of the timespan involved it is probable that all these roles and many more

were involved in the Palaeolithic stencils; and since the specific reasons for making such marks only seem to be remembered for a couple of generations[139] we shall never know why Palaeolithic people painted them.

For decades, despite the steadily growing number of 'Venus' figurines, it was believed that depictions of humans were rare in Palaeolithic art, and that they were badly done (even though many animal figures were equally incomplete or sketchy!). Now that at least 115 quite realistic human figures are known among the engravings of La Marche,[140] outnumbering all other species depicted at the site, it is clear that they are not so rare; their quality, together with that of the figures at nearby Angles-sur-l'Anglin, and that of the ivory heads from Brassempouy and Dolní Vestonice, shows that realistic images of people were by no means taboo, as had been supposed.

Nevertheless, it remains true that few depictions of humans can match the finest animal images in detail and beauty. More effort may have been put into the animal figures for some ritual purpose, but the answer might simply be lack of ability. Drawings of humans and animals require different skills, and it is noteworthy that Breuil, who learned much of his art from the 'Palaeolithic school', had a talent that was 'confined to portraits of animals. When he ventured to reproduce human figures ... his work was not on a higher level than that of any fairly efficient amateur'.[141] Breuil drew amusing caricatures of his teachers at school, and it is quite possible that some of the Palaeolithic portraits which look funny to our eyes are actually caricatures rather than attempts at serious portraits.

Genitalia are rarely depicted on Palaeolithic humans, even on figurines,[142] and pubic hair is never shown. On bodies drawn in profile, the phallus can be seen, but the vulva cannot. A recent study of over seventy depictions of supposed males found that about one third were ithyphallic.[143] On isolated heads, or where genitals are absent, males can still be confidently recognized from beards and moustaches, whereas breasts denote females (assuming that bearded ladies and hermaphrodites were as rare in the Palaeolithic as they are today, though it should be noted that the Brno male statuette has small breasts or exaggerated nipples). At La Marche, using these criteria, there are thirteen definite males (including eleven with beards).[144]

Of fifty-one bodies drawn at the site, four are definitely male, but only twenty-seven of the rest can confidently be seen as female: eight of these have breasts; the others have been identified on the basis of the size of hips and buttocks.[145]

Heads without facial hair have to be left 'neutral' – length of hair is not a sure guide. Hence the Brassempouy head is unsexed, although it is usually called a 'Venus' or 'Lady', and it has even been placed on a female body in one reconstruction.[146] Other scholars have attempted to identify males on the basis of nose-shape or faces that jut forward[147] – despite the fact that the two females engraved at Isturitz have equally jutting faces. There is a wide range of nose- and face-shapes at La Marche alone. It has also been claimed that Gönnersdorf has depictions of men with hairy legs, while one medical specialist has recently interpreted numerous unsexed figures as men on the basis of their general bodily proportions as well as their association with what he sees as weapons or hunting scenes (i.e. scenes of conflict, as opposed to supposed women who are found only in peaceful scenes).[148]

The presence of adornments can denote females, though it is not an infallible guide: the nine bodies at La Marche that seem to have bracelets and anklets are all female; analogous ornaments including 'necklaces' can be seen on the above-mentioned females from Isturitz (identified by their breasts, p. 173), as well as on the similar 'Femme au Renne' from Laugerie-Basse (with a vulva but no breasts), and on some female figurines from Russia.

Clothing is rarely clear; the 'Venus of Lespugue' seems to have a garment of some kind showing at the back, and belts are depicted occasionally (e.g. at La Marche and Kostenki). The bearded male, sculpted, engraved and painted, at Angles-sur-l'Anglin appears to be wearing a garment with a fur collar, and perhaps also some form of headgear.[149] However, the so-called 'Femme à l'anorak' at Gabillou is unsexed, and the 'hood' may simply be a hairline. Elaborate hairstyling is extremely rare (e.g. the Brassempouy head, the 'Venus' of Willendorf, or one female figurine from

Fig. 10.27 Tracing of engraved plaquette from La Marche (Vienne), and the face of a bearded man extracted from the mass. Magdalenian. (After Airvaux and Pradel)

Fig. 10.28 Tracing of an engraving of a bearded male profile on a small limestone plaque from Riparo di Vado all'Arancio (Grosseto, Italy). Final Epigravettian (c. 14,000 years old). It measures 40 mm high, 33 mm wide. (After Minellono)

Avdeevo). Using the strict criteria outlined above, it has been established that La Marche has thirteen male, twenty-seven female, and sixty-nine undetermined humans; a study of 410 parietal and portable human figures from Western Europe (including those of La Marche but not those of Gönnersdorf) produced a very similar picture: 10.2 per cent male, 24.8 per cent female, and 64.6 per cent neutral. However, seventy figures from Eastern Europe – all in portable art – proved to be 4.2 per cent male, 60 per cent female and only 35.8 per cent neutral.[150] The difference with the West is largely due to the fact that almost all the Eastern figures are statuettes, on which gender tends to be more easily recognizable.

Children are very rare in Palaeolithic art (perhaps even non-existent apart from the supposed baby of Gönnersdorf), though some specialists believe that a few of the humans of La Marche may be infants on the basis of head shape and bodily proportions.[151]

In Palaeolithic depictions of humans as a whole, details such as eyebrows, nostrils, navels and nipples are extremely rare. The legs are often too short (as in the art of many later cultures); the legs and/or feet of figurines are held together or slightly apart, and the female figurines of the Russian Plain usually have bent knees. Few figures have hands or fingers drawn in any detail (examples include La Marche, the 'Venuses' of Laussel and Willendorf, the man in the Lascaux shaft-scene, etc.). More than 75 per cent of female statuettes have arms depicted, usually held close to the body, for technical reasons (most are symmetrical, a few asymmetrical), and in more than 70 per cent of these the arms are directed to specific points rather than just hanging at the sides – only four have the arms directed to the breasts, while in twenty-eight the arms rest on the abdomen. Since it would be more normal to have the hands clasped at the pubic level, this has been seen as significant, and perhaps links these particular figurines to pregnancy (see below).[152] On other figures, the arms may be at the sides, raised horizontally (as on the Sous-Grand-Lac man), or up in the air – all these poses are represented at La Marche.[153]

Except for the two ivory heads mentioned

earlier, very few statuettes (apart from those of Siberia) have any kind of facial detail; their heads are held erect or tilt forward slightly (though one from Kostenki tilts upward). Many Palaeolithic humans are headless: in some cases, as in the female bas-reliefs of Angles-sur-l'Anglin, it is clear that they never had heads. The same is true of nine engraved humans at La Marche, but others at this site and elsewhere may have had their heads broken off, either purposely or accidentally.[154]

There are three heads seen full face at La Marche; all the others, together with 90 per cent of the bodies, are in profile, and it is therefore worth examining which way they face. Of fifty-seven isolated heads, thirty-five face right and twenty-two left; of fifty-one bodies, thirty-three face right and eighteen left. In short, of 108 people, sixty-eight (63 per cent) face right and forty (37 per cent) left. Once again, as with sexing (see above), the percentages are found to be consistent when the analysis is extended to 167 humans from other sites: 100 (60 per cent) of them face right, and sixty-seven (40 per cent) left.[155]

The depiction of women

Apart from a few probable males (e.g. Brno, Hohlenstein-Stadel – although some researchers believe the latter to be female), most carvings of humans are female. The 'Venus' figurines[156] have been presented in so many art histories and popular works on prehistory that they have come to characterize the period and its depiction of women; this is unfortunate, partly because such statuettes are rare when seen against the timescale of twenty-five millennia, and because the constant display of a few specimens with extreme proportions presents a distorted view.

The term 'Venus' was first used by the Marquis de Vibraye in the 1860s in connection with his 'Vénus impudique' from Laugerie-Basse (ironically, a very slim lady!, see p. 16), and was later adopted by Piette for the more corpulent figures from Brassempouy and elsewhere; subsequently, it seems to have become attached primarily to the obese statuettes.[157]

Despite the accepted view that Palaeolithic depictions of humans were badly done, early scholars nevertheless tried to use them as evidence of different races during the period. As with similar attempts involving horses and other species (see above), they were doomed to failure. Piette was among the first to have a try; apart from his erroneous belief that a Grimaldi carved head was negroid, his major mistake was to see the obese 'Venuses' as steatopygous: i.e. having the special fatty deposits which produce the massive, high and wide buttocks of some female Bushmen and Hottentots.[158]

In fact, he was confusing this phenomenon with the kind of buttock development which is found in all races. Only one statuette (the 'Polichinelle' of Grimaldi) could conceivably be steatopygous; the rest merely present proportions which one can see anytime anywhere on women who have produced lots of children. In short, 'Venus' figurines have no value whatsoever as indicators of race.[159] Only four (Lespugue, and one each from Willendorf, Grimaldi and Gagarino) have really extreme proportions, and only Lespugue has truly monumental breasts. On the whole, the obese carvings are not anatomically abnormal – they are simply bodies worn and altered by age and childbearing. They seem well nourished, with their adipose tissue concentrated in discrete areas rather than spread out as a continuous layer.[160]

One study which looked at 132 Palaeolithic 'Venuses', and tried to estimate their age group, divided them into the following categories: young (pre-reproductive, with a firm body, high breasts and a flat stomach); middle-aged (reproductive and potentially pregnant, with a fleshy body, big breasts and a protruding stomach); and old (post-reproductive, sagging all over). The result was thirty (23 per cent) young, twenty-three (17 per cent) middle-aged and apparently pregnant, fifty (38 per cent) middle-aged but not pregnant, and twenty-nine (22 per cent) old.[161] Despite its tentative and subjective nature, this exercise did at least suggest that the carvings probably represent women throughout their adult life. Unfortunately, it was done from pictures, and researchers working with the actual carvings found that the criteria used did not correspond with reality, and could not be determined on the originals – for example, on the Russian figurines one cannot estimate the age of the

females depicted, which can only be classed as women 'of mature shape'.[162] Specialists seem agreed that the carvings do represent a wide range of physiological conditions, often involving pregnancy or, in a few cases, childbirth (e.g. a figurine from Kostenki XIII). This would help explain why the breasts and abdomen are constantly emphasized, but not the thighs and buttocks (except in some West European examples), let alone the head and extremities. The small, slim Siberian figurines, on the other hand, are utterly different, with a total absence of any accentuation of the female anatomy.

Jean-Pierre Duhard, a male gynaecologist, sees the Palaeolithic female carvings as 'physiological portraits that depict morphological features linked to pregnancy', and feels able to assess whether they are pregnant or not, or whether they have had children in the past and even how many (and in the light of this interpretation, he sees the Laussel Venus's horn, with its parallel incisions, as an obstetrical cal-

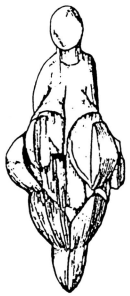
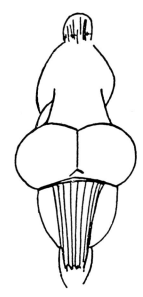

Fig. 10.29–31 The 'Venus' of Lespugue (Haute-Garonne) seen from the side. Ivory, probably Gravettian. Height: 14.7 cm.. Drawings show front and back views after Pales and Leroi-Gourhan respectively.

endar).[163] Most researchers are troubled by his somewhat subjective selection of which parts of the figures to 'take at face value' and which not. Duhard himself stresses that these are not anatomically realistic portraits, and points to the shortened arms and legs, the lack of proportion between head and body; but he also treats the breasts, abdomen, hips and buttocks as being relatively realistic, a fairly accurate guide to the subject's physiological history, how many children she has had, etc. However, one needs to bear in mind that the notion that each figure was meant to represent a female of a precise age and physical state is a theory, a possibility, not an accepted fact. Indeed some female researchers have cautioned against over-interpretation, and – as with animal imagery (see above) – treating the depictions as photographically accurate. Besides, it has been shown that these physical features can be found in non-pregnant women and in non-mothers: i.e. that one cannot distinguish mothers from non-mothers by shape, and that one cannot generalize in this way about the female form, such is its enormous physical variation and such is the stylization of these little carvings.[164]

As with bison and other species, some scholars have attempted to establish the measurements and conventions used in the construction of Palaeolithic human figures, but made the same mistake of using the images themselves, and not human anatomy, as the base of the study. Drawing lozenges around the Venuses merely shows that they have the same basic shape (as one would expect of human depictions), though with variations, and exaggerates the degree of their deformation. Anatomical studies, on the other hand, suggest that the figures' proportions are close to reality.[165]

One bizarre theory which was recently presented to explain the stylization of the female figurines claims that all such statuettes were actually produced as self-portraits by female carvers, many of them pregnant, who completely ignored the bodies of women around them and relied exclusively – for thousands of years – on the distorted views they could obtain by peering down at their own. There are numerous problems with such a view. First, it is as sexist to claim that all these images were

made by women as it is to assume that they were all produced by men (see below, pp. 188–9). It is claimed that the images show 'a realism in representation which sometimes approaches scientific exactitude', yet one wonders why artists of so long ago should have been concerned with such precision, which is an extremely modern feature. Moreover, the theory also falls down in its explanation of why the limbs are so abbreviated in these images – it is not difficult at any time to see one's arms and hands and to know their true shape, size and proportions. Similarly, when sitting down, one can see one's thighs, calves and feet extremely well, and even the most heavily pregnant woman must have remembered what her lower extremities looked like, even if she, like all the other artists, was totally ignoring the bodies of everyone around her. Besides, we know that the artists must have looked at other people, since they could not see or depict their own head or hair, and would need to be contortionists to obtain a detailed view of their own vulva![166]

As was mentioned earlier (see p. 100), 'Venuses' are not limited to the Gravettian and Magdalenian; there is good reason to believe that those of Brassempouy, at least, are Aurignacian; a possible 'rough' for one was reported in the Solutrean of Roc de Sers,[167] but is unconvincing. So far, only Angles-sur-l'Anglin has produced both parietal (four in bas-relief) and portable specimens (two statuettes), together with over a dozen other human depictions including the famous large and small 'versions' of a male 'portrait'.[168] However, the 'Venus' figurines were not found in any kind of privileged position. Very few such statuettes in western Europe have any stratigraphic provenance, though some (e.g. Lespugue, Courbet) seem to have been carefully hidden under rocks in caves and shelters.

In Eastern Europe, however, the excavation of open-air habitations has frequently encountered intact female figurines in pits in hut floors, particularly at Kostenki and Avdeevo, while others are found in the cultural layers:[169] for example, one from Kostenki I, found in 1983, and dating to about 23,000 years ago, was upright in a small pit, leaning against the wall and facing the centre of the living area and the hearths; the pit was filled with soil mixed with red ochre and was capped by a mammoth shoulder-blade.[170] At Avdeevo, a site dating to 20,000-21,000 years ago, twenty-five fragmentary or whole human figurines in ivory have been found, including fourteen intact specimens. Some have hair shown, one or two have faces, but none has a mouth or a vulva. In 1976/7, three were found together in a pit, along with normal flint and bone artifacts. One lay horizontally, face down, at the bottom; the other two were together, one horizontal and face down, the other on its side. 'Ebauches' — i.e. apparently unfinished figurines — and fragments are found in identical circumstances. Pits in the Russian sites may hold one, two or three figurines, of the same type or different; the pits themselves seem no different from others in the sites, and are not associated with hearths. In some cases, it is not even clear whether the figures were deliberately placed in the pits, or simply fell in from the cultural layer as the pit disintegrated. In any case, we do not know whether these pits were ritual or for storage.[171] The Eastern statuettes have often been interpreted by Russian scholars as a mother- or ancestor-figure, a mistress of the house.

The hidden figurines, both in Western and Eastern Europe, cast doubt on the theory that all Venuses were meant for public exhibition, and that those with a protuberance at the base (Tursac, Sireuil) or with peglike feet were hafted on to stakes or stuck in the ground for display.[172] Since the 'Venuses' cannot stand up, and it is unlikely that all were made to be stuck into the ground, one wonders if some were worn as pendants from the neck or waist, or carried around. Some may have been made for a single, specific use, others for posterity.

Finally, it is worth noting that no definite 'Venuses', either parietal or portable, have yet been found in the Iberian Peninsula; this is partly because bas-reliefs are limited to parts of France; but in view of the remarkable distribution of female statuettes from the western Pyrenees to Siberia, there seems no reason why Spain should not contain a few (Portugal has one or two very doubtful specimens), and no doubt future discoveries will alter the situation.

'Humanoids'
This category should comprise all those figures interpreted, but not positively identified, as being human. For example, there are many heads which could belong to animals but also resemble some of the more 'bestialized' heads on definite human figures: one, engraved on the wall at Comarque, represents a bear for some scholars, but others see it as human.[173] It cannot safely be assigned to either group. A number of rudimentary and sometimes grotesque heads on cave walls, seen full-face, are potentially human (e.g. Marsoulas, or Trois Frères), while others (such as 'phantoms') are best left undetermined. As mentioned above (see p. 156), the miniature figures grouped on some engraved bones have traditionally been interpreted as human (despite the lack of any distinguishing features) but may in fact be birds.

This is by no means a new problem. From the start, Breuil was well aware of the difficulties in separating 'elementary anthropomorphs' not only from animals but also from 'signs'. It occurred to him that the 'claviform' (club-like) signs might be derived from stylized females, but he dropped the idea because too many intermediate steps in the stylization were missing.[174] Today, however, we have far more 'stylized females', and the idea has re-emerged.

Such stylizations are particularly abundant in the Late Magdalenian, though they are not exclusive to the period. They occur both on cave walls (e.g. Fronsac)[175] and on slabs and plaquettes as at Gare de Couze and La Roche-Lalinde: Gönnersdorf alone has 224 such figures on eighty-seven engraved surfaces, as well as about a dozen figurines of similar type, in ivory, antler and schist; some of the figurines are perforated.[176]

The engraved figures are all stereotyped headless profiles (mostly facing right) with protruding buttocks; about a third can safely be identified as female humans because they appear to have breasts, as well as arms and other details. The other, more sketchy outlines from this site and elsewhere have been interpreted as females because of their resemblance to the more definite examples,[177] but they could be either young females or males (none is definitely male), and are best left as 'neutral': some of them would probably be classed as non-figurative if found out of context! The Gönnersdorf females have a wide range

Fig. 10.32 RIGHT: Stylized humans engraved on plaquette from Gönnersdorf (Germany): two female figures are depicted, 'face to face'. Magdalenian. The figure on the right is 11.3 cm long.

of proportions, and occur both singly and in groups of two, three and more, which have been dubbed as 'dances' for some reason. Some couples face each other, and one woman has what may be a baby on her back (Fig. 10.33).[178]

In view of the position of Venus statuettes in open-air habitations (see above), it is important to note that the intact figurines from Gönnersdorf were all found in pits.[179] It will be recalled that such statuettes are thought by some researchers to have been made for long-term use, while plaquettes had a short-term use before being discarded and/or broken (see pp. 58 and 90–91).

Various authors have proposed unilinear sequences of development from realistic to stylized, and even to 'signs'; Breuil did this for a number of species, and Leroi-Gourhan saw humans as being stylized into the headless profiles and ultimately into 'claviform' signs, which, like the profiles, occur singly and in groups.[180] Alas, all such schemes are subjective to some degree, and often ignore dating and other factors: for example, true claviform signs are exclusively parietal, probably Middle Magdalenian in date, and limited to the Pyrenees and Cantabria (and possibly also earlier at Lascaux); but the headless female profiles

Fig. 10.33 Engraved plaquette from Gönnersdorf (Germany), showing four schematized females with what may be a small child between them. Magdalenian. Width of plaquette: 8 cm.

are most abundant in Germany (and to a lesser extent in Dordogne), and seem to be Late Magdalenian! The claviform comprises a vertical line with a bulge at one side; but unlike the buttocks on the stylized females, the bulge is usually on the middle or the upper half of the 'sign'![181] It has been suggested that the 'upper bulge' on a claviform instead represents breasts, but this is based on supposedly stylized statuettes from Dolní Vestonice and elsewhere, comprising a rod with two little lumps near the top, and thus on analogies with sites even further away in space and time from the claviforms. In any case, stylization of humans or of any other theme can and probably did take place independently, in a number of phases and regions.[182]

The claviform question demonstrates the dangers of subjective impressions in these matters; a similar debate concerns some carvings from Mezin (Ukraine) which different scholars see as schematic images of women, as phallic symbols or as birds.[183] They may, in fact, be schematic female symbols (triangles, etc.) engraved on phallic birds! Clearly, in situations of this kind, the figures must be left undetermined, and speculation about what they depict must never be claimed as a firm identification.

Composites

We have just encountered some undiagnostic examples which could be either animal or human; but what of those figures which have clear and detailed elements of both? They have been called 'anthropozoomorphs' or 'therianthropes', but the term 'composites' is simpler, and does not give priority to either the human or the animal features.[184]

In the past, thanks to the dominance of the 'hunting magic' theories (see below, p. 171), all such figures were automatically and unjustifiably called 'sorcerers', and were assumed to be a shaman or medicine man in a mask or an animal costume.

But how can we differentiate people wearing masks, people with jutting, bestialized faces (see above) and humans with animal heads? Unfortunately we can't, and it is likely that all of them are represented in Palaeolithic images. Inevitably, interpretations have been heavily influenced by ethnographic parallels: for example, a bison-headed humanoid at Trois Frères (p. 178) bears some resemblance to North American Indians disguised as bison for a dance or hunt,[185] while another famous figure from the same cave – the 'Sorcerer', with his antlers and other animal parts – looks something like an eighteenth-century depiction of a Siberian shaman,[186] though in view of the vast distance in space and time separating the two images, and the often very inaccurate standards of depiction in the eighteenth century, it would be foolish to see any more than a superficial resemblance between the two. These Palaeolithic images may indeed be disguised humans, but it cannot be proved.

On the other hand, many scholars have noticed the strange resemblance between the Hohlenstein-Stadel statuette and Egyptian figures of a deity with a human body topped by a lion's head; in this case, therefore, the ethnographic analogy points towards an imaginary being – again without proof. Likewise, Breuil called the Trois Frères 'sorcerer' the 'God of the Cave' and considered him an imaginary figure, dominating the 'sanctuary'. Not all such 'sorcerer' figures, however, are in imposing posi-

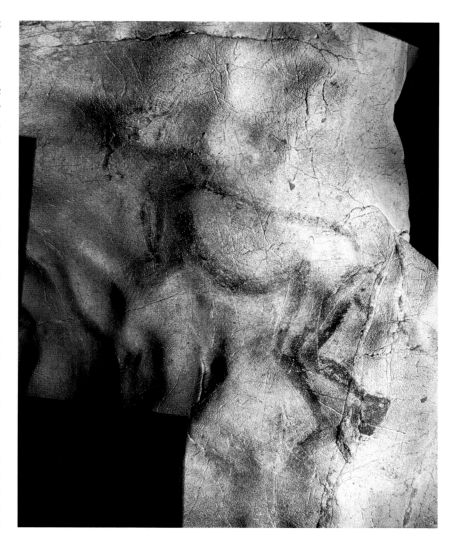

Fig. 10.34 & 10.35 Photomontage of the 'sorcerer' of Trois Frères (Ariège), together with Breuil's version of this painted and engraved figure. Probably Magdalenian. Length: 75 cm.

tions – indeed, some are hidden away among other images, and can be very hard to see.[187]

The Trois Frères figure is certainly strange: only the upright position and the legs and hands are really human. The rest is a mixture of different animals – the back and ears of a herbivore, the antlers of a reindeer, the tail of a horse, and the phallus in the position a feline would have it.[188] Very few such composite figures have horns or antlers – apart from the Trois Frères examples, Gabillou has clear horns (p. 180), as does a strange creature painted on a hanging rock at Chauvet Cave, which seems to be a bison upright on human-like legs.[189] Some researchers believe the vertical bison of Castillo (Fig. 8.5) likewise has human legs. Overall, however, composites or 'monsters' are extremely rare in parietal art – according to Leroi-Gourhan only about fifteen sites have them, with no more than half a dozen in each, and some of those he includes are probably not even humanoid.[190]

The distribution of human figures is interesting: as with parietal and portable art, there are many sites with none, some with one or two, while others like Gönnersdorf or La Marche have many. Pales' sample of 410 in Western Europe (including La Marche but excluding Gönnersdorf) comprised 187 realistic figures, 214 'humanoids', and nine composites[191] – the vast majority (c. 80 per cent) were engravings, with carvings accounting for almost all the others: painted humans are extremely rare.

Definite humans are scarce in parietal art, and there are few women resembling the Venus figurines or the stylized profiles (examples include Angles-sur-l'Anglin, Laussel, Fronsac). The distribution of sexed and unsexed humans within caves will be studied later (see p. 194).

Portable art accounts for over 75 per cent of Palaeolithic human depictions; as mentioned earlier, at Dolní Vestonice the human figures were not found in the same hut as those of herbivores, but were with carnivore figures. At Gönnersdorf they were scattered throughout the excavated area, unlike engravings of mammoths and birds: animals predominated on some slates at this site, while other slates were primarily devoted to stylized females; no intentional grouping of human figures with animals

or with symbols could be discerned.[192]

Non-figurative: 'signs'

The non-figurative category of Ice-Age markings was neglected until relatively recently, for the simple reason that it seemed uninteresting, or impossible to explain or define. Many lines, as mentioned earlier (p. 51), were simply ignored as *traits parasites*. Nowadays, however, thanks to the attention paid to the abstract 'signs' by Leroi-Gourhan, and to the discovery of similar non-figurative motifs in Australia and elsewhere (see above, p. 38), we have to come to terms with the possibility that these marks may have been of equal, if not greater, importance to Palaeolithic people than the 'recognizable' figures to which we have devoted so much attention. Certainly, it has been estimated that non-figurative marks are two or three times more abundant than figurative, and in some areas far more: for example, on 1200 engraved pieces of bone and antler from twenty-six Magdalenian sites in Cantabria, there are only seventy identifiable animals – all other motifs seem non-figurative.[193] In another study of 920 pieces of portable art from Cantabria, only 20 per cent have identifiable figures on them (140 with only figures, fifty more have figures together with 'signs'), while 80 per cent only have 'signs'.[194]

As already mentioned, it is hard to separate the non-figurative category from the two others; and it comprises a tremendously wide range of motifs, from a single dot or line to complex constructions, and to extensive panels of apparently unstructured linear marks. There are significant differences in content between portable and parietal examples, which make it extremely difficult to date parietal 'signs', particularly in cases where they constitute the only decoration and figurative images are absent, as at Santián and other sites.[195] Even in caves with both figurative and non-figurative images on the walls, 'signs' can be either totally isolated, clustered on their own panels or in their own chambers, or closely associated with the figurative – and sometimes all of these. Among 231 pieces of Magdalenian portable art from La Madeleine which bear non-figurative motifs, no two are identical, even where the simplest motifs are concerned, so this type of unstandardized decoration is very difficult to assess in terms of possible role or meaning.[196]

In the past, certain shapes were assumed to be narrative or pictographic: i.e. to represent schematized objects on the basis of ethnographic comparisons (see Chapter 11) or, more often, subjective assessment of what they 'looked like': hence, we have terms such as 'tectiforms' (huts), 'claviforms' (clubs), 'scutiforms' (shields), 'aviforms' (birds), 'penniforms' (feathers), etc. These are no longer taken literally, but, as they have entered the literature so decisively, are retained simply as rough guides to particular shapes. Even if these motifs were ultimately derived from real objects, they could now be purely abstract,[197] although it is impossible for us fully to separate them from 'realistic' figures: some signs may indeed be schematized or abbreviated (or unfinished!) versions of some object. Some authors now regard 'signs' as 'ideomorphs' — i.e. representations of ideas rather than objects,

but we simply don't know whether they are real or abstract or both.

We inevitably apply our understanding to what we see; our background, culture and art history come into play, and hence the early scholars tried to translate some signs as if they were hieroglyphics, in the vain hope of finding a Palaeolithic Rosetta Stone. Even if these signs were representational, their meaning would be far from straightforward without an informant: for example, among the Walbiri of Australia, a simple circle can mean a hill, tree, camp site, circular path, egg, breast, nipple, entrance into or exit from the ground and a host of other things; similarly, 'translation' of simple signs and pictographs in North America can often produce differing and unexpected results.[198] A mere resemblance of shape does not necessarily mean that a motif is an image of an object.

What is clear is that the simpler motifs are more abundant and widespread, as one would expect, since they could have been invented in

Fig. 10.36 'Signs' in the cave of Castillo (Santander), from 30 to 60 cm high.

many places and periods independently. The more complex forms, however, show extraordinary variability, and are more restricted in space and time, to the extent that they have been seen as 'ethnic markers', perhaps delineating Palaeolithic groups of some sort:[199] hence, 'quadrilateral' signs seem concentrated in Dordogne, particularly at Lascaux and Gabillou; a different kind of 'quadrilateral' is found in a Cantabrian group of caves (Castillo, La Pasiega and Las Chimeneas, all in the same hill, together with Altamira, some 20 km away); 'tectiforms' are found in the Les Eyzies region (Dordogne), in a small group of caves only a few kilometres apart, including Font de Gaume and Bernifal;[200] very similar 'aviforms' (now known as 'Placard signs') occur in painted form at Cougnac (eleven) and Pech Merle (three), 35 km apart, but also 165 km away, in the form of seven engraved specimens at Placard;[201] triangles are predominantly found in the centre of Spain, but are rare in Cantabria; perhaps most remarkable of all is the distribution of true 'claviforms', which are known in Ariège (Niaux, Trois Frères, Tuc d'Audoubert, Le Portel, Fontanet, Le Mas d'Azil) and in Cantabria (Pindal, La Cullalvera) some 500 km away (and perhaps also at Lascaux). Some of these motifs – such as the Placard signs – are too complex to be caused by independent invention and must be linked, but their bizarre distributions may argue against a role as ethnic markers, and be seen simply as proof of long-distance contacts, like many examples of portable art. Oddly, the Spanish Basque Country has an almost total absence of 'signs'.

As will be seen below (p. 193), Leroi-Gourhan divided 'signs' into two basic groups, the 'thin' and the 'full', which he linked to a sexual symbolism (phallic and vulvar); he later added a third group, that of the dots.[202] In view of the wide range of shapes on display, such a division was far too simplistic and schematic; using geometric criteria, other scholars have proposed seven or even a dozen classes of parietal sign, but still found some very hard to fit in.[203]

Similarly, the non-figurative motifs in portable art have been receiving an increasing amount of attention lately, and likewise divided into different groups.[204] Meg Conkey, for exam-ple, sought out the basic units of decorative systems in Cantabrian portable art, and their structural interrelations. In a sample of 1200 pieces, she found a set of about 200 distinct 'design elements', which she organized into fifty-seven 'classes'.[205] Her analysis — albeit based on a database that was poorly dated and provenanced — suggested that a core set of fifteen elements was used throughout the Magdalenian, and was widespread within and outside the region. As with the parietal signs, therefore, it will be necessary to focus on the more complex designs in order to identify regional variations in style (and possibly distinct social groups of some kind).

In both parietal and portable art, a full survey of the presence and interrelationships of different motifs, as well as of their association with each other and with other figures, is required, but will entail a more complete published corpus than we have, followed by computer analysis. It is the presence and absence of particular combinations which is revealing: in parietal signs, for example, very few binary combinations occur out of the range of possibilities, and only signs found in binary combinations also occur in 'triads'.[206] Clearly, these marks were not set down at random, but follow some set of rules and simple laws.

Can they therefore be seen as a primitive form of writing? Theories about this go back to the discovery of portable art, when a variety of enigmatic motifs associated with animal figures were seen as possible artists' signatures by a number of scholars including Lartet, Garrigou and Piette.[207] It is inevitable that, despite their wide variety of shapes, some of the Palaeolithic signs will resemble some of the simpler characters in certain early forms of writing; after all, the range of possible basic marks is somewhat limited. It has recently been claimed that some Palaeolithic signs have very close analogies with characters and letters in ancient written languages in the Mediterranean, the Indus Valley and China,[208] but, as with resemblances to objects, this does not necessarily prove anything. As far as parietal signs are concerned, it often appears to be their presence which was important, rather than their layout or orientation;[209] instead of forming a script, they were sometimes joined through superimposition, juxtaposition, or actual integration to form composites.

Where motifs on portable objects are concerned, however, it is still possible that they include some sort of 'pre-writing'; we simply don't know. What is almost certain is that the meaning of the signs and marks, no matter how abstract they may appear to us, must have been clear to the maker and to those who saw and/or used them. We can see this today with our road-signs and warnings: some have meanings obvious to everyone, others have to be learned, some are public and official, others are private and individual messages; but all are known to those who operate within that system and are therefore skilled at reading them.[210]

Colour schemes

A final factor in the content of Palaeolithic art which we have yet to consider is the use of different colours; we have already seen that the range was limited, and usually involved a straight choice between red and black. At first sight, it might seem possible to theorize that one of these represented males and the other females (for example in hand stencils), just as we have blue for boys and pink for girls; but the situation is far more complex.

Where signs are concerned, their colour seems linked to different regions and periods, but red usually dominates. As mentioned earlier (see p. 127), red is more visible in dark caves, so that whatever message — of warning, topographic guidance, instruction, etc. — the signs conveyed could readily be seen and understood. In the Iberian Peninsula, 23 per cent of the parietal signs are engraved, 27 per cent painted in black, and no less than 50 per cent painted in red, with a particularly high proportion of red signs in Asturias and western Cantabria.[211] In France too, about two-thirds of painted signs are red, and they are often the only (or almost the only) red figures in the cave.[212] This is particularly noticeable on friezes of black outline animals, such as Niaux's Salon Noir, or the frieze at Pech Merle (pp. 122–3).

The use of colour for both signs and animal figures alters significantly through space and time: a survey of thirty-eight caves in France and Spain found that black accounted for 48 per cent of figures and red for 43.7 per cent, only 5.3 per cent in yellow and brown, and mixtures (bichromes and polychromes) a mere 3 per cent[213] — a fact which underlines the fact that the latter are exceptional and unrepresentative of Palaeolithic art, despite their popularity in art-history books.

However, when the paintings were dated according to Leroi-Gourhan's styles (see above, p. 69), a difference was found: in Style III (Late Solutrean/Early Magdalenian), 56 per cent of figures were red, and 34.7 per cent black, while in Style IV (Middle and Late Magdalenian) these percentages were almost exactly reversed, with black animals becoming particularly dominant at the end of the period. Style IV appears to have a marked tendency for black animals and red signs.

Black and red, whether in signs or figures, are sometimes found in different parts of the same cave (e.g. Pech Merle, Castillo, Altamira, Chauvet), which may denote a chronological difference, or some purposeful colour scheme. Some signs are associated with a particular colour: about 94 per cent of claviforms are red, as are 100 per cent of tectiforms in Dordogne; most dots are red, and isolated dots are always red.[214]

There are also a few such associations discernible in animal figures, but these appear to be of regional rather than general significance: for example, in Spain 95 per cent of hinds are red, whereas stags can be either colour; Las Chimeneas only has stags, and they are all black, whereas all the seventeen hinds in Covalanas are red. At Castillo and La Pasiega, however, red dominates for deer no matter what their sex.[215]

All of the above should have made it clear that in Palaeolithic art there are no absolute rules, no certainties, and a very broad range of complex problems; but also that this is no random accumulation of pictures — on the contrary, it is characterized throughout by careful selection of surface, size, technique, colour, species, anatomical detail, degree of accuracy or stylization. It is now necessary to examine what sense different people have tried to make of all these facts over the last century, and in what directions current research is taking us.

11 | Reading the Messages

So far, we have looked at Palaeolithic art and asked questions about where, when, how and what; it is now time to grasp the nettle and tackle the most difficult topic of all: why?

Early theories: art for art's sake

The first and simplest theory put forward to explain the existence of art in this period was that it had no meaning; it was simply idle doodlings, graffiti, play activity: mindless decoration by hunters with time on their hands. Such a view, formulated most clearly by Lartet and Piette in the nineteenth century, reflected not only the anticlericalism of de Mortillet (see p. 18) who refused to believe that Palaeolithic people had any religion, but also the spirit of the age:[1] art was still seen in terms of the recent centuries, with their portraits, landscapes and narrative pictures. It was simply 'art', its sole function was to please and to decorate. In addition, Palaeolithic art indicated that the people of the period had plenty of leisure-time, implying that hunting was easy, there was lots to eat, and life was pleasant, calm and (in Piette's opinion) largely sedentary.

The 'art for art's sake' view arose from the first discoveries of Palaeolithic portable art in the 1860s and 1870s, and it must be admitted that there was little in the content and composition of these pieces to suggest any deeper purpose; indeed, it remains quite possible that some of the decoration of portable objects, and especially of utilitarian items, may simply be decoration – although more may be involved even here, from marks of ownership to the complex notations and seasonal images in Marshack's interpretations (see below, p. 205). Similarly, the newly discovered engravings in the open air do not seem as enigmatic as those inside caves; but it should be remembered that those of Fornols are all incomplete figures, not simple pictures of animals which happened to be passing; and in any case we know from Australian ethnography, for example, that open-air art can be just as taboo, fearful or filled with power and complex meaning as anything deep inside caves.

Of course, when parietal art began to be found, it rapidly became clear that something more was involved: as will be seen in this chapter, the restricted range of species depicted, their frequent inaccessibility and their associations in caves, the palimpsests and undecorated objects or panels, the enigmatic signs, the many figures which are purposely incomplete, ambiguous or broken, and the caves which were decorated but apparently not inhabited all combine to suggest that there is complex meaning behind both the subject matter and the location of Palaeolithic figures. There are patterns that require explanation, repeated patterns which suggest that individual artistic inspiration was subject to some more widespread system of thought. It is worth noting that, even in 1878, when only a few pieces of portable art were known, Cartailhac had already sensed that there was a purpose and a meaning behind this art, having probably been influenced by the writings of Edward Tylor.[2]

Nevertheless, despite all of the above evidence, the theory reared its head again in 1987,[3] with a claim that the art has no meaning, no religious, mythic, magical or practical purpose; it is simply play, mere aesthetic activity indicative of an inherent human desire to express ourselves artistically. This view resembles the assumption that someone is innocent until proven guilty: nothing in the periods before writing can be proved absolutely to be religious; but if Palaeolithic art simply comprised 'objects of contemplation', one would expect to find a pretty broad spectrum of subjects, accumulated more or less randomly on suitable surfaces. Yet the opposite is the case. Besides, countless studies of modern 'primitive' cultures make it clear that there is not necessarily a boundary between the sacred and the profane – everything can have a meaning for them, whether it be decoration, dances or mere gestures.[4] To compare Palaeolithic art with modern graffiti, or attribute it to some inherent human desire to leave a personal mark in places which are isolated or which stir the emotions is a projection of the present into the past.

It is undeniable that the artists must sometimes have derived great personal satisfaction, and perhaps enhanced prestige, from the best of their work, and certain 'realistic' figures are probably finer and more detailed than was strictly necessary to make their message intelligible – just as medieval craftsmen lovingly put minute details into stained glass which was placed so high up that nobody could see them without binoculars!

However, simple aesthetics explains nothing and has no bearing on meaning. An aesthetic interpretation 'reduces cultural phenomena to an innate tendency and directs explanation inward to mental states about which we can know nothing' – to infer a state of mind (pleasure) from the art, and then use it as an explanation is a circular argument![5] The theory tells us more about its proponents' reaction to Palaeolithic art than about the artists.

Hunting magic, hunting and shamanism

After the death of Piette in 1906, utilitarian theories took over, and primarily that of hunting magic. This was largely caused by the newly published (1899) account by Baldwin Spencer and F. J. Gillen of the life and beliefs of the Arunta of Central Australia, which inspired Salomon Reinach[6] to equate these 'primitive' users of stone tools with those of the Ice Age, and therefore assume that the same motivation lay behind the art of both cultures.

The Arunta were said to perform ceremonies in order to multiply the numbers of the animals, and for this purpose painted likenesses of the species on rocks. The analogy seemed perfect, and Palaeolithic art must therefore have had a magico-religious cause. For a few decades, all ideas and interpretations were subjectively chosen in this way from the ever-growing mass of ethnographic material from around the world. This is quite understandable – it appeared self-evident at the time that 'primitive' societies provided lots of precious information about the making and function of rock art in the past. Almost a third of the first monograph on cave art was therefore devoted to a review of the available ethnographic evidence: the present would teach us about the past, through the parallelism of the artistic activity of all primitive hunting peoples.[7]

'Sympathetic magic', including hunting magic, operates on the same basis as pins in a wax doll: i.e. the effect resembles the cause. The depictions of animals were produced in order to control or influence the real animals in some way: any injury to the image would likewise be inflicted on its subject. Desirable effects such as ensuring that the prey would be as fine and well fed as the image were also possible. In short, this kind of art is intended to exert influence at a distance, rather like modern advertisements; it is purely functional, and directly related to the need for food. This explanation had already been put forward for portable art in 1887 by Gustave Chauvet, who felt that the artists had engraved either a reindeer (their food) or a mammoth or bear (their terrible enemies), and then associated their arrows with the target.[8]

Breuil adopted the 'hunting magic' view because he felt that Palaeolithic art sprang from the hunters' anxieties about the availability of game. Most other scholars followed, particularly Bégouën who saw evidence for ritual and magic in almost every aspect of Palaeolithic art.[9] One feature which he constantly stressed was that some images seemed to have been 'killed' ritually with images of missiles (any shape marked on an animal body was either a missile or a wound), or even physically attacked with a weapon or by hand, presumably to improve the chances of the hunt, or (in the case of carnivore figures) to vanquish a fearsome enemy. Marks at the mouths or nostrils of animals were always interpreted as blood being vomited by a dying beast; marks on their bodies were always wounds or missiles, and dots were stones thrown at them: the famous bear of Trois Frères, covered in 189 small circles, was thus an animal stoned to death, with blood pouring out of its mouth.

The theory accounted for the broken and short-lived engravings on plaquettes: once the animal had been 'killed' and the ritual was over, the image had no further use or significance, and was discarded. It also had an explanation for why so many figures were drawn on the same pebble (as at La Colombière) or panel when so many equally good stones or surfaces were available but left blank: they were 'good hunting' pebbles or panels. If the first figure drawn there produced the desired result, they were used again and again.[10]

Such assumptions also influenced or dictated the interpretation of 'signs', which were thought to represent whatever object they looked like: the 'claviforms' were clearly clubs or throwing-sticks, and 'penniforms' were arrows, while the Dordogne 'tectiforms' were sometimes huts, but mostly detailed diagrams of pit-traps and gravity-traps, especially in cases where they were superimposed on a bison or mammoth (but why would only this tiny area depict such devices?). Other motifs were assumed to be lassos, nets, and so forth.[11] Occasionally, a more exotic ethnographic analogy was used, as where some signs were compared to the traps found in the East Indies that are intended for malignant spirits, for the souls of the dead, or for the spirit of game.[12]

Needless to say, the subjectivity, presupposition and wishful thinking which permeated this theory led to many errors: one of the most blatant concerns the clay bear of Montespan (p. 112) which, from the start, was presented as the epitome of the 'killed' image; the advocates of hunting magic claimed that they could see holes in the bear which meant that it had been stabbed with spears or struck with missiles in some sort of ritual attack. Despite the fact that these 'holes' were simply the natural texture of the clay, that the bear is only 1 metre from the cave wall, and that the ceiling descends to a very low level immediately beyond it, one still finds dramatic verbal and graphic reconstructions of the scene in popular books, with warriors dancing round a bear head/hide draped over a dummy, and throwing spears or firing arrows at it! These ridiculous flights of fancy are the legacy of the hunting magic theory, whose proponents saw what they wanted to see.[13]

Had they been a little more objective about the data, at Montespan and elsewhere, they would have realized that they were constantly stretching and adapting the theory to fit the evidence, or carefully selecting the facts to fit the theory. For example, faced with the rarity of carnivores in parietal art, they claimed that these were dangerous animals, difficult to hunt, and in any case would not have been considered very tasty. Yet one would think that hunting magic would be most useful for dangerous animals! In any case, carnivores are not so dangerous (they tend to keep well away from humans most of the time), and bulls or stags can be far more aggressive; lions are caught and eaten in Africa, and bears and dogs are considered delicacies in some parts of the world.[14] Besides, how can one reconcile that explanation with the similar rarity of reindeer or birds in parietal art, resources which were of great economic importance at times? Here, the proponents claimed that it was unnecessary to depict these species because they were in plentiful supply and easy to obtain! The inconsistencies in logic are blatant.

Overall, there are very few Palaeolithic animal figures – only about 3 or 4 per cent – with 'missiles' on or near them: bison seem to have the most, but less than 15 per cent of bison are marked in this way; many caves have no images

Fig. 11.1 'Tectiform' signs painted on a polychrome bison at Font de Gaume (Dordogne). Probably Magdalenian. Bison about 1.6 m long.

Fig. 11.2 Engraved bone from Isturitz (Pyrénées-Atlantiques): on one side there are two females (depicted either vertical or supine, judging from their breasts), one of whom has a barbed sign on her leg. The other side depicts a bison with a barbed sign on its side. Magdalenian. Total length: 10 cm

of this type at all.[15] Two caves are exceptional in this regard: Niaux, where about a quarter of the animals have 'missiles', and Cosquer, where twenty-eight figures out of 142 are marked with forty-eight long engraved lines.[16] One study before the discovery of the Cosquer Cave found a total of 302 'projectiles' on 138 animals, about half of which were bison, and only 20 per cent horses, with parietal depictions being rather more common than portable.[17] However, these marks could be all manner of things (see other theories below); even if they were missiles, we cannot tell whether they were drawn at the same time as the animal, or years or centuries afterwards (though pigment analysis, at Niaux for example, might shed some light on this). In some cases, such as a pseudo-arrow under the neck of a reindeer from La Colombière, the 'missile' was engraved before the animal![18] One of the most famous examples is a rhinoceros on a pebble from the same site which was thought to be pierced by a number of feathered projectiles: yet such missiles are weak and aero-dynamically unstable, and certainly could not penetrate rhino hide.[19]

Similarly, the marks at the mouths and/or nostrils of animal figures are open to many other interpretations such as voice, dying breath, or breath to show that the animals are very much alive (and see below, p. 181) – they have a variety of forms, and are mostly drawn on bison, bears and horses.[20] Supposedly 'falling' animals are often simply drawn on vertical supports, which dictated their orientation.

Besides, what is one to make of 'missiles' associated with human or humanoid figures? The classic example is the engraving of two women from Isturitz, one of whom has a harpoon-like motif on her thigh, which is identical to a mark on the bison engraved on the other side of the same bone! A little ibex in Niaux appears to have lances sticking out of it, but the same is true of the man at Sous-Grand-Lac, or the humanoids of Cougnac and Pech Merle (Fig. 11.3), as well as a very doubtful humanoid figure at Cosquer.[21] Perhaps they represent the more unpopular members of the tribe, or Palaeolithic 'wanted posters'!

Even if the figures marked with 'missiles' were symbolic killings, there are many possible reasons for them: they could be mythological, perhaps, or some sort of sacrifice. Sympathetic magic is just one possibility.

However, most scholars now dismiss the labelling of these motifs by what they look like to our eyes, and avoid words such as 'arrow' in favour of more objective terms, descriptive of

Fig. 11.3 Humanoid apparently 'pierced by missiles', in the cave of Pech Merle (Lot). Note the strange 'aviform' sign above, now known as a 'Placard sign'. Possibly Solutrean. The humanoid is 50 cm high.

shape (e.g. chevron with a median longitudinal line).

Breuil and others believed that the carnivores were drawn so that the artist might gain their strength, their skill in catching game, or perhaps be protected from them – a mixture of envy and fear; but in that case, why are carnivore images so rare except in the Aurignacian? Clearly, hypotheses of this kind are untestable. In the same way, opponents of hunting magic argued that many figures were far too detailed and perfect for the job, since presumably a simple outline would have done; but adherents of the theory countered that the better the portrait, the more effective the magic! The arguments go around in circles and lead nowhere.

But if hunting magic is no longer upheld as the primary motivation, Palaeolithic art is still seen as a 'hunter's art' by some scholars who, even today, take some of it at face value as a narrative. This is by no means a new concept: as we have seen, many signs were interpreted as objects associated with the hunt, and the scene of the wounded bison and the man in the Lascaux shaft (Fig. 11.8) has often been considered a depiction of a real event, to the extent that a search was made for the unfortunate hunter's grave nearby!

Proponents of 'hunting art' pick out the very few examples of what they consider to be hunting or stalking scenes, dangerous situations, and animals which are dead, wounded, alarmed or bolting, and present them as if they were typical of Palaeolithic art as a whole:[22] however, not only are these extremely rare, but their interpretation is by no means straightforward, and we have already seen (p. 156) that some of the 'groups of hunters' may really be birds, and that the 'dying' or 'wounded' bison of the Altamira ceiling may actually be rutting, giving birth or rolling in the dust. The art is considered to have been made by men, and to reflect their obsession with game animals and the hunt, rather like the endless pictures in modern American hunting magazines.[23]

A similar view, which follows on from the idea, mentioned earlier (see p. 147), that the deer figures at Lascaux and elsewhere are accurate records of prize stags, is to see the whole thing as 'trophyism': i.e. the hunting and display of trophies by men in order to impress and get the girls. The demonstration and advertisement of their hunting prowess, with the prey's attributes emphasized through twisted perspective, improved their status and thus their reproductive success.[24] Palaeolithic art was thus equivalent to a collection of scalps or medals. The idea is appealing, but does not stand up to scrutiny; it does not explain the enigmas in the art such as 'incomplete' or composite images, the large non-figurative component, and the layout in caves.

The suggestion that the organization of figures within the caves was arranged according to trophy values makes little sense: in that case, one would expect to find ferocious carnivores, bulls and prize stags dominating the principal panels, which is not the case except at Chauvet with its unique preponderance of rhinos, mammoths, big cats and bears. However, one critic of Leroi-Gourhan's statistical approach to cave layout (see below, p. 190) did confirm that there seem to be three groups of figures: the large herbivores (horse, bison, aurochs), the smaller herbivores (deer, ibex) and the dangerous animals (rhino, bear, feline), but rather than interpret them in terms of some metaphysical plan, he suggested that their distribution in the caves reflected their economic categorization by a hunter, based on the amount of meat they provided and the risk they involved.[25]

Another interesting hypothesis, based on a literal reading of the art, is that of corralling: once again, it is by no means a new idea, and the strange circular images drawn at La Pileta, with double finger-marks dotted around inside them, have often been interpreted somewhat tenuously as corrals with animals (or ruminant tracks) inside them[26] (even though identical marks are also found inside animal figures in the cave), while animals with signs painted on or near them, at Marsoulas, Castillo, Niaux and elsewhere, were thought by some authors to be in enclosures.[27]

Recently, however, Thomas Kehoe, an expert on animal drives in North America, has proposed that much of the painted decoration of Lascaux constitutes a diagram depicting an animal drive, with the rows of dots and the rectangles representing the drive-lines (made of small piles of rocks, or suchlike), blinds and corrals.[28] The barbed signs near some animals are brush barriers which help to funnel the animals into a corral. The little horses are heading for one enclosure, the 'jumping cow' is skidding to a halt in front of another, while the falling horse is going over the edge (Fig. 11.5). This is not the first time that the Lascaux signs have been interpreted in these terms: the quadrilateral images have previously been seen as Texas-style gates for shutting in cattle and horses;[29] but the new hypothesis treats the cave itself as a symbolic drive, forcing the animals down the narrow passages to the 'fall' at the end; the round Hall of the Bulls was perhaps conceptualized as a pound, and the cave itself could be a memory-device to preserve knowledge of how and where to place the cairns and corrals.

Similarly, the round chamber at Altamira is seen as a symbolic pound, with a drive depicted on the ceiling:[30] the curled-up bison at the centre are already dead, while those around them stand and face the hunters (the male humans engraved at the edge of the group). The Altamira ceiling has previously been interpreted as animals falling down a cliff,[31] but never before as a symbolic pound as well.

The hypothesis has the merit that it concerns a technique which is certainly the most efficient way to exploit large herbivores, and which was surely used throughout the Upper and probably the Middle Palaeolithic; it is also solidly based on ethnographic information, including the methods of depicting such drives in other cultures. Indeed, the interpretation seemed self-evident to Kehoe as soon as he saw these caves, so close were the analogies to ethnographic examples. It is not claimed that all the 'signs' are corrals or drive-lines, but their

Fig. 11.4 Carved antler (possibly a spearthrower) from Le Mas d'Azil (Ariège), showing three horse heads, one of them apparently skinned. Magdalenian. Length: 16.3 cm.

Fig. 11.5 The 'falling horse', painted in bistre and black at the end of the Diverticule Axial in Lascaux (Dordogne). Possibly Magdalenian. The horse is complete, painted around the rock, and measures 2 m from muzzle to tail.

close association with the animals at Lascaux certainly make his hypothesis very thought-provoking.

In the last three decades we have become so accustomed to esoteric interpretations of Palaeolithic art that we tend to forget that not all of it is necessarily mysterious and metaphysical: much of it may well be mundane and deal with everyday concerns. Certainly, some aspects of animal exploitation seem to be present: for example, there are one or two depictions of what may be stripped carcasses, or of animal hides spread out;[32] and the earliest art on blocks includes a series of what appear to be animal prints (probably a bear's front paw — a large cupmark with an arc of three to five little ones above);[33] other motifs which have traditionally been interpreted as vulvas may also be horse-hoof or bird-foot prints,[34] while the series of double incisions on the handle of the carved Lascaux lamp could be the prints of an ungulate such as a boar. In any case, it is certain that the observation and recognition of animal and bird tracks were of great importance in Palaeolithic life, and it is no surprise that they were occasionally reproduced.

However, even if hunting played a role in the production of some Palaeolithic art, for any or

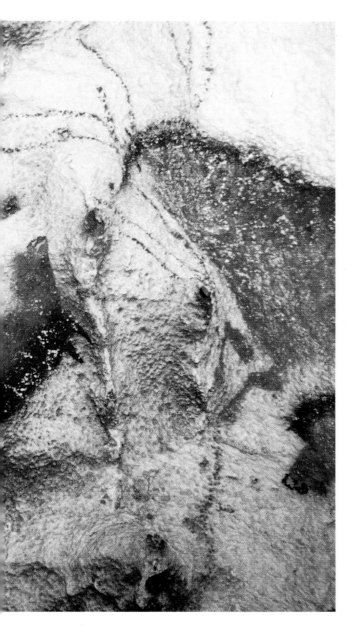

all of the reasons outlined above, the fact remains that the art's content is far more complex. We have already seen (Chapters 5, 10) that the species depicted cannot be taken as an accurate record of what was present in the environment; and it is equally true that they often bear little relation to the use people made of what was available: i.e. to the faunal remains in occupation sites.

It is often the case that we do not obtain an adequate picture of diet from the bones, because of differential preservation, or because the sample is small; and, as we have seen, it is often theoretically possible that the people who

ate the animals were not from the same period as the artists, though in most cases they probably were.

However, the results from many sites suggest that depicted species rarely correspond to eaten species in terms of quantity. This applies not only to parietal but also to portable art, and was noticed by some of the earliest specialists such as Cartailhac; in fact it was first observed in 1874 by Alexandre Bertrand when he compared the fauna consumed and the fauna depicted at the Swiss site of Kesslerloch. Since then, it has become an almost routine phenomenon. For example, at Gönnersdorf the mammoth is the second most frequently depicted animal (after the horse), but is very rare in the fauna, while the reindeer, whose antlers abound in the site, is rare in the art.[35] At Dolní Vestonice, the mammoth dominates the bones, but is rare in the art which is primarily devoted to carnivores.[36] At La Vache, the ibex is overwhelmingly predominant in the fauna but only accounts for about 18 per cent of the depictions, while the horse, which predominates in the art (over 25 per cent) represents only 0.01 per cent of the fauna![37] The bones from the Grotte des Eyzies and the Abri Morin are dominated by reindeer, but their portable art is primarily devoted to horses.[38] On the other hand, both the slabs and the bones of Limeuil are dominated by reindeer, and the horse dominates in both at Gönnersdorf.

In parietal art, the example always given is Lascaux, where reindeer account for 90 per cent of the bones but were drawn only once (0.16 per cent); as mentioned earlier (see p. 147), it is possible that many of the cave's stag figures may actually be reindeer, which would alter this disparity a little; but horses still account for 60 per cent of the figures. Bison, aurochs and ibex are in the cave's art but not in its bones; this situation is reversed for boar and roe deer.

The same picture is repeated elsewhere: the ibex dominates the art of Pair-non-Pair but is absent from its fauna; at Comarque over 75 per cent of the bones are reindeer, but the figures are mainly horses;[39] at Villars the people ate reindeer, but did not draw any; in many Ardèche caves, the mammoth predominates in the art (almost half of the figures), and the ibex

Fig. 11.6 & 7 Macro-photo of the head of a composite engraved figure at Trois Frères (Ariège), together with Breuil's tracing of the full figure. Probably Magdalenian. Total length: 30 cm. The head is about 3 cm across.

and reindeer are absent, whereas the bones show an abundance of the last two species and an absence of mammoth; at Gargas, there are no reindeer figures although over a third of the bones are of that species, but the mammoth, absent from the fauna, was depicted six times.

The situation recurs in Cantabria: at Altamira they drew bison and ate red deer; at Ekain the thirty-four horses are 57.6 per cent of the animal figures, but there are almost no horse bones in the fauna which is dominated by ibex and red deer; while at Tito Bustillo, the fauna is heavily dominated by red deer (up to 94 per cent), with ibex second, and very few

Fig. 11.8 Photomontage of the entire 'shaft scene' in Lascaux (Dordogne), showing rhinoceros, dots, apparently wounded bison, bird-headed ithy-phallic man and 'bird-on-stick'. Probably Magdalenian. Total width: *c.* 2.75 m.

horse bones, whereas the parietal art has twenty-seven horses (37.5 per cent), twenty-three red deer (31.9 per cent) and nine ibex (12.5 per cent).[40] On the other hand, the red deer dominates both the art and the bones at Covalanas and at La Pasiega.

In short, the relationship between depicted and eaten species ranges from non-existent to quite good; once again, there is no rule, no homogeneity. Nevertheless, it remains true that the restricted range of species depicted has no direct equivalent in any excavated level's sample of bones, and it is clear that the motivations behind the art were different to the environ-mental factors and economic choices which produced the faunal remains. A similar situation occurs in Gallo-Roman sites, where the animals depicted on pottery are over 75 per cent wild, with dogs dominating the other 25 per cent, whereas around 90 per cent of the bones come from domestic species.[41]

The disparities have been presented in the form of cumulative graphs which demonstrate, for example, that art and bones are far more different at Lascaux than at Gabillou.[42] Another study, however, which compared the combined percentages of depicted species in ninety French caves with those of bones from 151

Fig. 11.9 The 'sorcerer' engraved at the end of Gabillou Cave (Dordogne). Probably Magdalenian. 3.7 cm high.

other herbivores in flatter areas.

Although we can accept that the practice of sympathetic magic and the actual depiction of hunting may account for a bit of Palaeolithic art, the evidence for them is extremely scarce.[44] This has not, alas, prevented the recent resurrection of these old ideas, dressed up in the modern guise of 'transmitting information about hunting'. For example, all manner of figures and shapes have been interpreted utterly subjectively as depicting tracks, droppings and other visual 'cues' in an attempt to see the art as a means of communicating information about hunting – the most remarkable and idiosyncratic case is perhaps that of claviform signs, seen as depicting the marks left by deer rubbing antlers on tree-trunks!! It is blithely assumed that Palaeolithic art was primarily functional, useful in simple evolutionary terms, and that it is characterized by motifs linked with hunting information and 'mental preparation for the hunt'; corpuses of the art (often second- or third-hand sources, with some very poor illustrations) are then combed to find things which might conceivably represent what is required to support the theory, usually removing them from their context in the process.[45] This exercise in wishful thinking and total subjectivity leads nowhere. Even if one were to give these interpretations the benefit of the colossal doubt and accept them, the examples and 'visual cues' are so extremely scarce and unrepresentative that it is self-evident that the vast majority of Palaeolithic art has nothing whatsoever to do with hunting or teaching in such simple terms, regardless of date, location or technique.[46] Besides, if one wants to teach youngsters about animal tracks and spoors, it might just be more sensible to do so out of doors with the real thing, rather than in the depths of a cave with ambiguous imagery! Why is a databank necessary when one has oral traditions and full-time hunters at one's disposal?

There is a related phenomenon which could lie behind rather more of the art: shamanism. Many early scholars, as we have seen, interpreted the 'composite' figures as sorcerers or shamans in masks, basing themselves simply on similarities to ethnographic depictions and accounts. The theory has also persisted in connection with particular images: for example,

French sites spanning the entire Upper Palaeolithic, and those of thirty-five Spanish caves with the bones from thirty Spanish Upper Palaeolithic sites, was a complete waste of effort since it lumped all periods and vast regions together, thus completely ignoring crucial variations through space and time:[43] it only makes sense to compare an individual cave's figures with the bones from the same site, or at least from the same locality and period, since climate and local topography played a major role in economic choice – for example, ibex dominate the fauna in rugged ibex country, the

the famous Lascaux shaft 'scene' of the four-fingered bird-headed ithyphallic man with the apparently disembowelled bison and the 'bird-on-a-stick' has often been interpreted, quite subjectively, as a typical shamanistic séance, depicting the shaman, his spirit helper and a sacrificial animal; or as a shamanistic fight, a psychic conflict between two shamans (one in animal form) or between a shaman and a malevolent spirit.[47] Similarly, the enigmatic 'unicorn' figure in Lascaux, sometimes interpreted as men in a skin, has also been seen as a shaman driving the little horses before him.

The shaman is a very important figure, being a person with spiritual powers who combines the roles of healer, priest, magician, and artist as well as poet, actor and even psychotherapist! His most important function is to act as liaison between this world and the spirit world, a task sometimes performed by means of 'trances' and hallucinatory experiences — either symbolic, self-induced or using hallucinogenic fungi or other such substances. Cultures with shamans usually have a zoomorphic view of the world, and things are seen and experienced in animal form. Hence, from this perspective, Palaeolithic images are 'spirit animals', not copies of the real thing.[48]

In a theory of this kind, based on Siberian ethnography, the Abbé Glory suggested that many of the figures in Palaeolithic art were 'ongones', spirits which took the form of 'zoomorphs', 'anthropomorphs' and 'polymorphs', and which were asked to help in hunting, matters of health, and so on. Here, the lines emanating from animals' mouths would be an evil 'illness spirit' being exorcized, while damaged images will have been broken in anger when prayers were not answered.[49]

A closely related view, based on concepts widely held among modern hunting peoples, concerns the 'master-of-animals', usually a dead shaman; this humanoid or composite figure represents a third force mediating between living shamans and animals, and which constitutes the life-force of an animal or which can impart life-force to it. The living shamans derive their own life-force from the animals and then use it in the service of their clients.[50] According to this scenario, the artists were shamans maintaining links with the animals (with which they closely identified) through the master-of-animals; the power derived from the art came through the act of drawing, not from subsequent viewing. The lines at the mouths and nostrils of animal depictions represented the entrance or exit of life-forces.

It is worth noting that what appear to be lances or missiles hitting animals' bodies may likewise have an alternative explanation: for example, an Apache myth concerning the creation of the horse involves four whirlwinds penetrating the animal's flank, shoulders and hips, enabling it to breathe and move.[51]

Obviously, the shamanistic explanations, like the similar ones based on totemism in other cultures, involve ethnographic parallels; but from what we know about Upper Palaeolithic culture and life (particularly the Magdalenian), it seems likely that beliefs of this kind existed and thus played a role in the production of Palaeolithic art. Unfortunately, the latest attempt to apply such notions to the Ice Age[52] is based entirely on recent research on southern African rock art, involving hallucinations, trances and 'phosphene forms'.[53]

'Phosphene forms' are subconscious images, geometric shapes which seem to be present in the neural system and visual cortex of all human beings: some of the earliest art is thus thought to be an externalization of these images. This would explain the similarity between the finger tracings in Europe and Australia — they sprang from a common neural circuitry.[54]

These geometric forms are entoptic (i.e. can be seen by the eyes when the eyelids are shut). One way of inducing them is to alter the state of one's consciousness by trance or hallucination: psychologists have shown that the early stages of trance can produce such hallucinations in geometric forms. Hence, it has been argued that many simple Palaeolithic 'signs' are shapes of this type,[55] although any collection of non-figurative art is likely to include lots of marks which resemble entoptic phenomena, simply because there are very few basic shapes which one can draw, whatever the motivation or source — dots, lines, squiggles and basic geometric forms.[56] It has also been claimed, through analogy with supposedly shamanistic art in southern Africa, that in addition to the 'signs', many animal figures and, especially, the composites in Palaeolithic art may be hallucinatory images – i.e. they represent the insights and experiences of shamans in trance performances, and should thus be 'read' metaphorically.[57]

Of course, we have absolutely no evidence whatsoever to link any Ice Age art to something like shamanism except as a simple assumption. And even the existence of such practices in Palaeolithic society – which, as mentioned above, is quite likely, in no way proves how they were tied into the art, and leaves many questions unanswered: whether the shamans were artists, whether they were the only artists, what the pictures meant, how they were used, and what percentage of the art could conceivably be interpreted in this way – just that of the deep caves, or also that of the rock shelters, the open air, the portable objects?

Unfortunately, in the excitement at finding a possible key to some of the hitherto enigmatic motifs and features of the art, some researchers have predictably gone to extremes – just like those who misused ethnography in the early part of this century – and have not only made major assumptions about motive, content and meaning, but have also extended the hypothesis to ridiculous lengths, claiming, for example, that some of the 'signs' such as claviforms (and the 'butterfly' signs of Chauvet) are toxic mushrooms or hallucinogenic plants, others are magic drums, and handprints represent direct percussion on the cave wall.[58]

The new emphasis on 'altered states of consciousness' – associated with shamanic 'trance' – as the origin of numerous motifs is based on ethnographic accounts and neuropsychological studies which suggest that as shamans enter 'trance', they experience 'entoptics'. It is argued that people entering an altered state of consciousness go through three stages. In the first stage they see entoptics. In the second they elaborate these geometric images into objects with which they are familiar, and in the third stage they hallucinate the forms that are most commonly depicted in their rock art.

However, there are many problems with this approach to prehistoric art. First, a zigzag motif could easily be inspired by lightning, just as circles can be inspired by ripples in water.

Second, it does not require a shamanic trance to see entoptics – one can just as easily experience them through migraine or pressing on the eyeball, while altered states of consciousness can be induced through prolonged isolation, extreme boredom, nocturnal hallucinations, fervent praying, daydreaming, drowsiness, sleep deprivation and numerous other circumstances – even excessive masturbation! In fact hallucinogen- or trance-induced phosphenes account for a tiny fraction of such experiences. Hallucinations by older people with failing eyesight are very common and have been well documented since the eighteenth century. Even more pertinent to the topic of Palaeolithic art is that artists are often said to be in an 'artistic trance of creating', in another reality very much like a drug-induced state – for a profoundly creative mind, drugs are unnecessary to have a hyper-reality experience or the feeling of an out-of-body experience. Trance is merely a biological ability at which some people excel – it is a state of focused awareness, when one feels defocused from the outside world and often experiences a sense of falling, flying or whirling.

Different specialists disagree about how many of the 'entoptic' images derive from the wiring of the human nervous system rather than conscious recalls from things seen in trance. In any case, to seek the source of imagery only in altered states of consciousness and in universal neurophysiology has not produced a cross-cultural 'skeleton key' that unlocks the secrets of Palaeolithic art. In fact, it reveals no more than that the artists were human beings. It may pinpoint the ultimate source of the imagery, but tells one absolutely nothing of its meaning – even if one could be sure the imagery resulted from an altered state of consciousness, one would not know of what kind or in what context.[59]

In any case, the African research on which the new approach is founded is itself based on massive assumptions, wishful thinking, and extreme selectivity of image and of interpretation. In fact there is no hard ethnographic evidence to link any prehistoric southern African rock art with shamans, let alone with trance phenomena.[60] For example, the written ethnographic records contain no mention of trance whatsoever, nor of any practices, dreams or beliefs involving it. Yet it has been claimed that there is impeccable ethnographic evidence here for linking 'medicine men' and rainmakers with shamanism, and in turn linking the shamans with the rock art. Nobody denies that the San Bushmen sometimes went into trance, usually brought on by prolonged rhythmic dancing, but this was by no means restricted to medicine men; and besides, ritual and dancing among hunter-gatherers do not necessarily involve a trance state.

There is likewise no nineteenth-century ethnographic evidence of the medicine men or 'rain doctors' being the artists, nor of the artists being in a state of trance before or during the making of pictures. The records contain many stories from San folklore, but no information on the sex, age or status of the artists; they mention 'sorcerers' who, it was believed, could make rain, transform people into animals, and vice versa, perform healing, and so on, but there is absolutely nothing to indicate that they were more likely to have been painters than were other people. In short, there is nothing in the ethnographic record to prove any connection between shamanism and the execution of the art. None of the few ethnographic accounts of San artists at work refers to them as medicine men or being in any other than a normal state of consciousness. And even if one accepts fully the possibility that the creators of some of the art may have been influenced by hallucinations, that does not prove in any way that they were medicine men — unless one calls everyone who experiences an altered state of consciousness a shaman, in which case we are all shamans, and the word loses all meaning, becoming synonymous with human being!

Even the perfectly reasonable point that the images probably relate to the rock face as much as to each other has been taken to extremes, through the bald assumption that all the rockfaces were seen as interfaces between the world of daily life and the spiritual realm, so that any image placed on that interface necessarily had something, no matter how tenuous, to do with beliefs about relationships between the two worlds and with the activities of the shamans who mediated between them. Once again, such

absolute, blanket explanations leave no place for discussion. They cannot be tested or falsified, and tentative suggestions that maybe some images might have been placed on a rock which was not seen as an interface but which was simply a handy surface for the purpose, are dismissed as naive, simplistic or outmoded. In any case, such divisions between the real world and the spirit world scarcely existed in Bushman life, and so the supposed boundary is simply a construct of Western scholarship. One can certainly speculate that some of the animal figures might have 'pre-existed' and the artist simply liberated them from the rock, but that is hardly a profound or shamanistic insight, since many modern sculptors such as Michelangelo have said the same of blocks of marble! And there are certainly close ties between some animal figures and cracks or fissures in some caves (see below), but why should that necessarily be shamanistic?

Fig. 11.10 The sculpted frieze of La Chaire à Calvin (Charente), with sketch showing the two superimposed animals. Probably Solutrean. Total length: c. 3 m. The superimposed figures are 1.4 m long.

Fertility magic and sex

One of the most popular and durable explanations for much Ice Age art has been that it involves 'fertility magic': i.e. that the artists depicted animals in the hope that they would reproduce and flourish to provide food in the future. As mentioned earlier, such utilitarian theories about Palaeolithic art were largely

inspired by an early study of the Arunta of Australia, who, it was said, drew animals in order to make them multiply. There are certainly beliefs of this type in Australia even today: for example, it is believed by local Aborigines that 'Wandjina sites' (i.e. rock shelters in the Kimberley region containing images of ancestral creator spirits) are 'increase places'; and that if their images of goannas (lizards) and barramundi (fish) are retouched, then the Wandjina will increase the numbers of animals around the site.[61]

The proponents of this theory, like those of hunting magic (and they were often the same people), selected examples in Palaeolithic art which seemed to fit their ideas, and they often saw what they wanted to see: animals mating, and an emphasis on human sexuality too. Yet, as already mentioned, relatively few animals have their gender shown, and any genitalia depicted are almost always done so discreetly. There are no pictures of rampant stallions, and the only known animal with a blatant erection is a bison at Trois Frères which is probably a composite, and thus probably at least half human (though the phallus is thought to be that of a bison).[62]

As for copulation, in the whole of Palaeolithic iconography there are only a couple of possible examples, and they are dubious in the extreme. The best known is an alleged coupling of horses in the sculpted frieze at La Chaire à Calvin (Charente): there may be two

Fig. 11.11 Two superimposed humanoids on a plaquette from Enlène (Ariège), with a third at the other end of the stone. Note also the legs and belly of a bison. Magdalenian. Total length: 21 cm.

Fig. 11.12 Supposed copulation scene, of dubious date, from Los Casares (Guadalajara). (After Cabré.)

horse figures here, partly superimposed, but the 'stallion' is not in the correct position for copulation, and is smaller than the 'mare', and in any case its belly line may even be natural.[63] The other example, two bison at Altamira, is equally odd as they are claimed to be a female mounting a (larger) male in order to arouse him sexually![64]

Faced with an absence of copulation scenes, proponents of fertility magic have therefore interpreted many pairs of animal figures as being engaged in 'pre-copulatory activity': i.e. males advancing on females, or sniffing them.[65] Here too, a great deal of wishful thinking is involved: for example, at Teyjat we definitely have a male aurochs behind a female, and his muzzle is close to her rump — but they are part of a line of aurochs, and there is another cow in front of the first. Similarly, the two clay bison in the Tuc d'Audoubert have often been interpreted as a male about to mount a female: but neither has its sex shown clearly, so that one can only go by relative size; they are not close together, and the 'male' is actually behind and 15 cm beyond the 'female'; and, of course, the interpretation ignores the other two bison here, a small clay model and an engraving in the floor. There are occasional good examples — a portable engraving on antler from La Vache clearly shows a reindeer with a phallus and enormous antlers following another with udders, but at Angles-sur-l'Anglin there is a sculpted composition of two sexed bison, where the female is following the male![66]

The last resort of those seeking evidence of an interest in animal fertility was to see many figures, especially horses, as 'pregnant'. Yet even veterinarians cannot tell if a horse is pregnant from its profile alone (though they can do so from a back-view);[67] quite apart from possible stylistic conventions involved in giving horses big round bellies (like short legs, tiny heads, etc.), inflated stomachs can also be the result of eating quantities of wet grass; and some male horse figures with phallus shown look equally pregnant![68]

In a couple of cases, a small figure is superimposed on a large animal's body — most notably at Lascaux where a foal appears to be 'inside' a mare — but they are unlikely to be pictures of pregnancy. The Lascaux case has the

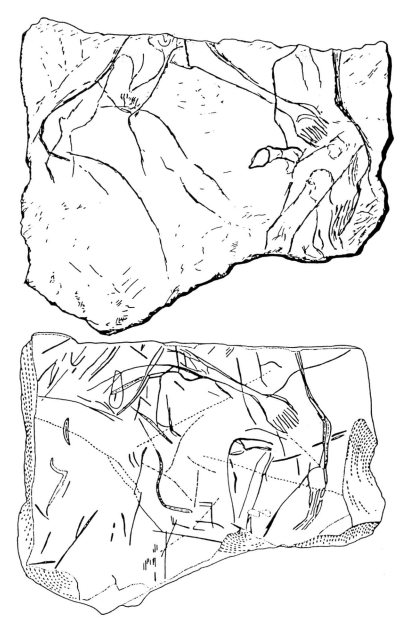

Fig. 11.13 Breuil's version and a more recent tracing (by C. Servelle) of a plaquette from Enlène (Ariège) showing the differences in content. Magdalenian. (After Bégouën et al.)

foal in the wrong position, and one of its limbs protrudes below the bigger figure's belly, so they are far more likely to be two superimposed images, or, according to another suggestion, a mare cantering alongside her foal.[69] Besides, as we have seen, there are very few young animals in Palaeolithic art.

So much for animal fertility, therefore. Surprisingly, more attention has been given to human sexuality, as if the artists were using fertility magic to produce their own children. Here again, however, much is in the eye of the beholder.

We have already seen that quite a few human

figures can be sexed, though relatively few have their genitalia marked: some are clearly female, fewer are clearly male, and most are 'neutral' — i.e. asexual. Since some of the definite males have erections, one might expect that intercourse would have been drawn from time to time.

There have been one or two claims for depictions of human copulation, but they are very sketchy, and it requires a good deal of goodwill and wishful thinking to see them as any such thing. For example, some of the 'fat ladies' of La Marche may actually be thinner ladies with men on top of them, and the 'Femme au Renne' of Laugerie-Basse also has faint lines engraved above her which may represent a person.[70] A large plaquette from Enlène (Fig. 11.11), with a bison as its principal figure, also has a couple of superimposed human/humanoid figures which have been interpreted as a male taking a female from behind[71] — yet apart from possible whiskers on the male the only indication of gender is a difference in size (the smaller figure's belt and long hair are insufficient evidence), and there is no explicit trace whatsoever of sexual activity.

The only plausible scene of human copulation in Palaeolithic art is to be found in the cave of Los Casares, where, according to a tracing by Cabré (Fig. 11.12), a male human has a huge phallus which is superimposed on the stomach of another human with prominent hips and abdomen, who is probably female.[72] Even here, however, the scene is not a realistic depiction of intercourse, and the date of the figures and the accuracy of the copy are in some doubt.

As with animal figures, the proponents of fertility magic therefore looked for scenes of 'pre-copulatory' activity: one example, on another engraved stone from Enlène (Fig. 11.13), has a male human with an erection, and facing him is a possible second figure with arm outstretched. In Breuil's original tracing, the second figure was 'completed' into a woman, with breasts and prominent buttocks, who was thus reaching to grasp the phallus in her hand; many of these features, however, are simply not there.[73] The 'woman' is extremely sketchy, and may not even be a figure, while the 'arm/hand' could equally well be an animal's tail with a tuft of hair at the end, and in any case 'touches' the man's stomach, not his sex. As the engraving is broken, the man's head is missing, but he may be a composite figure, since he appears to have a furry tail hanging at the back. Likewise, what Breuil described as a *scène intime* in Les Combarelles has proved to be nothing more than two anthropomorphous figures.[74]

Similar subjectivity influenced interpretation of the engraved rib from Isturitz depicting two humans (p. 173). They were seen as a male pursuing a female, the latter identified by her breasts and the former by a 'gesture' or a glint in the eye! Yet closer inspection reveals that the second figure, broken off at the chest, clearly had breasts too; moreover, the position of the breasts indicates that these are not two figures crawling along a horizontal surface, with one pursuing the other, but two vertical figures, one placed below the other,[75] or perhaps even two supine females.

What about pregnancies, then? As with horses, any fat female figure was interpreted as being pregnant; but recent re-evaluations of 'Venus' figurines (albeit controversial, see above, p. 161) have claimed that some could plausibly be interpreted in that way.[76] None is

Fig. 11.14 The 'playing-card' figure from Laussel (Dordogne). Probably Gravettian. Size: 20 cm.

definitely shown in the act of childbirth, although one stone figurine from Kostenki, different from all the others, appears to be a pregnant woman about to give birth in the Asiatic position (she is kneeling with legs apart, and her genitals are clearly marked); and gynaecologists have interpreted the Tursac figurine as giving birth, and its 'peg' as the emerging child.[77] Despite these possibilities, however, 'Venus' figurines as a whole do not seem to glorify maternity or fertility; they appear instead to be about womanhood. There are no known pictures of children, apart from the possible baby at Gönnersdorf and perhaps one or two infants at La Marche (see above, p. 60), and a few stencils of tiny hands.

Finally, an enigmatic carving from Laussel (Fig 11.14) which resembles a playing card, with two half figures joined at the waist, has often been claimed to be either a copulation scene or a woman giving birth, but it is so vague that it could be anything: indeed, some scholars see it as a full, symmetrical Venus figure like Lespugue, while another view argues that it is a person standing waist-deep in water![78]

The obsession with sexual interpretations, and particularly with female genitalia, in the art can be traced back eighty-six years. In 1911 the Abbé Breuil was consulted about certain deeply engraved motifs which had been found on some stone blocks in Early Upper Palaeolithic sites in Dordogne. Breuil declared these ovoid and subtriangular figures to be 'Pudendum muliebre', or vulvas.[79] This was a highly subjective interpretation, but ever since then most scholars have accepted it without question — even though a man of Breuil's profession should not perhaps be considered an expert on this particular motif — and it was certainly a major factor in Leroi-Gourhan's concept of the development and complementarity of male and female signs (see below, p. 193).

The 'identification' of so many examples of female genitalia led to ideas about a Palaeolithic obsession with sex; there is tautology in the reasoning — the figures are assumed to be vulvas, from which an obsession with sex is inferred, the evidence for which is the vulvas! Independent data are required to support the hypothesis; but when one searches, vulvas are remarkably hard to find in Palaeolithic art. If

Fig. 11.15 Tiny ivory pendant from Dolní Vestonice, about 22,400 years old, and only 2.5 cm high.

one allows that the only definite vulvas are those found in context, that is, in full female figures, then they are very few in number; of the motifs without context, only the vulva modelled in clay at Bédeilhac (see p. 110) can be identified with any confidence — all the rest are mere interpretations.

It is surprising to find that even among the 'Venus' figurines, often seen as proof of an intense interest in female sexuality, very few have the pubic triangle marked, and even fewer have the median cleft.[80] In fact, these figures accentuate the breasts, buttocks and hips; they may represent fecundity but they do not draw attention to the vulva, unlike the headless bas-reliefs of Angles-sur-l'Anglin which do seem to have this part of the body as their focus.

Where vulvas are depicted, nearly all are triangles, as one finds in female figures from many other periods and cultures. Yet the Early Upper Palaeolithic motifs have a wide variety of shapes, and scholars have needed a number of excuses in their desperate bid to fit all these shapes to the chosen interpretation: for example, there are 'incomplete vulvas', 'squared vulvas', 'broken, double vulvas', 'circular vulvas', 'relief vulvas', and even 'trousers vulvas'! Others have claimed unfinished vulvas, atypical vulvas, and even interpreted a single straight engraved line as an isolated vulvar cleft![81]

But in fact, as we know from rock art in Australia and elsewhere, very simple graphic designs can have a tremendous range of meanings; and in any case, if the Palaeolithic artists took pains to differentiate these shapes, it makes little sense for us to lump them all together into a single, subjectively chosen interpretation. Hence, more objective scholars have divided them into descriptive rather than interpretative categories: horned ovals, pear-shaped ovals, arched ovals, etc.[82]

Wishful thinking, therefore, permeates the search for human sexuality in Palaeolithic art. For example, the clay 'sausages' in the Tuc d'Audoubert were automatically seen as phalluses by believers in fertility magic, while others saw them as possible bison horns, but, as mentioned earlier (p. 110), they almost certainly represent the sculptor trying out the clay.[83] Similarly, one of the portable statuettes of Angles-sur-l'Anglin is a kneeling female, but Breuil originally thought it was a phallus.[84]

It is possible, of course, that there is purposeful ambiguity here, a visual pun; the same may apply to those 'vulvas' which look more like horse-hoof or bird-foot prints. We have seen that all boundaries and categories are blurred in Palaeolithic art, and it would not be surprising if the artists noticed and played with such ambiguities.[85] This kind of punning may also explain why they drew a red human outline around each of two little stalagmites, a few metres apart, sticking out of the wall at Le Portel, thus turning them into ithyphallic men.

A different example of ambiguity probably owes more to the preoccupations of the beholder than to the intentions of the artist: a series of little ivory carvings from Dolní Věstonice (Fig. 11.15) have always been seen by most male scholars as abstract female figurines with long necks and huge buttocks or breasts; the same applies to an ivory rod from the site, which has a pair of 'breasts' protruding from it, some way up. However, feminist scholars suggest that these carvings should be inverted, and clearly represent male genitalia![86]

Just as the 'hunting magic' theory has developed into a view of the art as reflecting a male preoccupation with hunting, so 'fertility magic'

has given way to a 'cheesecake' view of human depictions: i.e. this is early erotica made by men for their own pleasure. This is why the female figures have exaggerated breasts and buttocks, with no emphasis on the limbs or the face. They represent women's contours and were intended to arouse, to be touched, carried and fondled.[87] Similarly, the act of engraving the alleged 'vulva' carvings of the early phases reflected the 'digital activity of lovemaking around the labia'![88]

It is even suggested that the statuettes' heads are bowed in subservience, submission or shyness, while the enlarged rumps on the schematized females of Gönnersdorf and elsewhere are simply sexual lures. The rarity of clothing and personal adornments in the depictions, as well as of birds and other small resources, proves that it is an art which is limited to male preoccupations, and not concerned with a woman's side of life.[89] In this it resembles modern Eskimo art, comprising hunted animals and a few nude females.

Finally, some Palaeolithic depictions of women have been compared with pin-ups in *Playboy*, and it has been claimed that 'female figures often [*sic*] appear in sexually inviting attitudes, which may be quite the same as those in

Fig. 11.16 Annette Laming-Emperaire (1917–77) and André Leroi-Gourhan (1911–86) in the cave of Pech Merle, 1975.

the most brazen pornographic magazines. There are also anatomically detailed pictures of the vulva, showing the female sex organ sometimes frontally, sometimes inverted and from the back, open to penetration'![90]

This quotation reveals a lurid imagination. Apart from the bas-relief reclining females of La Magdeleine, there are no such 'poses' in Palaeolithic art. A few depictions do bear a superficial resemblance to the pin-ups selected, but the range of positions in which the female body can be depicted is pretty limited, as shown by the fact that all the comparative material was found in a single issue of the magazine! Another basic point is that we do not know the sex of the artists. The carvers of the 'vulvas' and figurines could just as easily have been female – it is known that women do sometimes produce rock art and sacred art in some 'primitive' cultures – and one can extend this argument to the whole of Palaeolithic art, invoking initiation ceremonies to explain menstruation, with lunar notation (see below, p. 204) as supporting evidence.[91] It is worth noting, too, that the very first novel about the Palaeolithic, Adrien Arcelin's *Solutré ou les chasseurs de rennes de la France centrale* (1872) featured a young woman who was both chief and artist!

In short, the theory that the art is about the male preoccupations of hunting, fighting and girls comes from twentieth-century male scholars. As mentioned above, feminist scholars have a different point of view about the supposedly sexual images (for the same reason, it would be interesting to see the reaction of a vegetarian culture to the animal figures). The macho view of Palaeolithic art is both simplistic and anachronistic. Some of the female figurines may have aroused men, but there is no reason to suppose that this was their intended function: after all, many of them may have been carefully hidden in pits. Moreover, images of female genitalia are far rarer than has been claimed in the past, and those which were depicted may well have been intended simply to indicate gender, rather than be erotic. The vast majority of Palaeolithic art is clearly not about sex, at least in an explicit sense. The next major theoretical advance, however, introduced the notion of a symbolic sexual element.

Laming-Emperaire and Leroi-Gourhan: the female and the male

Despite his stubborn opposition to parietal art, Emile Cartailhac was a very perceptive man; we have already mentioned (see above, p. 170) his early impression that 'art for art's sake' was an insufficient explanation. A further example occurred in 1902, when he stated that the cave of Pair-non-Pair appeared to contain a random accumulation of pictures, but that of La Mouthe had been decorated systematically.[92]

Thanks to the dominance of Breuil, who treated cave art as a collection of individual pictures, this pioneering insight was forgotten, and the concept did not re-emerge for half a century, until the art historian Max Raphael began work on Ice Age art. Raphael's contribution in its turn was forgotten for a long time, since it remained unfinished and largely unpublished, but it had a tremendous influence on two French scholars who arrived more or less independently at the same conclusion. Annette Laming-Emperaire and André Leroi-Gourhan, in a series of publications between 1957 and 1965, revolutionized the study of Palaeolithic art, which thus took its first leap forward since 1902.

Raphael insisted from the start that all lines should be traced, and he criticized Breuil's notion of *traits parasites* and his tendency to fill in lines (see Chapter 4). He also rejected the straight-line development from simple to complex, from abstract to naturalistic, and instead saw an infinitely complex expression of different ideas and content being developed over time. Raphael examined the imagery in caves in terms of panels and compositions, and, for the first time, envisaged a sexual symbolism: i.e. he saw 'signs' as male or female, and even foresaw the theory ascribing male symbolism to the bison and female to the horse. For example, he saw the Altamira ceiling as an opposition between the big hind figure and the group of bison, and thus as a confrontation between female and male, agility and solidity.[93]

One of the basic features of the approach adopted by Laming-Emperaire and Leroi-Gourhan was to ask new questions, while avoiding the intuition and the subjective use of ethnographic parallels which had characterized previous work and led to an impasse.

Interpretation was to be based on the total corpus of figures in a cave, not on one or two selected images. They too treated parietal art as a carefully laid-out composition within each cave; the animals were not portraits but symbols; they did not reflect explicit practices, but rather a complex metaphysical system or 'mythogram'. Previous concentration on the aesthetic and magical aspects was inadequate, as the ideas behind Palaeolithic art were far more complex and sophisticated. In short, they moved from analogy to observation, from naive and particularist explanations to the structures reflected in the subject matter and its associations. Although there was no hope of grasping the art's meaning, one might gain some insight into the richness of thought behind this symbol system.[94]

Although it was Laming-Emperaire who, following Raphael, first treated parietal figures as a composition in her study of the sculpted sites, as well as Lascaux and other caves, it was Leroi-Gourhan who confirmed the notion, extended it to all major caves and continued to develop the idea, while Laming-Emperaire eventually dropped it in favour of an even more complex explanation (see below).

An investigation of how many figures of each species existed in each cave, together with their associations and their location on the walls, led to a number of insights: Leroi-Gourhan eventually divided animal figures into four groups[95] based on the frequency of their depiction in parietal art as a whole. Group A was the horse, which constitutes about 30 per cent of all parietal animals; B was the bison and aurochs, which also account for about 30 per cent; C was animals such as deer, ibex and mammoth, representing another 30 per cent; D, the final 10 per cent, comprised the rarer animals such as bears, felines and rhinos.

He also divided caves into entrance zones, central zones, and side chambers and dark ends. It appeared that about 90 per cent of A and B were concentrated on the main panels in the central areas, the majority of C figures were near the entrance and on the peripheries of the central compositions, while D animals clustered in the more remote zones. He developed the concept of an ideal or 'standard' layout to which the topography of each cave was adapted as far as possible: these were organized sanctuaries, with repeated compositions separated by zones of transition marked with the appropriate

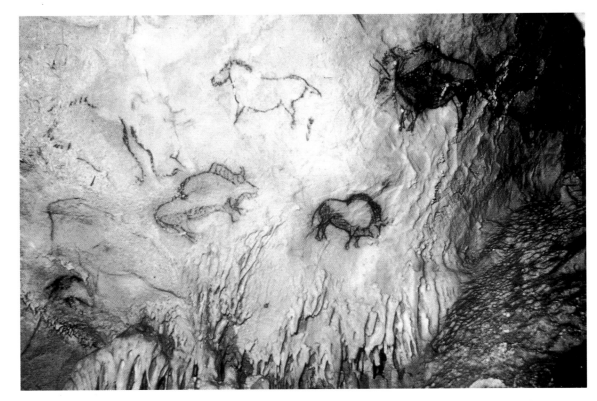

Fig. 11.17 The 'central composition' of black figures at Santimamiñe (Vizcaya), showing a horse surrounded by a number of bison, including two drawn vertically, back to back. Probably Magdalenian. Panel size: 2.5 m. The horse is 45 cm long.

animals or signs.[96] Different signs are found at the start and end of the sanctuary (mostly dots and lines), while more elaborate signs occur before and within the big animal compositions.

However, as Leroi-Gourhan was well aware, many caves are extremely complex topographically, and may have had different entrances during the Palaeolithic which are now blocked: one cannot simply assume that the present entry has always been the only one. Our doors may have been their windows! At La Pasiega, for instance, there are numerous possibilities for access to the cave through an abundance of different passages, so it is difficult to decide what was the centre or the back of this cave.[97] From the start, Leroi-Gourhan sometimes had to adopt bizarre itineraries within caves in order for his scheme to fit — even going to the wall furthest from the entrance to find the 'first series' of big signs, and then coming back again![98] He knew that there is no 'average' cave, but still believed that, though topography may have dictated layout to some extent, it was still the 'mythogram' that conditioned the choice of topographic adaptation;[99] i.e. he imagined that the artist who had been 'commissioned' to decorate the cave checked its topography and reliefs, and noted any features which appeared useful in accordance with the predetermined blueprint — hence the Altamira ceiling-bosses could only become curled-up bison, not felines or something else.[100]

His divisions into central and peripheral areas were somewhat tenuous (those in Gabillou, for example, are quite arbitrary — where do complex panels start and end?) and decorated caves are far more heterogeneous than the scheme suggests. It is not just a question of shallow caves and rock shelters versus simple corridors and complex systems of galleries, but a problem of content: as we have seen, some caves have a single figure or sign, others have animals and no signs or vice versa;[101] and how is one to fit the newly discovered open-air engravings into this concept? Leroi-Gourhan himself admitted that he omitted caves that were exceptions, incomprehensible or lacking the appropriate animals; and he conceded that there were different regional 'formulas', notably in Cantabria.[102] Far from presenting his theories as impeccable, he con-

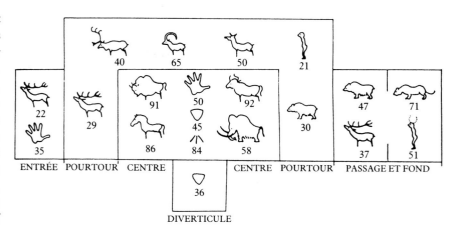

stantly pointed out the gaps and faults in them.

Even in caves with abundant animals and signs, the latter do not always begin or end the decoration, and the former often do not conform to the scheme: for example, bison are the last animals in the 'feline gallery' at the end of Lascaux, and among the last in Les Combarelles, while the clay bison of the Tuc d'Audoubert are at the far end of the cave. Leroi-Gourhan sometimes assumed changes in cave topography in order to explain 'anomalies' of this kind (e.g. wrongly claiming that the Tuc d'Audoubert was once joined to Trois Frères, so that the clay bison would not be at the end), or divided the caves up on the basis of the themes represented rather than topography, thus turning his scheme into a circular argument.[103] Even so, his own figures revealed that 60 per cent of bears and 66 per cent of rhinos are in the 'central' areas; indeed, the rhino is rarer in cave-ends than are ibex and deer! Ibex can be found in virtually any part of the cave. It is true, however, that the D group does not usually occur in entrance zones, and that felines tend to be in remote zones (a fact already noticed by Breuil). Chauvet Cave, of course, found long after his death, has a dominance of the 'rare' or 'marginal' animals — rhinos, mammoths, big cats — on its major panels, and does not correspond to the blueprint at all, except perhaps for having some bear figures tucked away in a small side chamber.

A further problem with Leroi-Gourhan's scheme was his use of quantitative data and percentages (which often contained small errors),[104] and especially the fact that the scheme worked on a presence/absence basis,

Fig. 11.18 Leroi-Gourhan's 'blueprint' for the ideal cave layout. (After Leroi-Gourhan)

not on abundance: hence a single horse figure is the equivalent of a mass of bison, and vice versa. Few scholars have been able to accept that abundance was irrelevant, quite apart from other variations such as colour, size, orientation, height above floor, visual impact, animation, use of rock-shape, technique and completeness. Leroi-Gourhan pointed out that, in Christian churches, images of Christ are often far outnumbered by the Madonna and saints, but admitted that the seventeen hinds of Covalanas probably had a more important role than the single (and dubious) aurochs head in the cave.[105] A number of other caves are similarly dominated by a single species.

Yet in his final presentation of the scheme he allowed A/B/C/D animals to be present in any proportions in a particular cave. Where the C group was concerned, only one of the species needed to be present in the formula. Caves with no A animals (Tête-du-Lion) or with no A or B (Cougnac) were seen as exceptions that prove the rule![106]

In fact, there are so many exceptions, contradictions and variations that one can seriously doubt the validity of the 'ideal layout'. Recent detailed studies, both of individual caves and of regional groups, constantly stress that each site is unique and has its own 'symbolic construction' adapted to its topography and concretions, and with multiple thematic linkages.[107] Few people would now uphold the universality of the layout, either in space or time. Leroi-Gourhan lumped all of Palaeolithic parietal art into his scheme, and believed that its structure remained homogeneous and continuous for 20,000 years – even in caves decorated in several phases, the later artists respected the first sanctuary's layout.[108] There is certainly a degree of thematic uniformity over this timespan: caves are decorated with the same fairly restricted range of animals in profile, and seem to represent variations on a theme: as Leroi-Gourhan pointed out, nobody drew a frieze of lions and storks surrounded by hyenas and eagles![109] He therefore thought that the animals depicted corresponded to a certain mythology.

The other foundation stone of the approach was the discovery of repeated 'associations' in the art, and the claim that there was a basic dualism. This, too, was a concept first developed by Laming-Emperaire after Raphael, and subsequently extended by Leroi-Gourhan. Having found bison 'associated' with both horses and women, Laming assumed that the horse might be in some way equivalent to female, and thus the bovids with the male. Leroi-Gourhan, on the other hand, saw the bison as equivalent to females; his view was based almost entirely on a couple of crude figures in Pech-Merle which he interpreted as human females derived from bisons with raised tails.[110] However, as animal images they are vague (they could even be horses!), and any similarity is superficial, being limited to the curve of the profile; some are definitely women, since breasts are shown. In short, the link between females and bison is so tenuous as to be non-existent.[111]

Laming also found a frequent association of horses with either bison or aurochs, a fact which Capitan and Bouyssonie had already noted in the portable art of Limeuil in 1924, though in caves it is of little surprise in view of their great abundance. In the new approach, the numerically dominant horses and bovids, concentrated in the central panels, were thought to represent a basic duality. One possibility was that this was sexual, though it was admitted that other explanations might be equally plausible (e.g. sun/moon, heaven/earth, good/evil, or something more complex). It was pointed out that horses live under the control of a stallion, and cattle under that of cows.[112]

Fig. 11.19 The 'bison-women' of Pech Merle (Lot). Probably Solutrean. Each about 10 cm long.

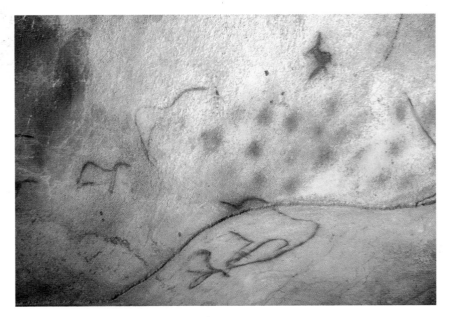

Having tentatively claimed a sexual duality, it was natural for Leroi-Gourhan to extend this to the abstract signs; his somewhat Freudian viewpoint was that they could be divided into male (phallic) and female (vulvar) groups, which he later dubbed 'thin' and 'full' – a third group was later added to include the dots. This view was inspired by two bear heads in portable art, one of which (La Madeleine) seemed to be associated with a phallus, and the other (Massat) with a 'penniform' sign.[113] Hence, where proponents of hunting magic saw missiles and wounds, Leroi-Gourhan saw phallic and vulvar symbols.

In the early years of his research, he felt certain that the two series of signs were of a sexual nature; but it was never explained why such a wide range of shapes should have only two basic meanings, even allowing for artistic licence. Besides, there were inconsistencies such as the claviform which is thin but, as we have seen (p. 164), was thought to be derived from a female silhouette – in the same way as the Coca-Cola bottle could be taken as a phallic symbol or a stylized female form according to one's preference!

Many scholars rejected the sexual theory: Breuil, for example, referred to it as a '*tentative déplorable*' and a '*perspective sexomaniaque*',[114] which was somewhat ironic from the man who unleashed the obsession with vulvas in 1911! However, most critics were confusing sexuality with genitalia; even if the division existed, it probably had nothing to do with eroticism, but simply reflected some sort of structure in society.

From these horse/bison and thin/full oppositions and associations, Leroi-Gourhan developed a view of the caves in which other animals were allotted sexual roles (e.g. ibex were male, mammoths female). Even clearly male bison were considered female symbols – in other words, animals were sexed by species and location, not by gender (except for stags and hinds, for some reason)! By 1982, however, he had dropped the emphasis on sexual symbolism, although he retained the entirely valid point that the cave itself probably constituted a female (womb) symbol, as it has in many subsequent cultures.[115]

Just as his scheme had fixed places for dif-

ferent species, with 'female' animals and signs clustering in central areas, and male elements both among them (horse) and around them, it also claimed that human males were mostly at the back of the caves, while females were in the well-lit entrance areas (as at La Magdeleine, etc.). Occasionally, he allowed the theory to dictate the evidence, to the extent that humanoids were identified, and gender attributed to unsexed humans/humanoids, on the basis of their location in the caves! For example, the 'Femme à l'anorak' of Gabillou was claimed to be a woman because of its position,

Fig. 11.20 The art in its setting: dots and a hand stencil in Pech Merle (Lot). Possibly Solutrean.

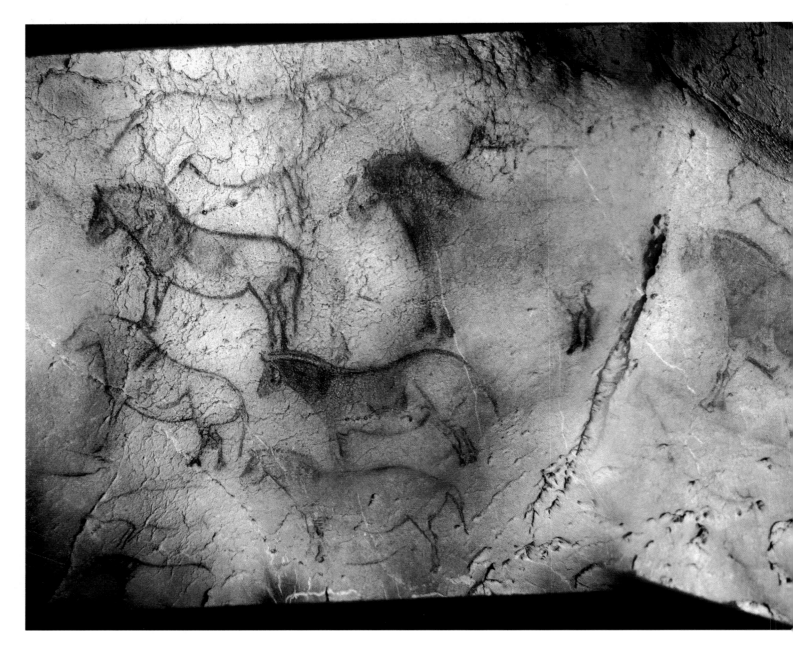

Fig. 11.21 Photomontage of the horse panel at Ekain (Guipúzcoa), c. 4 m wide. Note the red bison at top right, and the 'M' mark on the horse hides. Probably Magdalenian.

but he ignored the fact that the cave's clearest female is right at the very end.[116]

His claim that human males are always at cave ends or on the periphery of panels was untenable because he was including many unsexed heads, and he himself admitted that males can be found all over the place – in entrance zones, cave ends or around central panels.[117] In fact, there is a clear patterning of human/humanoid representations in caves, but it is somewhat different from that proposed by Leroi-Gourhan. Applying strict sexing criteria (see above, pp. 158–9), and dividing caves into

daylight, obscurity and total darkness, it has been found that the percentages of definite males, definite females and neutrals change dramatically: in daylight zones, 56.2 per cent of humans are female, 6.2 per cent male and 37.5 per cent neutral; in obscurity these figures have changed to 31.4 per cent/8.1 per cent/60.4 per cent, while in total darkness only 3.6 per cent of the humans are female, 16.3 per cent are male and a massive 80 per cent are neutral.[118]

Leroi-Gourhan felt his formula comprised a variety of binary oppositions, and that a symbol

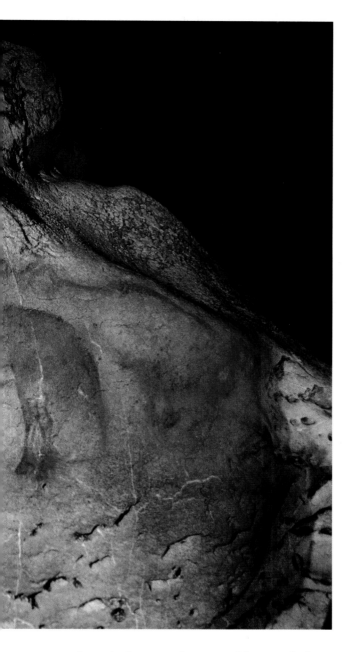

and aurochs are found with horses far more often than with any other species. In portable art, however, the same rules do not apply: at La Marche horses are found with practically every species depicted, while humans are found with almost every species except felines.[119]

The overwhelming parietal coexistence of horses and bison suggests that their duality is the basis of the whole figurative system, as Leroi-Gourhan believed, at least in the later phases of the Upper Palaeolithic; alternatively, these animals may each have a separate importance, making their frequent coexistence a matter of chance (the limited range of species depicted also increases the chances of their being found together). We simply do not know. However, a recent re-evaluation of parietal figures and their associations, using factor analysis, not only confirms the fundamental role and opposition of horses and bison, but also suggests that the bison is dominant, since it has a greater 'effect' on a panel centred on a horse than the presence of a horse does on a bison panel.[120]

It was also found that some 'associations' were rare or non-existent — such as bison and aurochs (this is equally true on plaquettes),[121] bison and stag, mammoth and hind or aurochs and reindeer; however, some species are rarely associated because of regional differences: as mentioned earlier (see p. 157), the majority of hinds occur in the Early Magdalenian of Cantabria, while mammoths are predominantly found in the Middle Magdalenian of Dordogne — this considerably lowers their chances of being found together. However, in those few caves where they do coexist, they are never together, whereas mammoths are sometimes found with stags.[122] In short, there seems to be a definite system or 'grammar' at work, but we do not know what it is; in the Magdalenian, at least, there appears to be a basic trinity of animals (horse, bison, ibex), though hinds and mammoths are also pretty fundamental. The system is not binary and is certainly far more complex than Leroi-Gourhan thought, and clearly changes over time (see above, p. 157), but discerning a structure tells one nothing about potential meaning.

Other scholars have run statistical tests on Leroi-Gourhan's figures, but found little evi-

such as the horse or bison could certainly have taken on or shed many associations in the course of the Upper Palaeolithic — he never saw the art as a rigid, fossilized system. Some of the associations which he and Laming discovered are certainly real, although the terms 'juxtaposition' or 'coexistence' might be preferable since they do not assume that the phenomenon is meaningful: for example, many hand stencils are found in close proximity to painted dots (e.g. Pech Merle, Castillo), and often near horses too; over half of parietal signs are with animal figures; in parietal art as a whole, bison dence to support the duality of signs or the male/female dichotomy in animal species, especially the homogeneity of male signs and animals, though the 'female animal' and 'dangerous animal' groups remained plausible.[123] Others claimed that Leroi-Gourhan's use of percentages of associations was 'idiosyncratic and fundamentally unsound' because he ignored group size and relied simply on presence/absence. A multivariate analysis of the animal figures in a number of sites revealed two small groups of neighbouring caves containing depictions of several species in similar proportions, but otherwise there was no overall pattern.[124]

The importance of Leroi-Gourhan's work is enormous, not so much for its results as for its impact on the field and on current and future research. It constituted a leap forward, and like all first attempts it oversimplified, went too far too soon, and made many mistakes. As with any new and powerful theory, there was an irresistible temptation to bend ambiguous facts to fit, while ignoring others. There was an excessive tendency to assume that Palaeolithic artists thought like twentieth-century French structuralists. Leroi-Gourhan felt that he had established the 'blueprint' for cave-decoration, the framework on which a variety of symbols and practices had been hung through the period. His results are too erroneous to be fully accepted, but also too revelatory to be dismissed: some of his conclusions, as we have seen, cannot be faulted. He found order and repeated 'associations', but not a universally applicable formula.

Inevitably, he modified his views over the years; Laming-Emperaire, however, who initiated this approach, eventually abandoned it in favour of a theory inspired by the work of Raphael and Lévi-Strauss, which virtually reverted to the totemism so beloved of early ethnographers, and which Reinach had presented (together with hunting magic) as an explanation of Palaeolithic art.[125] Breuil and Bégouën had opposed this view, since it meant that animals were mythical ancestors of human groups, and there would therefore have been taboos against killing or eating them — whereas they saw the depictions as killed animals.

Laming criticized the 'topographic deter-

minism' which had arisen around the sexual dualism theory and which, as shown above, meant figures and signs were 'sexed' by their location, in a perfectly circular argument.[126] She maintained that the horse and bison were probably fundamental, but did not necessarily constitute a sexual opposition. Instead she wondered if they might be an expression of exogamy practices: i.e. a decorated cave was a general model of a group's social organization, with animals of various sizes and ages representing either different generations or the mythic ancestors of certain clans. For example, a big bull and a horse at Lascaux would be clan ancestors. The male and female aurochs at Lascaux should not be seen as a bull and cow, but as two members of the bovid clan, such as a brother and sister. Two animals of different species and sex would be a matrimonial couple, while two of different species and the same sex would be an alliance.[127]

She believed that the main panels were origin myths, the principal animals were ancestors, and the secondary species were allied groups. The artists were not displaying a system or formula, but telling a story within a system. This approach was more flexible than that of Leroi-Gourhan because it did not expect everything to remain stable through time and space, but rather to be different in each cave. Unfortunately, Laming-Emperaire died before she could develop the idea further, although it is difficult to see how to do so without recourse to ethnographic analogies.

Whatever the faults and inadequacies of the work of these two scholars, they did succeed in changing irrevocably the way in which we think about Palaeolithic art. The images could no longer be seen as simple representations with an obvious and direct meaning; instead it was realized that they were full of conceptual ideas — they were the means by which the artists expressed some more metaphysical concept. Their work constitutes a very important transitional phase following the traditional approach of compilation and intuitive and ethnographic explanation. All current and future research will draw on their ideas and results, and modify or discard them. Breuil, Raphael, Laming-Emperaire and Leroi-Gourhan would all have agreed that, for work to progress, we must not

sit at their feet but stand on their shoulders.

It now remains, therefore, to look at a few of the varied directions in which current research is taking us.

Why is it where it is?

One of the most interesting new pieces of research has focused on the distribution of figures, and the shape of the wall beneath each one, in four neighbouring Cantabrian caves (Castillo, La Pasiega, Las Chimeneas, Las Monedas). It was found that in each of them the figures had been distributed according to the spatial characteristics of the cave, but there emerged clear differences between the content of 'concave' and 'convex' rock surfaces: no fewer than 88.8 per cent of horses were on concave surfaces, together with 93.1 per cent of hinds, 87.7 per cent of caprines, 100 per cent of stags and 100 per cent of hand stencils; on the other hand, 78.7 per cent of bison and 82.8 per cent of cattle occurred on convex surfaces.[128]

Moreover, the horse and bison were clearly the main representatives of the 'concave' and 'convex' groups respectively, and were usually bigger, more detailed and more complete than the rest of the animals. These 'hollow horses'

and 'bulging bovines' confirm the existence of a horse/bison dualism and the importance of these species in the art.

Of course there are exceptions to the concave/convex rule, even in these caves – almost every statement made about Palaeolithic art can be countered with exceptions – but a preliminary look at other caves suggests that there may be a degree of consistency: for example, in Covalanas, the only 'convex' figure is the bovid, while all the hinds and the horse are 'concave'; at Altamira, all the bison on the ceiling are 'convex', whereas the deer and horse are concave; at Font de Gaume the bison and reindeer seem to be on convex surfaces; the Tuc d'Audoubert clay bison are on convex, the Cap Blanc sculpted horses and the Tito Bustillo painted horses on concave surfaces; while almost all the hand stencils of Gargas, like those of Castillo, are on concave surfaces.[129]

Much more work will have to be done before we can assess how widespread is this careful selection of surface-shape. Doubtless many figures and caves will not conform, because no scheme can ever totally fit an art-form spanning so many millennia. A recent extension of this study – in fourteen Cantabrian caves, using the new direct dates

Fig. 11.22 The famous spotted horse panel in the cave of Pech Merle (Lot). The positioning of the animals was probably dictated or inspired by the natural horse-head shape on the right. The entire frieze – horses, hand stencils and dots – was produced by spitting pigment, and Michel Lorblanchet's recent re-creation of the panel by this method showed that the whole thing could be done in about thirty-two hours, although it probably comprises several artistic episodes. Charcoal in the right-hand horse has yielded a radiocarbon date of 24,640 years ago. This horse is about 2 m long.

Fig. 11.23 Bison drawn in the Réseau Clastres (Ariège) with concretions on top of it. Probably Magdalenian. Length: 1.14 m.

where available – has also begun to discern the 'decorative programme' of each cave, the stages in which its decoration accumulated, and how a particular location affected the content, technique and degree of perfection and finish of the images placed there.[130] Different areas of caves were often selected for either engraving or painting, with some areas apparently more important than others, or at least having a different value. The decoration is very far from random. In the terminal gallery of Altamira, for example, the engravings appear highly structured, with some areas of good wall left blank, and others crowded with figures. All six engraved bison and all the painted geometric images are on the right-hand wall. The bison are large, complete figures, mostly facing the entrance, while the majority of the deer on the opposite wall are simply heads and necks, mostly facing into the cave. The two animals are mutually exclusive in this zone.[131]

In several caves, a similar arrangement of figures around an entrance or fissure can be observed: in the Réseau Clastres, for example, the 'frieze' begins at left with the strange weasel-like figure and then the horse, both facing right towards a massive fissure; beyond the fissure are the three bison figures, with the first and only complete one facing left towards the opening. There is great symmetry here, with both the horse and the bison being about 4 m from the fissure, at either side, like sentinels.[132] One is reminded of the Hall of the Bulls at Lascaux, where the frieze likewise begins with a bizarre figure at left (the 'unicorn'), facing right, and followed by a whole series of animals grouped in a rotunda around the opening of a gallery. And in the recently discovered Covaciella, the main panel is again arranged symmetrically at either side of a large crack, with animals to its right facing left (including the principal bison images, Fig. 10.5), and vice versa.[133] Likewise, in Travers de Janoye (Tarn) the parietal art is clearly arranged around a large fissure.[134]

A different but doubtless hugely important factor in the selection of places to decorate and perhaps also in what to draw there is the long-neglected aural aspect: the acoustics. In recent years, detailed studies have been made of the acoustic properties of several caves, finding a frequent correlation between the locations of decoration and the areas of best resonance for men's voices.[135] This notion has also been extended to the content of cave art, with the suggestion that the artists ingeniously used the caves' acoustics and echoes to conjure up the sounds made, for example, by moving herds of hoofed animals. It has been found by yelling and clapping or striking stones together that in deep caves like Lascaux and Font de Gaume echoes in the painted chambers produced sound levels between twenty-three and thirty-one decibels. In contrast, deep cave walls decorated with stealthy cats, like Lascaux's 'Cabinet des Félins', produce sound levels of only one to seven decibels, whereas undecorated surfaces often are totally 'flat'.[136] Although some of the interpretations linking these findings to thundering ungulate herds or silent carnivores are less than convincing, and despite the fact that the acoustics of some caves and rock shelters must have altered somewhat since Palaeolithic times, studies of this type are nevertheless valuable in that they are trying to revive something that one might imagine gone for ever – the dimension of sound that accompanied whatever rituals may have been carried out in these sites. In view of the obvious intelligence of the artists, as well as the fact that they took full advantage of the morphology of caves and espe-cially of particular rock-shapes, it is extremely likely that they would have used any acoustic peculiarities to the full.

A different observation about the position of art in the caves concerns its visibility and accessibility. It is not always easy to judge whether the difficulties we experience in reaching certain galleries existed in Palaeolithic times: for example, Montespan's art can only be reached by wading upstream, with water up to one's waist, and, at one point, having to duck right under when the ceiling comes down to water-level; yet on very rare occasions, as in 1986, the cave is dry. Similarly, graffiti in the Tuc d'Audoubert (which now has to be entered by boat) show that it was visited in 1685 and 1702 by clerics in periods of great dryness.[137] One therefore cannot be sure whether the artists got their feet wet or not.

Visibility and access can also be affected by stalagmites and stalactites, and in some caves such as Cougnac these appear to have been broken by Palaeolithic people, thus making the figures more visible from a distance. In Candamo (Asturias), a number of concretions were broken to provide a theatrical setting for the drawing of a big black horse, 12 m up, which was thus visible from afar. In Pech Merle too, the black animals are visible from 10 m away, while the figures drawn with fingers on the ceiling were carefully placed so as to be visible from afar.[138] One can divide cave art into the clearly visible, the obscure, and the hidden – or, in other terms, the public and the private, made to be seen either by a human audience or by the supernatural.[139] The Hall of the Bulls at Lascaux or the spotted-horse panel at Pech Merle are other good examples of visible art in large chambers, whereas there are many examples of art tucked away, either in inaccessible galleries (such as that of Fronsac, 35 cm wide) or in nooks and crannies of large chambers (e.g. the signs hidden in the folds of the 'lithophone' at Nerja). Pergouset is a cave that has always been very difficult to enter, and, as mentioned earlier (see p. 124), contains a horse head engraved in a fissure, undoubtedly placed there by the artist to prevent it being seen easily. At Erberua, some engravings are located in a tiny chamber where one has to lie down.

It has been noticed that the parietal art is frequently associated with *bouches d'ombre*: i.e. chasms or entrances to lower galleries, many of which have water in them (e.g. at Tito Bustillo), and which may have been linked with myths concerning the Underworld, the cult of the Earth Mother, and 'chthonic deities':[140] for example, at Pergouset, the engraved 'monsters' seem to emerge from a mysterious *trou d'ombre*, while at Ekain the farthest frieze of horses is located above two holes which lead to the dark and mysterious depths of the cave. In all regions and at all times caves have been associated with placenta, with ideas relating to maternity and birth, and with the entrance to the underworld. Boundaries are areas of confusion to human beings, and, like the orifices of the body, are sacred places full of taboo; caves certainly constitute a boundary between the outside world and the underworld, and thus have a special significance in many mythologies. Water plays a major role in virtually every known religion, and running water in particular is frequently looked upon as potent, dangerous, and linked with spirits. It is therefore highly probable that where running water visibly crosses the boundary of the underworld, in either direction, it may take on special significance. This effect would be heightened where water enters the earth in an impressive setting (Le Mas d'Azil, Labastide), or where it emerges as a hot spring.

It has been suggested that these factors are likely to have played a role in whatever beliefs and rituals lie behind Palaeolithic parietal art — what might therefore be called the 'rites of springs': is it a coincidence, for example, that large clay models have only been found in two caves (Tuc d'Audoubert, Montespan) through which rivers flow, and that both contained a headless snake skeleton? A large number of decorated caves are located very close to springs, and there is a marked correspondence between certain parietal sites and thermal/mineral springs:[141] for example, the four caves in the Monte Castillo overlook the village of Puente Viesgo, through which a geological fault runs, responsible for a number of hot springs which have been found suitable for medicinal purposes; it would be unsurprising if in the last Ice Age very special significance was accorded to springs which did not freeze, and/or which were thought to have medicinal powers.

Recent work, therefore, has focused on the distribution of art within caves, and on that of particular decorated caves within the regions where they exist. For example, Pech Merle is in a dominant position, on the highest hill of its region, and is also the most important sanctuary of the area, surrounded by a dozen small ones within a radius of a few kilometres, in smaller, lower cavities.[142] The same is probably true of Chauvet Cave, easily the biggest cave of its region, far bigger than the other decorated cavities known in the Ardèche. Little attention has been paid to the problem of why the distribution of parietal art is so restricted in Europe, unlike that of portable art and despite the existence of many perfectly good caves in Germany, for example — though, as shown in Chapter 3, the present distribution of cave art may be partly an illusion, caused by geomorphological and taphonomic factors.

One theory is that southern France and northern Spain constituted 'refuges' into which people migrated in waves from other parts of Europe during periods of climatic deterioration from about 25,000 years ago onward: the 'cave-art regions' were more temperate, and had a greater abundance and diversity of fauna available, and people were able to live there by exploiting the plentiful reindeer and Atlantic salmon.[143] Salmon-fishing would have led to increased sedentism, and thus a greater need for elaborate procedures for resolving conflicts and marking territory: hence the cave-art sites were established as permanent ceremonial locations; while the portable art was perhaps linked to the more mobile life and seasonal aggregations associated with reindeer exploitation. In short, the art is seen as a manifestation of different social responses to processes of climatic and economic change and population movement.

Unfortunately, the theory does not stand up to scrutiny, simply because parietal art, as we have seen, is not limited to Franco-Cantabria, merely concentrated there (and in any case, the discovery of Palaeolithic engravings in the open air has rendered irrelevant most studies based on the known concentrations of parietal art). Moreover, reindeer were not the only staple resource in the Upper Palaeolithic — many sites, depending on their location and surrounding topography, had an economy based on horse, bison, red deer, ibex, or a mixture of several species. In addition, fish only began to be exploited on a significant and systematic scale at the very end of the period, long after the main phases of art production;[144] and of course, some pockets of parietal art such as the Rhône Valley were beyond the range of the Atlantic salmon.

Consequently, fish and reindeer exploitation can have had little or no bearing on the production of Palaeolithic art. It is still possible that France, Spain and the Mediterranean coast represented a refuge at times. However, we are unlikely ever to discover precisely why this art flourished where and when it did – it may constitute a kind of long-term Palaeolithic equivalent of Classical Greece or Renaissance Italy.

How many artists were there?

But was Palaeolithic art produced in a steady stream over the millennia, or does it represent a large but sporadic output by a relatively small number of gifted artists during this long period? Some scholars still hold the former view, but most now prefer the latter.[145]

As mentioned earlier (see p. 121), experiments suggest that many parietal and portable figures could each have been executed in a very short time by an artist with a modicum of talent and experience: all the black-outline animals of Pech Merle could have been done in less than three hours, and probably represent a single artistic episode: indeed, this cave is thought to have had no more than three or four such episodes – for example, analysis shows that its six black hand stencils were done with the same pigment and used the same hand.[146] Together with the archaeological and pollen evidence, these facts suggest that, far from being a great tribal sanctuary, Pech Merle was visited and used only very rarely, with long periods of neglect between the visits

Similarities between Palaeolithic images – primarily in portable art – have often, in the past, led to theories about 'schools' or workshops: Limeuil, for example, was seen as a centre of artistic and magic education. Images on stones and bones were thought to be artists' notebooks or sketches, and the 'mediocre'

specimens among the masterpieces were the work of pupils. While such notions are somewhat fanciful, and owe much to nineteenth-century views of art, some modern scholars have attempted to estimate and compare the percentages of 'masterpieces' and 'mediocre pictures' in different sites,[147] an exercise which is entirely subjective and takes no account of the artist's intentions.

Some clusters of images are so stylistically and technically similar that it is virtually certain that one artist or group is responsible. Hence, at La Marche, a particular style, technique and set of conventions (and errors!) have been recognized by those most familiar with its engravings.[148] Few would doubt that the Mas d'Azil and Bédeilhac fawn/bird spearthrowers (pp. 96–7) were by the same artist or, at least, two artists, one of whom had studied the other's work; the same could be said of the portable and engraved hind heads of Altamira and Castillo, some of the human profiles of La Marche and neighbouring Angles-sur-l'Anglin, the Labastide isard-heads and certain portable bison-head engravings from Isturitz.[149] Two perforated batons from Cualventi (Oréna) and Castillo – sites only about 15 km apart – are virtually identical.[150] On the other hand, other collections of portable art seem to be by a number of different artists: for example, the Castillo engraved shoulder-blades, or the slabs at Gönnersdorf where analysis of details suggests that at least a dozen different artists were responsible for the mammoths.[151]

The search for individual artists is by no means a new phenomenon in Palaeolithic art: Sanz de Sautuola himself saw the Altamira ceiling as a unified work, and in the Cartailhac/Breuil monograph on the cave there was some discussion of 'authors', and speculation that not all the cave's pictures were by the same hand. Until the end of his life Breuil insisted that one artist of genius could have done all the polychromes of Altamira.[152] Leroi-Gourhan, on the other hand, refused to accept that one could attribute even two adjacent images to a single artist.[153] He was obviously correct, in so far as it cannot be proved that more than one picture was done by a particular person; but there is a place here for intuitive reasoning: for example, Pales was certain that at least three of the ibex in Niaux's Salon Noir were by the same artist.[154]

One scholar, Juan-María Apellániz, has for some years been trying to establish firm criteria for recognizing the work of individual artists, to transform what have previously been intuitions into a reasoned method. He looks for an original way of drawing, a particular *tour de main*, the repetition of idiosyncrasies, peculiarities and details of technique and execution. There are variations, of course, since artists do not repeat themselves exactly, and may well have changed style slightly through the years: the same artist could certainly produce two very different figures, ten years apart, while, as we have seen, a change of tool can also affect the hand. Only complete figures can be used.

It is advisable to work directly from the original images, rather than from tracings and copies by other people which, as we have seen (Chapter 4), can exaggerate the similarities or differences between figures. Occasionally Apellániz has attempted to work from other people's tracings,[155] but more often uses direct observation, sometimes backed up with photographs and copies, as in the case of the Altamira ceiling, where he confirmed earlier scholars' intuitions that the ceiling was largely the work of one 'master'.[156] It is relatively easy to compare the standing and curled-up bison figures, since, as mentioned earlier (see p. 140), they share many characteristics; however, it is much harder to extend the comparison to animals of different species such as the horse and hind on the ceiling.

Studies of this type can be particularly easy on portable objects, where one can be pretty sure adjacent images are contemporaneous, and where similar figures often form friezes and were almost certainly the work of one artist (e.g. on the Torre bird bone, p. 95, all the horns have their ends drawn in the same way, probably by the same hand); the same applies to objects such as bone discs with decoration on both sides: on that from Le Mas d'Azil with a calf on one side and a grown bovid on the other, or that from Laugerie-Basse with a chamois on each side, the two figures are identical in their use of space, their proportions, composition, silhouette and technique.[157]

In his work on parietal art, Apellániz has produced mixed results. At Altxerri and Ekain,

he has 'identified' different artists or groups working in different parts of the caves.[158] At Las Chimeneas he believes that all the black stags are probably by the same hand, and at Lascaux the same applies to the four huge black bulls, whereas the cows seem to have a number of authors.[159] One of the clearest sets of similar figures is found in a group of caves: in Covalanas and, 150 m away, La Haza; in Arenaza, 45 km to the east, and La Pasiega, 60 km to the west. The fidelity to the general 'canon' of composition in hind and horse figures, the dotted outlines, the similarity in details such as eyes and extremities of limbs are such that he claims this group of caves can be seen as a 'school': i.e. the 'master' or principal artist of Covalanas had disciples or companions who worked in the other caves.[160] However, he does not believe an artist worked outside a particular cave: each cave may have formed part of a 'school' of this type.

How valid is this approach? The criteria presented as objective and reasoned are still rather subjective and intuitive; he is assessing the degree of similarity between figures which may be by one hand, but which may also simply display the accepted canons and conventions of a particular period and culture.[161]

In an effort to counteract these problems, Apellániz is developing statistical techniques, using a series of variables and measurements of various parts of the animal outlines (like those made by other scholars on bison figures, see p. 143), and subjecting them to factor analysis in order to assess their degree of similarity: so far this method has confirmed that the Chimeneas stags, for example, are extremely similar in the treatment of the neck, back and rump.[162] It will never be possible to prove that the same hand was responsible, but this approach has certainly established a very high degree of probability in some cases.

We can only speculate about how Palaeolithic art's 'rules' and 'canons' were passed through space and time so faithfully. It is likely that most decorated caves were visited only rarely in the Upper Palaeolithic, and in any case many of their figures are hidden. Portable art, however, was probably circulating over great distances, and people (including artists, presumably) were often moving seasonally, and coming together annually or periodically in aggregation sites when much could be taught and learned.[163] Finally, there must have been a great deal of art in perishable materials, and, as we have seen (Chapter 9), open-air art was probably abundant and widespread. It seems certain, therefore, that Palaeolithic people were surrounded by and familiar with many forms of art, and so there is really no mystery about how its styles and techniques were passed down.

Art as information, art for survival

Equally, it is almost certain that some of the art itself constituted a means of storing and transmitting information of various kinds: this is particularly likely for some non-figurative motifs. Once again, such notions are by no means new: Lartet interpreted some sets of marks on bone and antler as 'hunting tallies' and, as mentioned earlier (p. 168), other motifs were thought to be artists' signatures or proto-writing.

A different kind of role for parietal non-figurative marks was first suggested by Breuil in 1911: he saw that some panels at Niaux, covered in dots and lines, were located just where the main passage divided, and felt that the marks might therefore be topographic guides. There do seem to be occasional links between 'signs' and important places in a cave: at Lascaux, for example, some lines of dots are located at points of topographic transition, and similarly in other sites signs are positioned where passages turn, become narrow or branch off.[164]

In an extension of this view, and of the similar idea that certain caves might have been decorated as symbolic animal drives/pounds (see p. 175), some scholars have even suggested that the artists took advantage of the topography of cave walls to provide or store information about the surrounding landscape: for example, Leroi-Gourhan speculated that horse and bison were primarily in the spacious, central parts of caves as these represented the wide-open pastures, while figures of felines clustered in the dark and narrow depths because these represented their lairs.[165] Others believe that some cave art might be a model of the local Palaeolithic hunting territory: it has been

pointed out that the axes of the main cave of Ekain parallel exactly the line of flow of the streams in the area, while a conical rock in the cave is like the hill housing the cave itself.[166] It is an interesting idea, but very difficult to put forward in more than a handful of cases, particularly as we do not know how certain features of the landscape (such as stream-flow) may have changed. Nevertheless, it is quite possible that some of the art acted as a mnemonic device in this way, to instruct the younger members of the group (see below).

A few items of portable art have been interpreted as actual maps: a stone from Limeuil with engraved meanders on it was thought to be a possible river map; and an apparently abstract design engraved on a fragment of mammoth tusk from Pavlov has also been seen as a map. However, the best known is a fragment of mammoth tusk from Mezhirich (Ukraine), dating to the twelfth millennium BC, which is engraved with narrow bands, transverse lines and some geometric forms in the middle. It has been claimed to represent a map of a river with four dwellings along it and some fishing nets across it;[167] but in fact the object was published upside-down in terms of the direction in which the marks were made, and the central geometric forms are totally unlike the Palaeolithic huts of this region – in any case, the engraving seems to have been accumulated rather than produced as a single composition.[168]

Fig. 11.24 Tracing of the so-called engraved 'map' on the ivory plaque from Mezhirich (Ukraine). (After Gladkih et al.)

Fig. 11.25 The Taï plaque, a rib fragment 8.8 cm long, with its complex, continuous serpentine accumulation of marks.

The 'hunting tally' theory about sets of marks has been revised by a number of scholars in recent years: some have used ethnographic analogy to interpret groups of dots and lines on bones as being connected to games of chance for telling fortunes;[169] others prefer to search for evidence of Palaeolithic proto-mathematics. Absolon used to look for a system based on five and ten, but since this is now considered to be imposing our modern norms on the past, other scholars such as Boris Frolov claim to have found repeated multiples of almost every number from one to ten on portable objects from the former USSR and elsewhere.[170] There is said to be a particular frequency of five and seven, with three and four as 'supplementary number sets'. Frolov sees this numbering system as the basis of an incipient Palaeolithic science.

He has been vigorously opposed by Alexander Marshack, who points out[171] that these numbers are quite random, and the discovery of any repetitions in Palaeolithic engravings is more indicative of a researcher's ability to count than of any number system in the Palaeolithic, and that claims about isolated sets of marks like these cannot be tested. There is no reason to see them as numbers (let alone arithmetic) when, in Marshack's view, they are more likely to be 'notations' – i.e. sets of marks accumulated as a sequence, perhaps from observation of astronomical regularities. He has tested linear sets of such marks (most notably a serpentine set on an Aurignacian object from Abri Blanchard) for 'internal periodicities and regularities', particularly as possible examples of lunar notation, with varying degrees of success,[172] though he believes that there are also non-lunar notations. However, he has stressed

that of the thousands of Ice Age objects he has studied which are incised with sets of marks, only a few dozen bore what he considers to be possible notations, the most impressive and complex of which is a series of more than 1000 short incisions on an Upper Palaeolithic bone from the Grotte du Taï in eastern France.[173]

Marshack's claims have been tested by means of new technology in recent years, with Francesco d'Errico, in particular, striving through use of the Scanning Electron Microscope to discover reliable ways to establish how incisions were produced. By incising bones with different techniques and tools, he produced firm criteria for recognizing how such marks are made and whether with one or several tools. When he applied these criteria to a Late Upper Palaeolithic bone from Tossal de la Roca, Spain, which has four series of parallel lines on each face, he concluded that each set of its incisions was made by a different tool. He also detected changes in the technique and direction of tool use between sets of incisions, which strongly suggests that these markings were accumulated over time, and thus may well be a system of notation; while in a recent technical re-analysis of an engraved antler from La Marche, using these same criteria derived from experiments, d'Errico concluded that the piece was an artificial memory system with a complex code based on the morphology and spatial distribution of the engraved marks.[174]

Whatever the validity of their results, these studies have at least focused attention on a type of marking which previously was ignored or dismissed as random; it is now clear that they are coherent and ordered, and were carefully made over a period of time. Similarly, a study of the painted pebbles from the Azilian, the very end of the Ice Age, has revealed that the sixteen signs drawn on them were found in only forty-one of the 246 possible binary combinations (indicating that some sort of 'syntax' was employed) and that there was a predominance of groups numbering from twenty-one to twenty-nine or their multiples, which may again have some connection with lunar phases or lunations.[175]

The phases of the moon would certainly have been the principal means available to Palaeolithic people for measuring the passage of time, and other scholars have likewise sought evidence of lunar observation, but of a different kind: an ivory plaquette from Mal'ta (Siberia) has hundreds of pits engraved on it in spirals; these have long been interpreted as symbols of moon worship. Frolov claims that there are seven spirals, the biggest of which has 243 pits in seven turns, and the others 122 (243 + 122 = 365). By analogy with similar motifs found among modern Siberians, he interprets it as calendrical ornamentation (243 days being the gestation period of the reindeer, the staple food at Mal'ta, as well as the length of the winter, while the summer lasts about 122 days). Unfortunately, this view is far from reliable, since about a quarter of the original plaquette's surface is missing, and this area has been filled with wax and then engraved so as to complete its patterning![176]

A different method of noting the passing of time, of course, is by means of the seasons: those who uphold the 'hunting art' theory point out that there are very few 'spring scenes' (emaciated animals, pregnancies, births) and a preponderance of autumn images (fat animals, bouts, pre-copulatory behaviour), while Marshack has interpreted many groups of figures, especially in portable art, as representing particular seasons – though some of his interpretations such as the 'moulting animals' of La Colombière or Lascaux are by no means definite.[177] We have already seen (Chapter 10) that little reliable zoological and ethological information can be extracted from Palaeolithic images, though a few details such as the hook on a salmon jaw are certainly indicative of a specific time of year.

Marshack has developed the concept of the 'time-factored' symbol to explain the periodic use of images or accumulations of notation. They were images of seasons, relevant to economic, ritual or social life, and made to be used at the correct time and place. The distinctive features of a season were depicted (e.g. on the Montgaudier baton) to provide a reference point in the calendar. Hence, where previous scholars saw the 'penniforms' around the 'Chinese horse' at Lascaux as missiles or phallic symbols, depending on their pet theory, Marshack would see this as a horse in summer coat running among plants or ferns.

Marshack's importance lies not only in devising new techniques for investigating and documenting the art (see Chapter 4), but also in asking new questions about it, and in postulating that many images were built up and reused over long periods. He believes that we are faced with a very complex tradition of symbolic accumulations which were regionally specialized and differentiated. The notations are merely one type of sign system: many other 'marking strategies' were in use simultaneously. He has focused on what seem to be non-figurative, geometric markings found on non-utilitarian artifacts, always studying the sequence and process of accumulation, rather than just the finished images.

In this way, he has found motifs of different types recurring throughout the Upper Palaeolithic, such as zigzags, 'ladders', double-lines, and (particularly in the former USSR) fish-scale or netlike patterns. One of the most fundamental is the 'band motif' on portable objects, which may be equivalent to the finger-tracings on cave walls, and which, he has speculated, may represent one aspect of a water-related symbolism[178] – whereas, predictably, Breuil saw the parietal 'meanders' as serpents, and Leroi-Gourhan saw them as perhaps phallic. Marshack does not claim that they actually represent water, but that they were 'iconographic acts of participation in which a water symbolism or mythology played a part', a theory which fits well with the possible connection (see p. 200) between certain caves and water entering or coming out of the earth.

These marks are no longer seen as random accumulations but instead as a sophisticated, sequential, additive image system; no longer

Fig. 11.26 Tracing of some painted and engraved horses in the Panneau de l'Empreinte at Lascaux (Dordogne), showing extra eyes, ears, muzzles and legs. Probably Magdalenian. The 'multiple' horse covered in arrow-like marks is about 1 m in length. (After Glory)

dismissed as idle doodlings, but considered to be possibly the most complex element in Palaeolithic iconography. The act of image-making was the important factor, and it was not always done in a public or participatory setting: as we have seen, a great deal of parietal art is hidden away in inaccessible recesses and tiny chambers, such as the horse and 84 'P' signs in a narrow passage of the Tuc d'Audoubert; the little chamber at La Pileta with its animals and double finger-marks;[179] the magnificent bison that was engraved high in an inaccessible shaft in Les Combarelles II by an artist precariously balanced several metres above the ground; or the mammoth painted high up a 'chimney' in Bernifal. It is assumed that different people squeezed into these crannies at different times to perform some symbolic activity.

Some images, both portable and parietal, both hidden and public, were clearly 'touched up' at times, such as the 'bison-women' of Pech Merle, or the Cougnac megaloceros figures (see above, p. 115). Opinions still differ about what kind of behaviour lies behind this phenomenon. Marshack firmly believes that these images were used and reused, and has published many examples such as engraved animals on pebbles and bones from Polesini, Italy, which had legs or muzzles added at some point after they were originally drawn.[180] As we have seen (p. 89), some of his claims that each of a series of marks was made by a different point or hand are simply not verifiable — a variety of lines or marks could have been done by the same tool or hand, and one cannot take this as proof of long-term retouching.[181]

Many theories have previously been put forward to explain this phenomenon which seems to be most frequent in the Magdalenian of Franco-Cantabria. A proponent of hunting magic such as Bégouën saw it as the renewal of an image which had 'worked' before, the act of drawing thus being more important than the finished picture (conversely, the images would be broken if results did not comply with the prayers). Other scholars believe that some figures, as at Lascaux, were simply improved or rectified over a short period (they show a unity of style), the artist experimenting until satisfied with the figure's position and quality; while others suggest that the multiple hooves, muz-

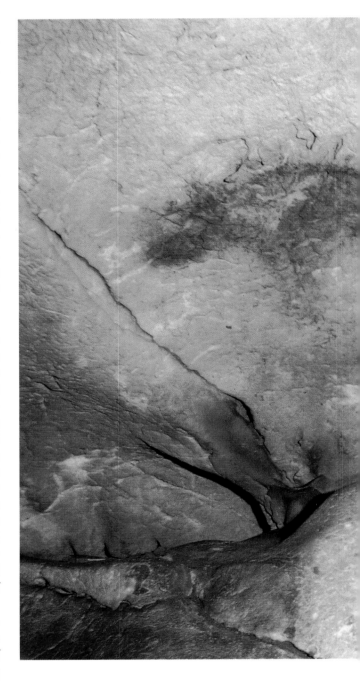

zles, etc. (e.g. on reindeer at Ste Eulalie, Lot), may be an attempt to depict movement.[182] If so, this technique goes back to the Aurignacian, since a bison in Chauvet Cave appears to have seven or eight legs and multiple outlines, thought to represent either perspective or movement.[183]

The fact that we cannot assess the timespan involved prevents us establishing which explanation is correct; where carvings are concerned, Marshack's view seems more plausible,

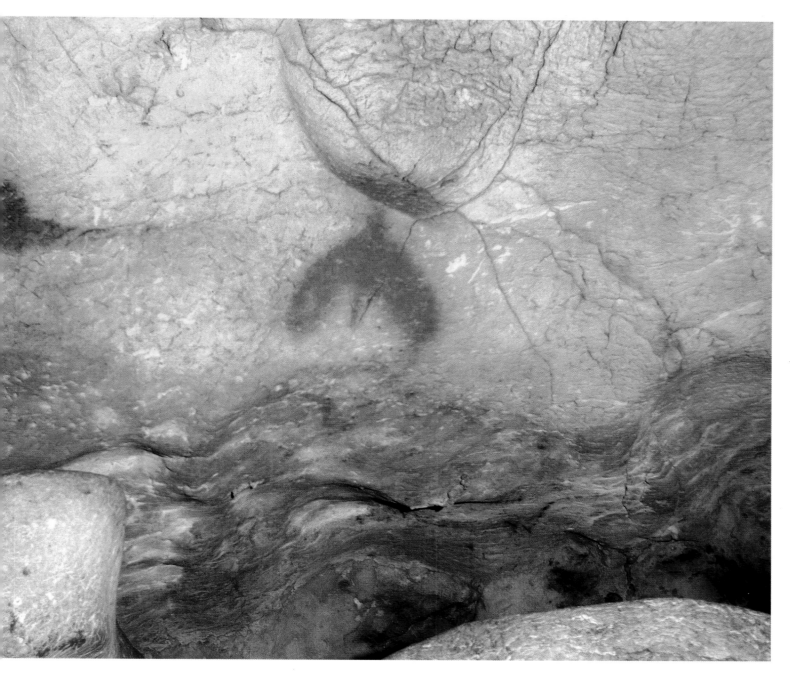

since some of them – such as those of Vogelherd, or even the Tata plaque – are worn and polished from long handling, and the Vogelherd figurines seem to have had marks added to them at times.[184] However, it has been pointed out that some of the Vogelherd figurines – the horse, mammoth and a feline – have perforations at the front and hind legs, and may therefore be pendants, which would also account for the wear and polish.

Even though some marking activity may have been private, it is nevertheless true that much of it is on open view, both in caves and on objects, and must therefore contain messages of various types.

Interpretations of Palaeolithic art have been largely governed by the knowledge and pre-occupations of academics in different decades. Just as 'art for art's sake' was the theory which best suited the end of the last century, and 'sympathetic magic' was taken from the ethnography of the early years of this century,

Fig. 11.27 The enigmatic 'spade' sign at Le Portel (Ariège) with a painted horse head nearby. The man painted around a stalagmite (see Fig. 8.3) is below the sign. Probably Magdalenian. The sign is 45 cm across.

so it was no coincidence that the structural and sexual interpretation arose during the 1950s and 1960s (the era of the sexual revolution, and of Claude Lévi-Strauss), that ideas about astronomical observation were developed during the Space Age, or that an obsession with hallucinogenic phenomena and altered states of consciousness has arisen in recent decades and our 'New Age'. As with Stonehenge, each generation re-invents the evidence in its own image. Just as our computer age produced grossly exaggerated views of that monument as an accurate eclipse predictor, the new technology has inevitably led to some novel ideas about the purpose and meaning of Ice-Age art.

A great deal of research has been done in recent years into information and its processing, and into the nature of memory and how we can use our memory centres more efficiently. One of the major aspects of this work concerns how art can be employed to transmit information and imprint it on the mind.

One such approach has been an attempt to apply 'Information Theory' to cave art; it was thought a useful technique since it ignores the potential meaning of the figures, and concentrates solely on their 'quantity of information'. First applied to language in 1945, the theory defines a message's quantity of information as a function of the number of letters of which it is formed and the language in which it is written. The unit of information is called a 'bit' (an abbreviation of binary digit).

As we have seen, it is a fair assumption that in the last Ice Age most of the cave art contained a 'message' which was not aimed at us, and which we cannot understand clearly. This should not matter to Information Theory, but there is a more crucial obstacle in the fact that we do not know what kind of classificatory system should be employed in our analysis. This is important, because the frequency of each 'symbol' directly affects its quantity of information; the more examples of a particular motif in a sample, the fewer 'bits' are allotted to each of them. But how should we divide up the figures? Should bison be separated from aurochs, or stags from hinds? And what about all the other factors such as size, number, colour, orientation, technique, age, sex, posture, etc.?

In short, the applicability of Information Theory to cave art is severely limited, and the few results achieved so far could equally well have been produced by common-sense:[185] it is obvious that in a panel of similar figures, such as the horses of Ekain or the bison-ceiling of Altamira, each individual figure is of less information value than a single, unique figure such as the bizarre 'sign' in the cave of Le Portel (Fig. 11.27) which resembles a 'spade' from a playing-card, and which has also been seen as a hut, a vulva or a bird in flight. But it does not mean that we are any closer to understanding what that information might be!

Exercises of this type are akin to a Martian coming to Earth and trying to calculate the information contained in an incomprehensible English message by lumping it together with some Polish, Basque and Latin in order to establish the probabilities of the different symbols used. They involve an attempt to measure the degree of complexity of a structure or a message, but do not take into account the context; and, as we have seen, with cave art, the context may be even more important than the content.

In this computer age, it was perhaps inevitable that a cave with figures on its walls should be interpreted as a great 'floppy disk', a storehouse of information for long-term retention and recall. Such an explanation does have the merit of taking into account the position and distribution of figures within a cave, and sees the blank spaces between the groups as part of the message. Indeed, the art itself is only part of the experience: the task of reaching it, and the strange sensations and occasional dangers involved, are all part of the process of imprinting the message indelibly. It has long been assumed that cave art, at least, has much to do with the initiation of young members of the group, and the 'information' and 'survival' theories have again brought this to the fore.

We have seen (Chapter 9) that rock art was done out of doors in the same period, but caves are where parietal art has survived most successfully, and the secrecy and darkness involved in a visit to them must have been a memorable experience for any initiate — especially in cases like Pergouset whose access was physically very difficult. Quite apart from the physical difficulties encountered in some caves, whether rock formations, or flowing water or holes to crawl through, there is the utter blackness, the total silence, the loss of sense of direction in the often labyrinthine passages, the change of temperature, and the frequent sense of claustrophobia.[186]

The fear of being abandoned, lost and alone in the dark, would have concentrated the mind wonderfully, and prepared the apprehensive initiate for anything. Similar techniques are still in use today for more sinister purposes, such as brainwashing or debriefing: victims are taken to alien, unpleasant or unfamiliar surroundings, subjected to physical and mental discomfort, made confused and uncertain about what is to happen; their entire frame of reference is removed, the everyday world is completely undermined – a phenomenon which Henry Miller called 'moving day for the soul'. The technique of sensory deprivation also concentrates the mind, and a cave would certainly affect all the senses of scared initiates quite radically, and leave them vulnerable to indoctrination.

The advocates of this idea do not assume that Ice Age people were aware of all these factors – no doubt they used instinct as much as planning – but the experience of centuries will have taught them which techniques were most successful; and success was crucial since the tribe's very survival would depend on its accumulated learning and rules being transmitted and imprinted effectively and indelibly. Hence the careful positioning of certain figures in niches, or in places where they could be effectively and suddenly revealed by cunning lighting – these were the rudimentary 'special effects' of the time which must have had considerable impact on impressionable and expectant minds. One can speculate that the mass of unsexed human figures in the dark cave-depths represent these adolescents and their ambiguous, transitional status.

But how exactly could these images transmit information? Clearly, we cannot know precisely what the different animals or signs were 'metaphors' or mnemonics for, which rules or events they were connected with. But they must have been vivid, overwhelming visual impressions, especially by the flickering flames of small lamps which, as mentioned earlier (see

p. 127) make the animals seem to move; and pictures of this sort are extremely powerful conveyors of information: the cliché that 'a picture paints a thousand words' is all too true here, and cave pictures still shout down the millennia, though now their language is alien to us.[187]

We know from modern studies that the human mind can 'take in' up to half a dozen items at a glance, and these convey far more information to the mind when arranged in familiar patterns: for example, if one glanced at the above cliché, and at six unconnected words, the cliché would be perceived more clearly and remembered. The same is probably true of Palaeolithic art, which certainly comprises a 'vocabulary' of symbols, some of which must have had considerable information value, and certain combinations of which may have had special significance.

One further idea concerning mnemonics involves not the thinking of the computer age, but methods developed by the Ancient Greeks and Romans, and maintained in medieval Europe. These entailed memorizing vast quantities of information through a technique of impressing 'places' and 'images' on the memory. Before the invention of printing, a trained memory was often of vital importance; people would therefore imagine, in their 'mind's eye', a large deserted place such as a great cathedral or a whole series of rooms. By walking around it mentally, and memorizing thousands of places, each associated with something they wished to remember, they could later repeat the 'journey' and recall everything perfectly. The Greek poet Simonides, who wrote the first text about this method, called it 'inner writing'; and some Roman scholars used it to commit long texts to memory – Cicero's speeches, legal points, or Virgil's entire *Aeneid* (9700 lines, which could even be recited in reverse order!).[188]

Committing information to memory in this way is second nature to Australian Aborigines who use vast stretches of desert, containing thousands of special places (water-holes, hollows in trees, etc.) which they can associate with the Dreamtime wanderings and adventures of their ancestors. In this way, a huge amount of tribal lore, vital to survival, is preserved and imparted little by little to the next generation during initiations and festivals. Art plays a crucial role here; and it is quite possible that similar methods were employed during the Upper Palaeolithic period, both in the open air and in the caves. The latter are suitably quiet, deserted settings for a 'mental walk', containing numerous special places such as niches and concretions by means of which information could readily be committed to memory.

In short, the concept of computerized information has combined with the ancient art of mnemonics to produce a new view of early art, worthy of the modern age. Naturally, little evidence survives to support the idea, apart from the obvious fact that some of the art is in caves, and carefully set out within them. We have no proof of Palaeolithic initiation ceremonies, nor of how information was transmitted or imprinted.

Some scholars believe that parietal art constitutes a system of communicating information, though we need to learn much more before we can feel confident that it is a semiological system capable of transmitting coded messages.[189] Others claim that portable objects are more likely than parietal art to be involved in the storage and rapid transmission of information, for the simple reason that they are portable and not all of them are decorated. Certainly, their different motifs do display interesting distribution patterns, as, for example, in the Magdalenian of Cantabria, where there is distinct regional variation: some motifs are widespread, while others are unique to major centres ('aggregation sites') such as Altamira, or to particular areas, and may thus mark group boundaries.[190]

It is believed that having a set of motifs or a common art-style, different from those of one's neighbours, helps to maintain a group's identity and cohesion (an adherence to norms is the basis of a culture's stability), while motifs shared with neighbours would foster intergroup solidarity[191] – 'a portable art version' of Leroi-Gourhan's view of complex signs as ethnic markers; it has even been claimed that some objects such as 'Venus figurines' were display items indicative of a system of communication and information-exchange designed to maintain contact and alliances between the participants in mating networks which had become strained through climatic deterioration and expansion of hunting territories![192]

Unfortunately, there is no evidence whatsoever to support these fanciful ideas – we have seen, for example, that 'Venus' figurines are by no means a homogeneous group limited to two short periods, and thus there is nothing to link them with 'alliance networks' – indeed, it has been pointed out that we have no idea whether the Gravettian statuettes of France, Moravia and Russia had any more cultural links than the ancient Egyptians, Maya and Peruvians, who all built pyramids but were centuries and huge distances apart.[193] Besides, the environmental deteriorations thought to have triggered this supposed crisis (and the above-mentioned migrations of people into Franco-Cantabria) have been assumed, not demonstrated. There were many climatic fluctuations in the course of the Upper Palaeolithic.

Like all the other views outlined above, these theories merely reflect one of archaeology's current preoccupations – prehistoric social systems, the delineation of group boundaries, and the links between style and information content; and like the preceding theories they may well contain some truth. Time will tell, but meanwhile at least one of the proponents has decided that the 'art as adaptive information' idea is really a truism and based on some very weak assumptions[194] – and in any case should not be assumed to apply to all of Palaeolithic art.

Other recent ideas are simply too vague to be useful: for example, the view has been expressed that group cooperation was required for large-scale communal hunts (another truism), and that cave art may represent the rituals involved in reinforcing the authority of the hunt coordinators; alas, there was no suggestion as to how this was done.[195]

All in the mind?

Finally, as mentioned above, the latest approach to Palaeolithic art has focused on the origin of the oldest motifs, and drawn on current research in neuropsychology. This view may well account for some of the more disturbing cross-cultural parallels in images that are far more complex than meanders or simple geometric shapes. For example, the Lascaux man, with his erection and his 'bird on a stick', is remarkably like a petroglyph in Petrified Forest, Arizona, depicting an ithyphallic man with a bird on a stick.[196] And what are we to make of the fawn/bird spearthrower of Le Mas d'Azil (see p. 98) which is identical (apart from the turd!) to a pose struck by Disney's Bambi in a film made before the object was discovered? There is also a striking analogy between one Palaeolithic scene and a Greek amphora of the eighth century BC: the Magdalenian baton from Lortet shows a fish that seems to be leaping up towards the genitals of a deer, and above are two lozenges with dots inside; the amphora has a very similar composition, except that the deer is replaced by a horse, and there are three lozenges.[197]

Are these simply remarkable coincidences? Or are they indicative of some widespread phenomenon such as 'universal myths' of very remote antiquity, or perhaps something akin to the 'entoptics' and originating in the human mind? Future research may help us decide.

The realization that motifs and motives in Palaeolithic art are not easily recognizable has meant that researchers have found it ever harder to move beyond detailed descriptions and well-meant speculations. What it comes down to, basically, is whether one is content to work with the art as a body of markings that cannot be read, or whether one wants to have stories made up about it! Since proof simply does not apply in attempts to explain the meaning of prehistoric art, it has been argued forcefully that the use of ethnographic analogy to achieve a 'best fit hypothesis' – as in the case of shamanism, for example – is a viable and valid enterprise. However, in view of the problems outlined above, it would seem that more caution and rigour are needed to avoid the abuses of ethnography seen earlier this century, as well as the simplistic wholesale transfer of specific interpretations from one body of evidence to the other in what has been called 'ethnographic snap'. The obsession with 'trance' phenomena will in twenty or thirty years' time, like hunting magic or structuralism before it, be seen as just another stage in the history of Palaeolithic art studies, but probably not a decisive one

(though certainly a simplistic and ill-informed one, since the basic analogy of 'trance' with shamans is utterly rejected by current scholarship on shamanism).[198] In all such cases one is left with a few nuggets of useful and sound interpretation, but the rest becomes so much verbiage, of only historical interest.

The fundamental problem with such attempts at interpretation is that what begin as signposts become marked tracks, then tramlines, and finally circular roundabouts that take us nowhere. Certainly it is healthy for all paths to be explored thoroughly, and each one contributes some new piece of the jigsaw, but it is vital to avoid asserting that the jigsaw only has one piece! It is unhelpful to resort to universalistic concepts of supposed primitive mentality: to simply assert, for instance, that shamanism or sympathetic magic or a fear of the Underworld must have existed in the Upper Palaeolithic tells us nothing specific about that culture, and to simply claim the existence of such beliefs does not even test the hypothesis. This is why Leroi-Gourhan rejected the use of simplistic parallels of this kind. One of the supreme difficulties in using ethnographic parallels, however cautiously it is done, and however many 'ifs' and 'buts' are inserted, is that human cultures are enormously diverse, with an inherent tendency to produce unpredictable variations in behaviour. In short, there is absolutely no reason to suppose that Palaeolithic people drew pictures for the same reasons as very recent Bushmen or Australians – on the contrary, the thousands of years and miles separating them render such a belief ridiculous. We have no idea, and no means of telling, how much change there has been in the cognitive structures of non-literate societies over all these millennia. Many researchers find it irritating that no interpretation of Ice Age art now meets with widespread agreement, and criticize the growing number of specialists who prefer to concentrate on other aspects of its study – those outlined in the first ten chapters of this book – and abandon any serious attempt at interpretation as a waste of effort. But, as Robert Layton has said quite rightly, 'looking for specific meanings in Palaeolithic motifs is frustrating and relatively unproductive'.[199]

12 | CONCLUSION

Readers may by now be wondering what exactly we know about Palaeolithic art, and the answer is both a great deal and virtually nothing. For a start, although we have about 300 decorated sites in Europe alone, and thousands of pieces of portable art, they probably represent a tiny fraction of what existed originally – most outdoor art has gone, so has anything in perishable materials, and we do not know how many decorated caves remain undiscovered. Even in the caves we have, untold quantities of figures may have been weathered or washed away, trampled into the floor or covered by stalagmite and clay – in rich sites such as Castillo or Trois Frères new figures are still being discovered today; while a single new discovery such as Cosquer and especially Chauvet can have the most profound impact on our concept of Ice Age art's content and development.

There is a vast literature on the subject, and hundreds of tracings of figures; but, as we have seen, many tracings, cave-plans and studies are inaccurate, incomplete or biased, and everything needs to be checked, improved and amplified. Much portable art lies unknown and unstudied in museums and private collections, while some major parietal sites still await adequate publication.

It is only with full and accurate facts at their disposal that prehistorians can hope to draw valid conclusions – but, as shown in Chapter 10, it can be difficult enough to identify the figures, let alone assess what they are doing or what they might mean: indeed, some would argue that we should abandon the attempt to extract zoological or ethological information from the art, whereas others see these topics as crucial to our understanding of the art's message.

Thanks in large measure to the work of Léon Pales, rigorous and objective anatomical criteria have been applied to a large number of figures, and many doubtful cases have been weeded out. Currently, much effort is being devoted to distinguishing between secure identification and mere interpretation, between what is definite and what is only probable: in future, prehistorians working on Palaeolithic depictions will have to be far more careful about presenting probabilities as certainties, or mere possibilities as probabilities.

As we have seen, the current emphasis is on total recording of the art: not just all the marks present but also their context. In a cave this entails study of its topography, sediments, climate, archaeological contents, and its use between the Palaeolithic and the art's discovery. Moreover, not only the art's location, date and subject matter are investigated but also the technique of execution, and the time taken to produce it. Such topics may seem less fascinating than the art's meaning, but they are fundamental to any theorizing, and it is no coincidence that Breuil devoted more attention to the question of 'when', and Leroi-Gourhan to 'how', in the latter part of their lives.

We need these solid facts. As Sherlock Holmes said, in *A Scandal in Bohemia*, 'It is a capital mistake to theorize before one has data. Insensibly, one begins to wish facts to suit theories, instead of theories to suit facts.' On the other hand, facts do not speak for themselves: you have to choose a theory at some point, and test it against the data. Unfortunately, this has been done badly in the past: as we have seen, all-embracing theories were presented, and facts bent or ignored in order to make them fit. It is unlikely that anyone well acquainted with Palaeolithic art will ever try this again.

The reason is simple: we are faced here with two-thirds of known art-history, covering at least twenty-five millennia and a vast area of the world. It is ironic that so many books on art begin with a token picture of Lascaux, since it stands at the half-way mark in the history of art! – indeed, probably far closer to us than half-way, since Palaeolithic art only makes sense if it was preceded by an extremely long period of development (doubtless primarily on perishable materials). What we are trying to investigate ranges from simple scratches on bones and stones to sophisticated statuettes, and from figures on blocks and rocks in the open air to complex signs hidden in the inaccessible crannies of deep caverns. Almost every basic artistic technique is represented, together with an often simultaneous use of different conventions such as realism, schematization, stylization and abstraction. Most scholars would now agree that it is futile to try to encompass all of this within one theory.

Not all of it is necessarily mysterious or religious, although some parietal art is almost certainly tied to ritual and ceremony. Some is private, some public; some figures are discreet while others attract attention to themselves.

There is a certain thematic unity — profile views of the same restricted range of animals are present throughout in some areas, together with a mass of apparently non-figurative motifs — but beyond this any homogeneity is more assumed than real. As we have seen, the most recent studies emphasize that every cave has its own unique decorative programme, and it is becoming ever harder to establish regularities in the midst of such wide variation.

Similarly, portable art is marked by great regional variation, both in forms and motifs; it is a parallel, perhaps complementary art-form, reflecting different preoccupations, and it is extremely rare (as with the Altamira/Castillo engravings) that one artist seems to have tackled both.

One can therefore study the complex phenomenon of Palaeolithic art at three principal levels: the individual sites (either a cave, an object, or a group of objects); regional groupings (of motifs, complex signs, dominant animal species, styles, etc.); and as a whole, seeking homogeneity and regularity. As the last option loses favour, efforts are being focused on the other two levels of study, as well as on how things changed through the period. As our database becomes more complete and objective, it will become easier to identify regional and chronological groupings and characteristics, and hence theorize more securely about social systems and information exchange. In the past, we have tried to 'run' before we can 'walk', trying to extract more information than

was available, simply because we dislike unintelligible things. But Palaeolithic art is perverse – every new 'piece of the jigsaw' raises fresh questions, and casts doubt on the conclusions already reached. There are no absolute rules, there are always exceptions.

We have seen that the current 'hit-parade' of theories involve a cosmological structure, the marking of group boundaries, the fostering of cooperation, the recording of seasonal phenomena, the maintenance of mating networks, the initiation of the young, and shamanistic trance images. 'Golden oldies' include art-as-pleasure, hunting and fertility magic, totemism, and a macho preoccupation with hunting and girls. In view of the timespan and vast area through which Palaeolithic art is spread, it is probable that all these theories contain some truth, and that there were many other motivations which may never be known.

It is generally agreed that Palaeolithic art contains messages, no doubt of many kinds (signatures, ownership, warnings, exhortations, demarcations, commemorations, narratives, myths and metaphors); their basic function was probably to affect the knowledge or the behaviour of those who could read them. We, alas, do not know how to read them. We can analyse their content, their execution, their location and their 'associations'; but the way everything was combined into an experience has gone for ever. Without informants, we can only use the art as a source of hypotheses, never as a confirmation of them; without the artists our chances of correctly interpreting the content and meaning of a decorated cave are very slight – and how would we know if we were right or not?

Nevertheless, we are constantly learning more about the art and the people who produced it; every new discovery, every new idea adds to the picture we are building up. Despite its frustrations, the study of Palaeolithic art is extremely rewarding, not only through the beauty of the figures and the exhilarating experience of visiting the caves – the very places where the artists worked – but also because it still represents our most direct contact with the beliefs and preoccupations of our ancestors, and therefore constitutes one of the most fascinating episodes of prehistory.

Notes

Preface
1. Bahn 1997, intro. For discussions and references see Davis 1986, 1987; Bednarik 1986; Pfeiffer 1982; Stoliar 1977/8, 1981; Ucko 1987; Zhurov 1977/8, 1981/2.
2. Sieveking 1979.
3. Pfeiffer 1982.
4. Lasheras, Múzquiz and Saura 1995.

Introduction: Cave Life
1. Andrieux 1974, and pers. comm. See also Pales and de St Péreuse 1976, pp. 129, 133; 1979, p. 140; Pales *et al.* 1976.
2. *L'Art des Cavernes* 1984, pp. 433-7; Vialou 1986.
3. Bahn 1996, p. 8.
4. Arias *et al.* 1996.
5. Bahn 1987b; Tyldesley and Bahn 1983; Thieme 1996.
6. Arl. Leroi-Gourhan 1983.
7. Vallois 1961; Dastugue and de Lumley 1976, pp. 616/17.
8. Bachechi *et al.* 1997.
9. Roper 1969, p. 448. The often-cited deep cut at the front of the skull of a woman from Cro-Magnon is now known to have been caused by the pickaxe of an excavator in 1868 (Delluc 1989, p. 397)!
10. Duhard 1991a, p. 154.

1 The Discovery of Ice Age Art
1. Lartet 1861, p. 214; Sacchi 1990, p. 13.
2. Pittard 1929.
3. Faure 1978, p. 75.
4. Contrary to most textbook accounts, the Chaffaud bone was not found in 1834 (see de Mathurin 1971; Worsaae 1869).
5. Lartet 1861. For his later finds, see Lartet & Christy 1864, 1875.
6. Bouvier 1977, pp. 54–7.
7. Cook & Welté 1995, p. 90.
8. Kühn 1971, p. 14.
9. Simonnet 1980, p. 13.
10. Contrary to the often-repeated claims (e.g. Nougier & Robert 1957) that the 'paintings' of Rouffignac were first mentioned by François de Belleforest in a publication of 1575, and occasionally by later authors, there is not a single mention by anyone of drawings in that cave before the mid-twentieth century (see de Saint Mathurin 1958; Delluc & Delluc 1981).
11. Molard 1908.
12. Chabredier 1975, p. 12; Ollier de Marichard 1973, p. 28.
13. Cartailhac 1908, p. 522; Ollier de Marichard 1973, p. 28.
14. *Découvertes* 1984.
15. Richard 1993, p. 63.
16. Kühn 1955, pp. 45/6.
17. Maria gave this account in conversation with Herbert Kühn in 1923 (see note 16). She claimed she was five years old at the time of the discovery; other informed sources place her age at eight (García Guinea 1979, p. viii) or nine (*ibid.* p. 19).
18. Contrary to popular accounts, Maria would not have used the word *toros* for the animal figures she saw (García Guinea 1979, pp. viii and 20).
19. Kühn 1955, p. 46.
20. Sanz de Sautuola 1880.
21. Kühn 1955, p. 48.
22. Madariaga 1972, p. 83: in a letter to Cartailhac dated 19 May 1881, de Mortillet claimed that merely looking at sketches of the Altamira figures made it self-evident that this was 'a farce, a simple caricature, made and displayed to the world to make credulous palaeontologists and pre-

historians look ridiculous'.
23. Madariaga 1972, p. 86; García Guinea 1979, pp. 424. On de Mortillet's anticlericalism, see Reinach 1899 and Bahn 1992.
24. Minvielle 1972, p. 169.
25. Harlé 1881.
26. The earliest known find of a Palaeolithic lamp occurred in 1854 at La Chaire à Calvin (Charente), and several others were also discovered before Altamira, but they were either contested or ignored, while some remained unpublished for years because of the controversy over the existence of a Palaeolithic lighting system (de Beaune 1987, p. 12).
27. Bégouën 1947, pp. 494.
28. Madariaga 1972, p. 28.
29. Kühn 1955, p. 49.
30. García Guinea 1979, p. 42; Madariaga de la Campa 1981, p. 302.
31. Breuil 1964, p. 11.
32. Piette 1889, p. 206.
33. Jagor 1882.
34. Mallery 1893, vol. 1, p. 177.
35. Roussot 1965; Delluc & Delluc 1973, p. 205.
36. As mentioned in note 26, early finds of lamps had met with resistance or ridicule; Rivière himself had originally failed to take seriously the claims of a M. Detrieux who informed him in 1896 that he had found a lamp near Le Moustier (de Beaune 1987, p. 12). In short, the history of the discovery and acceptance of Palaeolithic lamps closely mirrors that of cave art itself.
37. Menéndez Fernández 1994, p. 343.
38. Roussot 1972/3.
39. Richard 1993, p. 65.
40. Bégouën 1932, p. 8; 1952, p. 289.
41. Cartailhac & Breuil 1905.
42. Cartailhac 1902.
43. Brodrick 1963, p. 63.
44. Cartailhac 1908, p. 520.
45. *Ibid.*, p. 512.

2. The Oldest 'Art' in the World
1. Bahn 1997, intro.
2. Bednarik 1992.
3. Dart 1974; Oakley 1981, pp. 205–6.
4. Oakley 1981, pp. 208–10.
5. See Marshack 1981 and Wreschner 1975, 1980 for references.
6. Crémades 1996.
7. Bordes 1969; Marshack 1977, pp. 289–92; 1976, p. 140; 1976a, pp. 278–9.
8. Mania and Mania 1988; Bednarik 1995; Müller-Beck and Albrecht 1987, p. 68. Feustel (in Müller-Beck and Albrecht 1987, p. 60) even claims that marks on one Bilzingsleben bone may be the depiction of a large animal, although this view has not been widely adopted.
9. Bednarik 1993.
10. Marshack 1997.
11. Marshack 1988, 1988a, 1990.
12. de Barandiarán 1980, pp. 217/57.
13. Marshack 1996.
14. Leonardi 1983, 1988, 1989 pp. 63–7; and in Clottes 1990, vol. 1, p. 129.
15. Stepanchuk 1993.
16. Baldeon 1993, pp. 25/6 – though the perforations may not all be man made.
17. Marshack 1976, p. 139; 1976a; Freeman and González Echegaray 1983; Crémades 1996.
18. Marshack 1976, pp. 135–41; 1976a, p. 277; Kozlowski 1992.
19. Frolov 1981, pp. 63, 77. For a sceptical view, see

Kozlowski 1992, p. 36.
20. Bordes 1969; Marshack 1988, 1988a, 1990; Bednarik 1992a.
21. Hublin *et al.* 1996; Marshack 1988, 1988a, 1990; Poplin 1983, p. 62.
22. Hayden 1993, pp. 123–4.
23. Demars 1992; Bordes 1952.
24. See note 5; Mészáros and Vértes 1955.
25. Crémades 1996, p. 494.
26. Bednarik 1992a, 1995.

3. A Worldwide Phenomenon
1. Bahn 1987, 1991.
2. Aveleyra 1965; Lorenzo 1968, pp. 284/5; Messmacher 1981, p. 94. A mastodon bone, found at Valsequillo, Mexico, dating to perhaps 22,000 years ago, has often been claimed (e.g. see Canby 1979, p. 350) to bear engravings of a mastodon and a large cat. However, careful study has shown that it has only cutmarks, not engraved figures (Lorenzo 1968, pp. 286–8).
3. Messmacher 1981, p. 84.
4. Purdy 1996, p. 10; 1996a, pp. 4–5.
5. Whitley and Dorn 1987, 1993.
6. Whitley *et al.* 1996.
7. Julien and Lavallée 1987, pp. 49–50; see also Linares 1988.
8. Aschero and Podestá 1986, pp. 40–43.
9. Cardich 1987, p. 110.
10. *Ibid.*, p. 112.
11. Crivelli Montero and Fernández 1996.
12. Roosevelt *et al.* 1996, pp. 378–80.
13. Guidon and Delibrias 1986. Another spall from a layer of *c.* 32,000 years BC bears a vague red patch, but it cannot be assumed to be humanly made, and might be natural.
14. Chaffee, Hyman and Rowe 1993.
15. Bahn 1991, p. 92.
16. Bednarik 1989.
17. Prous 1994, pp. 90/1.
18. On Zimbabwe, see Walker 1987, p. 142; 1996, pp. 11–13. Mori (e.g. 1974), like Frobenius before him, has proposed that the earliest rock-engravings of the Sahara may date back to the late Pleistocene. However, other scholars such as Muzzolini (1986, pp. 312–14) point out that this reasoning is based on a few isolated, and possibly anomalous, radiocarbon dates from occupations at the foot of decorated rocks, and which may have no association at all with the engravings. The question therefore remains open. New research in the Saharan Atlas Mountains, however, is suggesting that some large naturalist engravings may well date back to the late Pleistocene after all (Hachid 1992).
19. Leakey 1983, pp. 21–2. Anati (1985, p. 54; 1986, pp. 789/91) believes that some rock art in Tanzania may be 40,000 years old, partly because levels of that date have been excavated in decorated sites, but also because some parietal figures seem to resemble those on the Apollo 11 Cave's stones.
20. Wendt 1974, 1976.
21. Butzer *et al.* 1979, p. 1212; Bahn 1991.
22. Beaumont 1973, p. 44, note 44.
23. Singer and Wymer 1982, pp. 116, 149.
24. Van Noten 1977, p. 38.
25. Saxon 1976.
26. Hachi 1985, and pers. comm. 1996.
27. Otte *et al.* 1995.
28. Kökten 1955.
29. Hovers 1990.
30. Weinstein-Evron and Belfer-Cohen 1993.
31. Belfer-Cohen and Bar-Yosef 1981, p. 35.

32. Ronen and Barton 1981.
33. Stordeur and Jammous 1995; Stordeur et al. 1996.
34. Anati 1986, pp. 789/92.
35. Wakankar 1984, 1985; Anati 1986, pp. 789/93.
36. Kumar et al. 1988.
37. Bednarik in The Artefact 18, 1995, p. 87.
38. Kumar 1996.
39. Marshack 1972b, p. 66; the other object is a flat frag-ment of much harder stone, 6 by 7 cm, containing four series of engraved notches along two of its very straight, right-angled edges. A more delicate linear marking on the faces of the plaque is much deteriorated, however.
40. Bednarik and You 1991; Bednarik 1994a.
41. Sohn Pow-Key 1981.
42. Sohn Pow-Key 1974, p. 11.
43. Aikens and Higuchi 1982, p. 107.
44. Okladnikov and Myedvyedyev 1983.
45. Okladnikov 1972, p. 75; Novgorodova 1980, pp. 38–51; for a contrary view, Formozov 1983.
46. For a critical review of claims for Pleistocene art in Australia, see Rosenfeld 1993.
47. Wright 1971; Gallus pp. 56–9 in Bednarik 1986.
48. Rosenfeld et al. 1981.
49. Brown 1987, p. 59; Brewster 1986; McGowan et al. 1993.
50. Cosgrove and Jones 1989; McGowan et al. 1993.
51. Loy et al. 1990.
52. Watchman 1993.
53. Cole, Watchman and Morwood 1995.
54. Watchman and Hatte 1996.
55. O'Connor 1995.
56. Chaloupka 1993.
57. Chaloupka 1984; Murray and Chaloupka 1983/4. For arguments against, and a reply by Chaloupka and Murray, see Archaeology in Oceania 21, 1986, 140–47. On the prob-lem of identifying species in Australian rock art, see Chapter 10, p. 135).
58. Lewis 1988.
59. Loy et al. 1990; Nelson 1993, and reply by Loy 1994.
60. Dortch 1979, 1984; Bednarik 1991, p. 27; 1994, p. 33.
61. Vanderwal and Fullagar 1989.
62. Morse 1993.
63. Lorblanchet 1988, p. 286.
64. Dragovich 1986.
65. Nobbs and Dorn 1988; Dorn et al. 1988.
66. Nobbs and Dorn 1993.
67. Fullagar, Price and Head 1996.
68. Hallam 1971.
69. Aslin and Bednarik 1984a/b, 1985; Bednarik 1986.
70. For this subject in general, see Clottes 1990. For regional studies, see Barandiarán 1973, Moure Romanillo 1985, Corchón 1986 (N. Spain); Chollot 1964, 1980, de Saint Périer 1965, Piette 1907 (France); L'Art Préhistorique 1996 (Pyrenees/N. Spain); Lejeune 1987, Twiesselmann 1951 (Belgium); Adam and Kurz 1980, Bosinski 1982 (Germany and Switzerland); Eppel 1972 (Austria); Graziosi 1973, Zampetti 1987, Leonardi 1988 (Italy); Freund 1987, Hennig 1960, Valoch 1970, Müller-Beck and Albrecht 1987, Jelínek 1988 (Central Europe); Kozlowski 1992 (Eastern Europe); Abramova 1995, Praslov 1985 (former USSR).
71. Campbell 1977, vol. 2, figs 102, 105, 143. A brief account, without illustrations, can be found in Sieveking 1972. For the Sherborne bone, see Chapter 6.
72. See, for example, Bahn and Cole 1986.
73. Chollot-Varagnac 1980, p. 457.
74. de Saint-Périer 1965.
75. Bouet et al. 1986/7; Delporte et al. 1986; a comput-erized databank has also been set up for the French pari-etal sites (Djindjian and Pinçon 1986).
76. Villaverde 1994; Fortea 1978; Pericot 1942.
77. Pales 1969; Pales and de St Péreuse 1965.

78. Bosinski and Fischer 1974, 1980.
79. Bégouën and Clottes 1987; Bahn 1983.
80. Graindor and Martin 1972; Martin 1973.
81. Breuil thought some red streaks in the Welsh coastal cave of Bacon Hole might be Palaeolithic (1952, p. 25), but these have now faded away, and are thought to have been either a natural phenomenon or the traces left by a workman having cleaned a paintbrush on the wall in 1894! The claims in 1981 for art on the walls of a cave in the Wye Valley (see Illustrated London News Jan. 1981, pp. 31–4; Current Anthropology 22 (5), Oct. 1981, pp. 501–2) proved to be based on a mixture of natural features and wishful thinking (see Antiquity July 1981, pp. 81–2, 123–5; I.L.N. May 1981, p. 24; Current Anthropology 23 (5), Oct. 1982, pp. 567–9).
82. Hahn 1986, p. 253; 1991; Kozlowski 1992, p. 41.
83. Delluc and Delluc 1991, pp. 206–11 and 130.
84. Kozlowski 1992, pp. 41–2.
85. Maringer and Bandi 1953, p. 23.
86. Bosinski 1982, p. 6; Breuil 1952, p. 24; Freund 1957, p. 55.
87. Kohl and Burgstaller 1992; Bednarik 1996.
88. Kozlowski 1992, pp. 41–2.
89. Oliva 1996, pp. 120, 129.
90. Abramova 1995, pp. 109–10; Bader 1965, p. 25; Maringer and Bandi 1953, p. 27; Shimkin 1978, p. 280. Abramova mentions rectilinear engraved lines on the walls of Mgvimevi cave in western Georgia, and engraved or pecked handprints near rock shelters on the River Gubs, as well as the painted Chichkino figures.
91. Bader 1965, 1967; Shimkin 1978, pp. 280–82.
92. Shchelinsky 1989.
93. Petrin and Shirokov 1991; Petrin 1992.
94. For example, the lone engraved bison of Ségriès (Alpes de Haute Provence) is thought by most (though not all) specialists to be a fake, and was therefore not included in L'Art des Cavernes: see de Lumley 1968. For the doubts, see Pales and de St Péreuse 1981, pp. 75–86: the figure is quite different in its proportions, technique and style from unquestionably authentic bison, and is isolated in a region empty of Palaeolithic art (though this latter point has lost its force in recent years through the discoveries of deco-rated caves and objects in parts of France, Spain and Portugal which previously were blank spaces on the map).
95. Aujoulat et al. 1984.
96. GRAPP 1993.
97. Arte Rupestre en España 1987 is still the most up-to-date synthesis of the Spanish caves, together with papers in Chapa and Menéndez 1994. See also Straus 1987.
98. Graziosi 1973; Naber et al. 1976.
99. Araújo and Lejeune 1995. For the open-air engravings, see Chapter 9.
100. Basler 1979; Kujundzic 1989. The cave of Pecina has only non-figurative engravings, attributed to the Mesolithic.
101. Cârciumaru and Bitiri 1983.
102. For distribution maps, see L'Art des Cavernes 1984, and notes 96, 97.
103. Delluc and Delluc 1973; L'Art des Cavernes 1984, pp. 208/9; Fortea 1981. For other examples of caves with sin-gle figures, see Jordá 1979, p. 332; 1985.
104. L'Art des Cavernes 1984, pp. 376–7, 347–9.
105. I., pp. 378–80.
106. Leroi-Gourhan and Allain 1979.
107. Vialou 1986, p. 187. It is noteworthy that Vialou (p. 166) counts clusters of the same motif as a single unit: thus the 189 circles engraved on a bear figure at Trois Frères are counted as one sign. If a different system were adopted, the resulting figures and percentages would be totally modified.
108. Bégouën and Clottes 1987.
109. Vialou 1986, p. 351.
110. Tosello 1983. Moure Romanillo (1988, p. 79) has

since noted that, of twenty-six sites in Cantabria with fig-urative portable art, eighteen also have parietal art of the same age.
111. Bahn 1982, 1984.
112. Conkey 1980.

4. MAKING A RECORD

1. Delporte 1993, pp. 61/2.
2. See, for example, Rivenq 1976, which compares mod-ern photographs of a frieze in Montespan with those taken in 1926; Bégouën 1980, p. 689, which does the same for Trois Frères; or Omnès 1982, pl. XIV, which does the same in Labastide.
3. Brodrick 1963, p. 53; Ripoll 1994.
4. Breuil, letter to Glyn Daniel, 20 August 1955, in author's collection; see also Aujoulat 1987, p. 22.
5. Bégouën & Clottes 1987.
6. Boyle et al. 1963, p. 15.
7. Delluc & Delluc 1984, p. 57; 1991, p. 65. Non-existent circles also appear on some of Breuil's versions of portable engravings from Enlène – see Bégouën et al. 1984/5, pp. 2, 40; this article (pp. 71/2) shows that Bouyssonie's copies of portable art are often more reliable than those of Breuil which feature omissions, additions and a certain amount of wishful thinking.
8. Bégouën 1980, p. 684; Vertut 1980, p. 674.
9. Breuil 1952a, p. 11.
10. Cartailhac & Breuil 1906; Breuil & Obermaier 1935.
11. Breuil 1957, p. 238.
12. Breuil 1949.
13. Breuil 1957, p. 237.
14. For unquestioning acceptance of Breuil's tracings, see, for example, Leason 1939, p. 51: 'In verifying his theory the Author has had recourse to published sources, but, coming from the distinguished hand ... of Professor Henri Breuil, their accuracy can hardly be called into question.'
15. Pales 1969, pp. 105/6.
16. This was a problem, for example, in a recent re-exam-ination of the heads of the Trois Frères felines – see Bégouën & Clottes 1986/7; and comment by Clottes in Marshack 1985, pp. 175–7.
17. Bégouën 1942, p. 6.
18. Archambeau & Archambeau 1991, p. 68.
19. Freeman et al. 1987, pp. 206–8, 233/4.
20. Clottes, in Clottes 1990, vol. 1, p. 203.
21. Garcia 1979; Martin 1974; Sauvet in GRAPP 1993, pp. 359–62; Sacchi in GRAPP 1993, pp. 363–4.
22. Bégouën & Clottes 1987.
23. Pales 1969, p. 40; Airvaux & Pradel 1984, p. 213; Faure 1978.
24. Marshack 1972 (1991).
25. d'Errico 1987.
26. Bégouën 1980, p. 685; 1984, p. 79; Vertut 1980.
27. Vertut 1980, p. 667.
28. Airvaux et al. 1991; and see INORA 1, 1992.
29. Marshack 1975.
30. Vialou 1982; 1981, p. 81.
31. Marshack 1975; see also Marshack 1985, pp. 68, 77–9 (and pp. 63/4 for criticism by Lorblanchet).
32. Vertut 1980, p. 672.
33. Girard et al. 1992.
34. Lorblanchet 1984; see also Aujoulat 1987.
35. Lorblanchet 1993, p. 122.
36. Ruspoli 1987, pp. 177–86.
37. Bégouën 1984, p. 79.
38. Delluc & Delluc 1984a.
39. Lasheras et al. 1995.
40. Clottes 1996.
41. See 'Prehistoric art leaps back to life' in New Scientist, 31 August 1996, p. 11.
42. Layton 1991, pp. 26–30.
43. Laurent 1963, 1971.
44. Tosello 1983, p. 285.

45. Faure 1978, pp. 68/9.
46. Roussot 1984c.
47. Bouvier 1977, pp. 54–7.
48. See Pales 1970, which puts an end to an error of this kind.
49. Lorblanchet 1984.
50. Bégouën 1942, p. 9.
51. Pales & de St Péreuse 1965, p. 221.
52. Pales 1964, p. 12; Pales & de St Péreuse 1965, p. 230.
53. Pales 1969, p. 26; Pales & de St Péreuse 1976, 1981 and 1989.
54. Bednarik 1986, p. 35; Meylan 1986, p. 47.
55. Meylan 1986, pp. 46–50; Lhote 1972, p. 323.
56. González Morales 1989, p. 272.
57. Ucko 1977, p. 8.
58. Rivenq 1984.
59. Lorblanchet 1984.
60. Vialou 1982.
61. GRAPP 1993, pp. 315–85.

5. How Old is the Art?
1. Barandiarán 1988; 1994, pp. 47/8.
2. Airvaux and Pradel 1984; Bégouën and Clottes 1979, p. 19.
3. Piette 1894, p. 129.
4. Bahn 1984, pp. 164/5.
5. Berenguer 1986, p. 669.
6. Faure 1978.
7. Crémades 1989, p. 470.
8. Bahn 1982.
9. Combier 1984.
10. The development of Piette's views is explained in Breuil 1909; and see Breuil 1959, p. 15; Delporte 1987.
11. Sacchi in GRAPP 1993 pp. 311–14.
12. Lister and Bahn 1994.
13. González Echegaray 1968; 1974, pp. 39–42.
14. González Echegaray 1972. A similar attempt to date caves from the dominant species depicted was made by Jordá, quoted in Altuna 1983, p. 236.
15. Bégouën and Clottes 1985, p. 43.
16. Rouzaud 1978, p. 32.
17. Liger 1995.
18. Clottes, Duport and Féruglio 1990 (and also in *Bull. Soc. Ariège-Pyrénées* 46, 1991, pp. 119–32).
19. Delluc and Delluc 1991.
20. Laming-Emperaire 1959, pp. 32/4; Roussot 1984a.
21. Carcauzon 1986.
22. Gaussen 1964.
23. Straus 1982, p. 78.
24. Combier et al. 1958; Combier 1984.
25. Moure and González in *International Newsletter On Rock Art* 3, 1992.
26. Girard et al. in *International Newsletter On Rock Art* 12, 1995.
27. Clottes et al. 1992.
28. Arl. Leroi-Gourhan 1983, and in Leroi-Gourhan and Allain 1979.
29. Cartailhac and Breuil 1906; see also Barandiarán 1973, pp. 302–7.
30. For Labastide, see Omnès 1982, pp. 182, 187; for Gargas, Barrière 1976, pp. 406–9.
31. de Saint Mathurin 1973, 1975, 1978.
32. Almagro 1976.
33. Utrilla 1979; see also Almagro 1976, p. 99; Straus 1982.
34. Barandiarán 1988, 1994; Moure et al. 1996.
35. Jordá Cerdá 1972, 1980.
36. Cartailhac and Breuil 1905.
37. de Balbín Behrmann and Moure Romanillo 1982; Moure Romanillo 1986.
38. Glory 1964.
39. Leroi-Gourhan and Allain 1979.
40. Bahn 1994.

41. For example, Breuil 1952, p. 149.
42. González García 1993; Delluc and Delluc 1991, pp. 316/17.
43. Lorblanchet 1993, p. 122.
44. Stoliar 1977/8, 1981, 1985, 1995; similarly, Hornblower (1951) assigned these models to the Gravettian, but there is no evidence for this either. For a reply, see Zhurov 1977/8, 1981/2; Bahn 1984, p. 343; 1987a.
45. See Altuna 1975, p. 104/5, or Altuna and Apellániz 1978, p. 22 for an illustration of the difference between a side and a front view of a living red deer's antlers.
46. Rousseau 1984; Pales and de St Péreuse 1981, pp. 94–7.
47. Guthrie 1984, p. 58; Mithen (1988) also adopted this view.
48. Riddell 1940, pp. 158/9.
49. Breuil 1977.
50. de Saint Mathurin 1973, 1975.
51. Breuil 1977, pp. 57/8.
52. Ripoll 1964.
53. Chauvet et al. 1995.
54. Breuil 1912, 1952.
55. Breuil and Lantier 1959, p. 225.
56. Laming-Emperaire 1962, 1959.
57. Jordá Cerdá 1964, 1964a; see also Altuna and Apellániz 1976, pp. 158–63; Barandiarán 1973, p. 310.
58. Leroi-Gourhan 1971.
59. Leroi-Gourhan 1962.
60. Delluc and Delluc 1978, p. 386.
61. Moure Romanillo 1985, pp. 124/5; Barandiarán 1973.
62. Delluc and Delluc 1983, pp. 32/3.
63. See Ucko and Rosenfeld 1967, p. 78 for a discussion of these points.
64. Delluc and Delluc 1991, p. 348.
65. Apellániz 1991, p. 15.
66. Clottes 1990a.
67. See, for example, Bahn 1984 for Pyrenean cultures, or Soffer 1985 for those of the Central Russian Plain.
68. Lorblanchet 1977.
69. Guthrie 1984, p. 72.
70. See papers in Lorblanchet and Bahn 1993.
71. Almagro et al. 1972, p. 469.
72. Bégouën and Clottes 1985, p. 49; 1979a, p. 60.
73. Altuna and Apellániz 1976, pp. 157, 165; Apellániz 1982, p. 92; Almagro 1976.
74. Lorblanchet 1995, pp. 277/8.
75. Pericot 1942; Jordá 1964; Lorblanchet 1974, pp. 82/3; see also Straus 1982.
76. Airvaux et al. 1983.
77. Bednarik 1994b.
78. Clottes et al. 1995.
79. Clottes in Chauvet et al. 1995, p. 118.
80. Clottes et al. 1992, 1996.
81. Valladas et al. 1992; Clottes et al. 1992.
82. Lorblanchet 1993a.
83. Moure et al. 1996.
84. *Ibid.*
85. *Ibid.* Igler et al. 1994.
86. Lorblanchet et al. 1995.
87. Lorblanchet 1995, pp. 279/80.

6. Fakes and Forgeries
1. Bahn 1993. On criteria of authenticity of Palaeolithic art, see Barandiarán 1995.
2. Vayson de Pradenne 1932 [1993]., p. 276.
3. *Ibid.*, pp. 142–50.
4. *Ibid.*, pp. 278–9.
5. *Ibid.*, pp. 278–9, 500.
6. Delporte 1993, pp. 2131.
7. Niedhorn 1990.
8. de Mortillet 1898; Chollot-Legoux 1964, pp. 446–59; Leonardi in Clottes 1990, vol. 1, p. 132; Delporte 1993,

p. 98; 1995, p. 58; and especially Pales 1972, pp. 238–50.
9. Bisson and Bolduc 1994. As Marshack has pointed out (*Science* 264, 1994, p. 329), all these figurines were offered for sale to the American Museum of Natural History in 1939; the museum contacted Hallam Movius, who recommended the acquisition of only one; it is certainly noteworthy that Movius did not recommend that the museum should also snap up all of the others which, since their re-emergence in 1994, have been trumpeted as being of major importance! Movius was America's leading Palaeolithic specialist, and it is not every day that Palaeolithic figurines from Grimaldi or anywhere else are available for sale, so it seems likely that he smelt a rat.
10. Delluc and Delluc 1991, p. 209.
11. Marshack 1988b; Delporte 1993, pp. 144/5.
12. Schuster and Carpenter 1988, p. 1363. Cook (in McDermott 1996, p. 251) reports that the head was offered for sale to the British Museum in 1948, but evidence was found that it was made recently from ancient ivory.
13. Couraud 1985; Bahn 1984a; Bahn and Couraud 1984.
14. Kraft and Thomas 1976; for the case against, see Meltzer and Sturtevant 1983 (as well as Sturtevant and Meltzer in *Science* 227, 1985, p. 242).
15. Griffin et al. 1988; Bahn 1990.
16. d'Errico et al. 1997; Stringer et al. 1995.
17. Madariaga 1972, p. 45.
18. Jordá 1970; Moure 1993, pp. 175, 180.
19. Clottes, Menu and Walter 1990, p. 183.
20. Altuna et al. 1992; Apellániz 1995; Barandiarán 1995, pp. 25–9.
21. Bahn 1993, pp. 54–8.

7. Portable Art
1. Marshack 1987.
2. Buisson 1990; Fages and Mourer-Chauviré 1983; Absolon 1937. It should be noted that the so-called 'Aurignacian flute from the Abri Blanchard, located in the Musée des Antiquités Nationales' mentioned by White (1986, p. 109) is actually a specimen of unknown date from La Roque-St-Christophe (Dordogne) and housed in the British Museum! There is no flute from the Abri Blanchard. (See also Delluc and Delluc, in Clottes 1990, vol. 2, p. 63).
3. Pfeiffer 1982, pp. 181/2.
4. Turk, Dirjec and Kavur 1995. A previously reported specimen from a pre-Aurignacian layer at Haua Fteah, North Africa, is now known not to be a flute.
5. Allain 1950.
6. Dauvois 1989, 1994, pp. 13/14; Harrison 1978.
7. Roussot 1970, pp. 9/10.
8. Bibikov 1981; Lister and Bahn 1994, p. 108.
9. Huyge 1990 (and see discussion in *Rock Art Research* 8, 1991, pp. 61–4), 1991; Dauvois 1994, p. 18.
10. Dams 1984, 1985, 1987a, pp. 60, 193–5; Glory 1968, pp. 55/6.
11. Dauvois and Boutillon 1990.
12. Philibert 1994.
13. Audouin and Plisson 1982.
14. Péquart 1960/3, pp. 211–14.
15. Knight, Power and Watts 1995, pp. 85–9.
16. Taylor 1996, pp. 99–107.
17. Kozlowski 1982.
18. White 1992, and in a whole series of similar papers.
19. Marshack 1991, p. 49; Bednarik 1995, 1995c.
20. See, for example, Ladier and Welté 1994.
21. Hahn 1972; Soffer 1985.
22. Bader 1967 (1976).
23. Praslov 1985.
24. Taborin 1982, pp. 50/51.
25. Poplin 1983.
26. Le Mort 1985; see also Bégouën, Bégouën and Vallois 1937.

27. Taborin 1977.
28. Taborin 1985, 1993; Desbrosse *et al.* 1976.
29. Imaz 1990, pp. 270/4.
30. Soffer 1985a.
31. Bahn 1977, pp. 252/3.
32. Taborin 1985.
33. Bahn 1982, 1977.
34. Bader 1967 (1976); Soffer 1985a, p. 259.
35. Pond 1925; for a more detailed and illustrated account of Aurignacian production of ivory beads, see Otte 1974; and several papers on ivory beads in Hahn *et al.* 1995.
36. Taborin 1982.
37. Jia Lanpo 1980, p. 52; *Atlas of Primitive Man in China* 1980, pp. 114/15.
38. Leroy-Prost 1984, p. 45.
39. Frolov 1981, p. 63, 77; 1977/9, pp. 85, 153/4; for Temnata, see Crémades *et al.* 1995.
40. Marshack 1972, pp. 255–8; Faure 1978, p. 65.
41. Omnès 1982, p. 185; Bosinski 1973, p. 39; Clottes 1990, vol. 2, p. 216. Marshack has admitted (1985, pp. 72/3) that his own experiments showed that a line's cross-section altered as the line/tool changed direction; it is the changes of direction and of rhythm which suggest the presence of different tools or hands to him.
42. Bosinski and Fischer 1974, pp. 5/6; Faure 1978, p. 65; Marshack 1985, pp. 72/3.
43. d'Errico 1987; 1994.
44. Bégouën and Bégouën 1934; L. Bégouën 1939, p. 298; Bégouën *et al.* 1984/5. See also de Saint Périer 1930, p. 81; Capitan and Bouyssonie 1924, p. 40; and Pales and de St Péreuse 1976, p. 17.
45. For Ferrassie and Abri Durif, see Pales and de St Péreuse 1979, pp. 137/8.
46. Bégouën *et al.* 1984, p. 145; many of the La Marche slabs are also engraved on both sides – see Pales and de St Péreuse 1981, p. 22.
47. Omnès 1982, p. 185; Bahn 1983.
48. Fritz 1990.
49. Péquart 1960/3, p. 222; see also Pales and de St Péreuse 1981, p. 22, for the situation at La Marche.
50. Omnès 1982, p. 184.
51. Bégouën and Clottes 1981, p. 42; 1983; 1985.
52. Moure Romanillo 1985, p. 103.
53. Rozoy 1985, 1988, 1990.
54. Bosinski 1973; 1984, p. 318.
55. Moure Romanillo 1985, p. 103.
56. Pales and de St Péreuse 1979, p. 141; 1976, p. 17.
57. Couraud 1985, 1982; Bahn and Couraud 1984; Bahn 1984a.
58. Broglio 1992.
59. Thévenin 1983, 1989; Couraud 1985; d'Errico 1994.
60. Rozoy 1992.
61. Bosinski 1973, p. 41; Rozoy 1988, p. 176; Russell 1989, 1989a, pp. 242–7.
62. Crémades 1989, p. 60.
63. Delporte and Mons 1980.
64. Bégouën and Clottes 1985, p. 45; Pales and de St Péreuse 1979, p. 141/2.
65. For painted bones, see Gladkih *et al.* 1984, p. 142; and Shimkin 1978, pp. 275, 283; for paste inside engraved lines, see Marshack 1981; 1979, p. 288.
66. Barandiarán 1994, p. 58.
67. Péquart 1939/40, p. 451; Almagro 1976. For Le Mas d'Azil, see Pales 1970; Delporte and Mons 1975. For a brief review of the topic, without illustrations, see Sieveking 1983. For plaquettes as a whole, see Sieveking 1987.
68. de Sonneville-Bordes 1986, p. 633.
69. Almagro 1976, p. 68 note.
70. Delporte and Mons 1980, p. 43 – this paper also gives references to some of the earlier experimental work by these authors. For the nineteenth-century experiments, see Leguay 1877, 1882.

71. Crémades 1989, p. 60.
72. *Ibid.*, p. 481.
73. Barandiarán 1971.
74. Breuil 1909, p. 380.
75. Bahn 1984, p. 116.
76. Baulois 1980.
77. Allain and Rigaud 1986; Praslov 1985, p. 187.
78. de Sonneville-Bordes 1986, pp. 634–6.
79. Noiret 1991.
80. Bellier 1981/2, 1984; Péré 1988.
81. Barandiarán 1968; Sieveking 1971.
82. Jelínek 1975, pp. 458/9.
83. Passemard and Breuil 1928.
84. St Périer 1930, p. 97; 1936, pp. 116–20; Bellier 1984, p. 26.
85. Buisson *et al.* 1996.
86. For the new Spanish specimens, see, for example, Fortea 1981; for Labastide, see Simonnet 1952; Omnès 1982, p. 192.
87. Fritz and Simonnet 1996.
88. Bellier 1984, p. 30.
89. For the French and Swiss type, see Bandi and Delporte 1984; for the relationship with antler shape, see Leroi-Gourhan 1971, figs. 34/5. Not all compositions of this type, however, were determined by the antler's shape – see Pales and de St Péreuse 1965, pp. 227/8.
90. Cattelain 1977/8.
91. Bahn 1982, pp. 257/8; 1984; Robert *et al.* 1953; Simonnet 1991. Bandi has claimed (1988) that the animals are not defecating but giving birth; but he is then unable to explain the presence of what he accepts are birds …
92. Péquart 1960/3, p. 299.
93. See Bahn 1978; Bahn and Otte 1985; Péquart 1960/3, p. 219; Vasil'ev 1985.
94. Klíma 1984, 1983, 1982.
95. Neugebauer-Maresch 1996, pp. 102–3; Delporte 1990, p. 126.
96. Vandiver *et al.* 1989; Soffer *et al.* 1993.
97. For Isturitz, see St Périer 1930, pp. 109–16; 1936, pp. 126–33; Mons 1986. For Bédeilhac, see Bahn and Cole 1986. Bégouën *et al.* (1987) report a sandstone sculpture of a bison from Enlène which was deliberately destroyed by being sawn into at least three pieces.
98. St Périer 1936, pp. 123/5.
99. For a fine example, see Pales and de St Péreuse 1979, pp. 118/19; on the topic as a whole, see Chollot-Varagnac 1980, p. 452.
100. Gvozdover 1995.
101. Arambourou 1978, pp. 116–24.
102. Eppel 1972, p 78.
103. Praslov 1993; Gvozdover 1995; Abramova 1995.
104. Neugebauer-Maresch 1995.
105. Lister and Bahn 1994; Abramova 1995.
106. Christensen 1996; on ivory working in general, see Hahn *et al.* 1995.
107. Hahn 1979, 1982, 1986.
108. Delporte 1993.
109. de Saint Mathurin 1978.
110. Breuil 1959, pp. 15–17 and 109; Breuil and Lantier 1959, pp. 193/4. In Breuil 1909, p. 399, he stated that Piette's 'Vallinfernalien' layer (= Gravettian) was above the 'Eburnéen ancien' layer containing the ivory sculptures, which must therefore be pre-Gravettian. And in 1957, p. 233, he stated that the Brassempouy figures were at the very base of the Aurignacian. Yet Delporte (1993) claims that there is no evidence of any French statuette being Aurignacian! Other Breuil references to this topic are cited by Pales 1972, p. 250.
111. Marshack 1984; 1976; 1976a; Bosinski 1982; Hahn 1984.
112. Hahn 1971; Bosinski 1982; Marshack 1987, 1987a.
113. Dauvois 1977; for other observations on how female statuettes of ivory and stone were carved, incised and pol-

ished, see Delporte 1993.
114. Jelínek 1975, p. 411.
115. Praslov 1985; Abramova 1995.
116. Tarassov 1971; Delporte 1993.
117. Delporte 1993.
118. Gamble 1982 is the principal example of such an approach; it has been demolished by Soffer 1987, pp. 335–9.
119. Soffer 1985; Gladkih *et al.* 1984; Jelínek 1975; Lister and Bahn 1994.
120. Bosinski 1984, p. 315; Bosinski and Fischer 1980, pp. 47, 99, 119 and pl. 101.
121. Leroi-Gourhan 1978; for other examples of distortion to fit available space, see Pales 1976/7, pp. 90/1, and Pales and de St Péreuse 1976, p. 53, and 1981, pp. 128–31.

8. BLOCKS, ROCK SHELTERS AND CAVES

1. Berenguer 1987, p. 665.
2. Delluc and Delluc 1978, p. 393.
3. *Ibid.*, pp. 428, 391.
4. Delluc and Delluc 1984b; 1978, pp. 215–21.
5. Lejeune 1981; Eastham 1979, pp. 367–70; see Vézian 1956 for a variety of examples in one cave. For the Pech Merle bison, see Lorblanchet 1995, p. 6.
6. Bednarik 1986, pp. 44/5; see also Breuil 1926, p. 366.
7. Marshack 1977. For Gargas, see Barrière 1976; for Pech Merle, see *L'Art des Cavernes* pp. 470/71; and Lorblanchet 1992.
8. Bednarik 1986, pp. 43/4.
9. Delluc and Delluc 1983a; 1985, p. 60; Bednarik 1986.
10. Bednarik 1994; Clottes and Simonnet 1990, pp. 71–4.
11. Vertut 1980.
12. Delluc and Delluc 1983, pp. 58/9.
13. Bégouën and Clottes 1987, 1980.
14. Allain and Rigaud, in Leroi-Gourhan and Allain 1979, pp. 103, 106. For experiments in parietal engraving, see Lorblanchet *et al.* 1973; Lorblanchet 1995, p. 221; on parietal engraving in general, Féruglio in GRAPP 1993, pp. 265–74.
15. Martin 1989.
16. Bégouën 1925/6, p. 508; Bouillon, quoted in Ucko and Rosenfeld 1967, p. 241. The idea was first put forward by Max Bégouën, and used in his prehistoric novel *Les Bisons d'Argile* (Fayard: Paris, 1925).
17. d'Errico 1994.
18. Eastham and Eastham 1979, pp. 375/6. On the subject of light sources and visibility of engraving, see also Pales and de St Péreuse 1965, pp. 227/8.
19. Bégouën 1931.
20. Almagro 1981, 1980, 1976.
21. For Fontanet, see *L'Art des Cavernes* 1984, p. 435; for Montespan, see Rivenq 1976, 1984.
22. Breuil 1952a, p. 12; Beltrán *et al.* 1973.
23. Beltrán *et al.* 1967.
24. Beasley 1986; Bégouën, Clottes and Delporte 1977.
25. Bégouën, Casteret and Capitan 1923; for the clay-head theory, Garcia, cited in Chapa and Menéndez 1994, p. 222.
26. Roussot 1984.
27. Delporte 1973.
28. Roussot 1984.
29. Delluc and Delluc 1978, p. 389; a picture of the Cap Blanc tools can be seen, although with no indication of size, in White 1986, p. 135.
30. Roussot 1981; Laming-Emperaire 1962, p. 186. See also Simek 1986, p. 406. On Palaeolithic sculpture in general, see Barrière in GRAPP 1993, pp. 275–80.
31. Regnault 1906, p. 332.
32. Bader 1965, p. 28; Palchik, in Petrin 1992, pp. 163/4; Cârciumaru and Bitiri 1983. For a list of pigment analyses since 1950 (mostly at Lascaux and Altamira) see Couraud 1983, p. 108. See also Martin in GRAPP 1993, pp. 261–4.

33. Cartailhac and Breuil 1908, p. 28; Lemozi 1929, p. 147.

34. Martí 1977; Cabrera Garrido 1978.

35. For Niaux, see Brunet 1981, 1982; for Las Monedas, see Ripoll 1972, p. 53.

36. Demailly 1990, p. 109.

37. Judson 1959.

38. Couraud 1982, p. 4.

39. Couraud 1983, p. 107; 1983a, p. 6.

40. Couraud and Laming-Emperaire, in Leroi-Gourhan and Allain 1979, pp. 153–70.

41. Couraud 1982, 1988. On Arcy, see Couraud 1991.

42. Ballet et al. in Leroi-Gourhan and Allain 1979, pp. 171–4.

43. Vandiver 1983.

44. Lorblanchet 1990, 1993a, 1995; Lorblanchet et al. 1990.

45. Clottes, Menu and Walter 1990, 1990a; Clottes 1993; Menu et al. 1993.

46. Buisson et al. 1989.

47. Lorblanchet 1995, p. 147.

48. Palchik, in Petrin 1992, pp. 163/4.

49. Couraud 1982, p. 4 and in Leroi-Gourhan and Allain 1979, pp. 163–4. Couraud's experiments thus confirmed the opinions of Rottländer (1965), and have recently found further support in experiments by Vaquero (1995, pp. 72–5).

50. Cabrera Garrido 1978.

51. Pepe et al. 1991.

52. For Baume-Latrone see L'Art des Cavernes 1984, pp. 333–9, and Bednarik 1986, p. 36; for Pileta, see Dams 1978, pp. 86/7.

53. Marshack 1985, pp. 78,81; Lorblanchet 1990, 1993a, 1995.

54. Girard, Baffier and Chevalier 1995.

55. Perpère 1984, p. 43.

56. Couraud 1982, p. 4; and in Leroi-Gourhan and Allain 1979, p. 165. See also Vaquero 1995 pp. 92–3.

57. Leroi-Gourhan 1981, p. 30; 1982, p. 13; Vaquero (1995, pp. 122–6) believes that these Lascaux animals were sprayed, with hide cutouts, held against the wall by assistants, delineating the image.

58. Plenier 1971, pp. 69/70.

59. Marshack 1985, pp. 77–9.

60. Bohigas and Sarabia 1988; Moure Romanillo et al.1984/5; Almagro 1969.

61. Clot, Menu and Walter 1995.

62. Couraud 1982, p. 4; Barrière 1976, p. 75; Lorblanchet 1995, 1980, p. 34; Groenen 1988.

63. Pfeiffer 1982, pp. 241/2; Wildgoose et al. 1982; Vaquero (1995, p. 110) claims that spraying with tubes can produce a hand stencil in less than half a minute.

64. Leroi-Gourhan 1967. See also Moore 1977; Walsh 1979; Wright 1985.

65. Lorblanchet 1995, 1980, pp. 35, 37; Pfeiffer 1982, pp. 241/3; Groenen 1988.

66. Suères 1991; Couraud 1982, p. 4. The experiments reported by Wildgoose et al. produced somewhat equivocal results on this point and others.

67. Sahly 1963; Barrière and Sahly 1964. The medical view of the Gargas hands was presented earlier by Janssens 1957. For a review of incomplete hand stencils, see Pradel 1975; Clottes and Courtin 1994, pp.67–77. For the most recent studies arguing for missing fingers and against bent fingers, see Suères 1991; Barrière and Suères 1993.

68. Tardos 1987, 1993. It should be noted that, among the 55 hand stencils in the Spanish cave of Maltravieso, those which used to be thought to have the little finger missing (Almagro 1969) have now been shown by ultra-violet light to be complete hands with that finger painted out for some reason (Ripoll et al. 1997).

69. Lorblanchet 1980, pp. 37/8.

70. Lorblanchet 1991, 1995, pp. 209–23, 236–9. Also

Lorblanchet in GRAPP 1993, pp. 257–60.

71. Couraud 1982, p. 5.

72. Glory 1964; Couraud 1982, p. 5; and in Leroi-Gourhan and Allain 1979, p. 166.

73. Marshack 1985, p. 68.

74. J. Bégouën 1939, p. 284.

75. Altuna and Apellániz 1976.

76. Chauvet et al. 1995.

77. Plenier 1971, pp. 69/70; de Balbín Behrmann and Moure Romanillo 1982, pp. 62/3.

78. Lorblanchet 1995, p. 128.

79. Aujoulat 1985; see also debate in Bull. Soc. Préhist. Ariège-Pyrénées 47. 1992, pp. 95–106; and Aujoulat in GRAPP 1993, pp. 281–8.

80. Clottes and Simonnet 1990, p. 66.

81. Rouzaud 1978, pp. 126/7.

82. Breuil and Lantier 1959, p. 223.

83. Sauvet 1983, p. 52.

84. For Roucadour, see L'Art des Cavernes 1984, pp. 511–13; for Baume-Latrone, see Bednarik 1986, p. 34; for Chauvet, see Clottes in Chauvet et al. 1995, p. 90.

85. Delluc and Delluc, in Leroi-Gourhan and Allain 1979, pp. 175–84; Barrière and Sahly 1964, p. 179.

86. L'Art des Cavernes 1984, pp. 596–9.

87. de Beaune 1987, 1987a.

88. Roussot 1995, p. 321.

89. Girard, Baffier and Chevalier 1996.

90. de Beaune 1987, 1987a, 1984; Delluc and Delluc, in Leroi-Gourhan and Allain 1979, pp. 121–42; Ruspoli 1987, pp. 28–30; Vaquero 1995, chapter 2.

91. Roussot 1984, p. 76.

92. de Beaune 1987, 1987a, 1984.

92. Omnès 1982, pp. 105–7.

94. Delluc and Delluc 1991, pp. 89–92.

9. Art in the Open Air

1. Bahn 1992a, 1995.

2. Fortea 1981.

3. Jorge et al. 1981, 1982; Jorge 1987; Jorge et al. 1981/2.

4. Martín and Moure 1981; de Balbín and Moure 1988. For La Griega, see Sauvet 1983.

5. Ripoll and Municio 1992, 1994; Ripoll et al. 1994.

6. Sacchi 1987; Sacchi, Abelanet and Brulé 1987; Sacchi et al. 1988; Bahn 1985.

7. Martínez García 1986/7, 1992.

8. de Balbín et al. 1991, 1995; de Balbín and Alcolea 1994; de Balbín, Alcolea and Cruz 1995; de Balbín, Alcolea and Santonja 1995.

9. Clottes 1995.

10. Bahn 1995; Clottes 1995; Lorblanchet 1995, 1995a; Zilhão 1995.

11. Bednarik 1995a, 1995b; Dorn 1997; Phillips et al. 1997; Watchman 1995, 1996.

12. Bednarik 1995b; Zilhão 1995a, 1995b.

13. Villaverde 1994.

10. What Was Depicted?

1. Fortea et al. 1990, p. 240; Chauvet et al. 1995, pp. 111, 115.

2. Examples cited in Bahn 1985a, from pp. 166, 283/4 and 328 of Bandi et al. 1984.

3. See Rosenfeld 1984, and Macintosh 1977; Macintosh compared his identifications with the answers later given by an Aboriginal informant, and found that out of twenty-two items he had been wrong about fifteen, and only superficially right about the other seven!

4. Cited by Clottes 1989.

5. Examples given by Clottes 1989, and by Pales 1969, p. 52.

6. Rousseau adopted 'valable' and 'probable' categories in his 1967 study of felines, but Pales (1969, p. 52) disagreed with some of his assessments, finding him too ready to accept examples into both groups.

7. Pales 1969; Pales & de St Péreuse 1976, 1981, and 1989.

8. Delluc & Delluc 1991, p. 110.

9. Lorblanchet 1977.

10. Lorblanchet 1989, p. 118. On unidentifiable animals, see Clottes in GRAPP 1993, pp. 193–6; on imaginary animals, see Bégouën in GRAPP 1993, pp. 207–10.

11. Leroi-Gourhan 1983, p. 258.

12. For Pindal, see Leroi-Gourhan 1982, p. 49; 1983, p. 258; for Chauvet, Chauvet et al. 1995, p. 83; La Marche, Pales & de St Péreuse 1981, nos. 66/7, pls. 84–7; 1976, pl. 52; Lascaux & Pergouset, Lorblanchet 1995, pp. 127/8. On composite animals in general, see Bégouën in GRAPP 1993, pp. 201–5. Recently, Soubeyran (1995) has even suggested that the famous rhino in the Lascaux shaft (see p. 178) is a bison with rhino horns!

13. Rousseau 1967, p. 76; 1974; Pales 1969, p. 64.

14. Lhote 1968; Leroi-Gourhan 1958b, pp. 520/1; for examples of headless animals, see Leroi-Gourhan 1971, p. 471; and Ripoll 1972, pp. 56–60.

15. G., L. & R. Simonnet 1984, p. 29; also illustrated in Clottes 1989.

16. Bahn & Cole 1986, p. 144 and fig. 28.

17. Lorblanchet 1974, p. 70; 1977, p. 46; 1989, p. 115.

18. Bahn 1986, p. 117, and p. 107 (comment by Clottes); Davis 1986, p. 516.

19. Ucko & Layton 1984.

20. Rousseau 1984a; Barandiarán 1972, pp. 351–8.

21. Guthrie 1984, p. 38; Mithen (1988) prefers to see these features as teaching aids for hunters.

22. For Morin, see de Sonneville-Bordes 1986, p. 632; for the fawns, see Lorblanchet in Marshack 1985, pp. 62/3; for the Mas d'Azil bone disc, see Chollot-Varagnac 1980, p. 408 (many scholars, including Graziosi 1960, Barandiarán 1968, Leroi-Gourhan 1971 and Apellániz 1982, 1986, have wrongly attributed this piece to Laugerie-Basse).

23. See examples in Lhote 1968.

24. Baffier 1984, pp. 143. 150.

25. Baffier 1984.

26. Pales & de St Péreuse 1981, p. 87.

27. Clottes, Garner & Maury 1994; Fortea 1995, p. 270.

28. Clottes, Garner & Maury 1994, p. 69.

29. Leason 1939; Souriau 1971 likewise believes that the deer on the Lortet baton are stretched dead on the ground, one with head bent back, whereas most scholars have seen them as galloping and bellowing.

30. Riddell 1940.

31. Pales & de St Péreuse 1965, p. 229; 1981, p. 96; Pales 1969, p. 115.

32. Delluc & Delluc 1989.

33. Soubeyran 1991.

34. Guthrie 1984, p. 46; Riddell 1940, p. 160. On seasonal imagery, see also Dubourg 1994.

35. Vojkffy, quoted in Bandi 1968, p. 16; Bandi et al. 1984, p. 29.

36. Soubeyran 1991, p. 534.

37. Vojkffy, quoted in Giedion 1965, p. 193.

38. Bandi 1968, p. 17; Breuil & Obermaier 1935, p. 31 and pl. 27; Leroi-Gourhan 1984, p. 77; see also Jordá 1980, p. 282.

39. Baffier 1984, p. 153.

40. Freeman 1978; 1984, p. 222; 1987.

41. Leroi-Gourhan 1984, p. 76.

42. Azema 1992. On animation in general, see Crémades in GRAPP 1993, pp. 289–96.

43. Graziosi 1960, pp. 91/2 and pl. 90; Barandiarán 1971a; 1973, pp. 187/8 and pl. 45, no. 1; Breuil 1936/7, p. 6; Bahn 1982, pp. 257/8. Three of the pieces were also studied in Sieveking 1978. For the demolition of this claim, see Barandiarán 1993, 1994.

44. Delporte 1984, pp. 112–14; see also Caralp et al. 1973.

45. For a general survey of the fauna and its depictions, see Powers & Stringer 1975.

46. Leroi-Gourhan 1971, p. 463.

47. For example, Piette 1887.

48. Capitan, Breuil & Peyrony 1924; Bourdelle 1956; Blanchard 1964; for a summary of the problems, see Barandiarán 1972, p. 346; Pales & de St Péreuse 1981, p. 58; Prat 1986; and de Sonneville-Bordes & Laurent 1986/7, p. 72. On equids in Palaeolithic parietal art, see also Aujoulat in GRAPP 1993, pp. 97–108; and Clottes & Courtin 1994, pp. 87–98.

49. Lión 1971; Altuna & Apellániz 1978, pp. 107–9, 113–19; Bosinski & Fischer 1980; Pales & de St Péreuse 1981.

50. Pales & de St Péreuse 1981, pp. 26/7.

51. *Ibid.*, p. 38.

52. Leroi-Gourhan 1984, pp. 76–8.

53. Schmid 1984, p. 157.

54. Altuna 1975, pp. 112/13; Altuna & Apellániz 1978; see also Barandiarán 1972, pp. 365-8; and Bahn 1982a, p. 23.

55. Rousseau 1973; also 1984a, pp. 187–9; Altuna & Apellániz 1978; and Barandiarán 1972, p. 373.

56. Bahn 1982a, p. 21.

57. Bahn 1983a.

58. Arambourou 1978, p. 49; Bahn 1982a, pp. 22/3.

59. Piette 1906; Pales & de St Péreuse 1966, & 1981, pl. 1; Bahn 1978a; Jordá 1987. For Erberua & Oxocelhaya, see Larribau & Prudhomme 1989, p. 490.

60. Paillet 1993; Züchner 1975, p. 23; Barrière in GRAPP 1993, pp. 113–19; Clottes & Courtin 1994, pp. 99–104; for an earlier study of bison figures, see Capitan, Breuil & Peyrony 1910.

61. Pales & de St Péreuse 1981, pp. 63–106; 1965.

62. de Lumley 1968; for objections, see Pales & de St Péreuse 1981, pp. 75–8.

63. Pales & de St Péreuse 1981, p. 72.

64. Leroi-Gourhan 1984, pp. 77–80.

65. See Barrière in GRAPP 1993, pp. 109–13; Clottes & Courtin 1994, pp. 99–104.

66. It was primarily Breuil & Laming-Emperaire who said the small bovids were *Bos longifrons*; the sexual dimporphism solution was put forward by Zeuner (1953) and Koby (1954).

67. Koby 1968.

68. On deer in general, see Crémades in GRAPP 1993, pp. 137–50; Clottes & Courtin 1994, pp. 114–19.

69. Barandiarán 1969; Ripoll 1972, and 1984, p. 272; see also Olivié 1984/5 for cervids in Cantabrian art.

70. On La Colombière, see Faure 1978, pp. 47, 50, 76; for Covalanas and La Haza, see González Morales 1989, p. 278.

71. Vialou 1984, and in Leroi-Gourhan & Allain 1979.

72. Kehoe 1990, p. 189; and letter to T.F. Kehoe from Danish zoologist M. Meldgaard, 30 March 1987. Other possible examples of 'swimming' animals may include the deer on the Lortet baton, with fish between their legs and 'above' their backs; and some horses at Lascaux are said to hold their heads just as if they were crossing a river (according to Gunn, in Marshack 1979, p. 299).

73. Blanchard 1964a; Breuil 1957, p. 237.

74. Vialou 1984, p. 214.

75. Bouchud 1966.

76. Pales & de St Péreuse 1989.

77. Schmid, in Bandi et al. 1984, p. 90.

78. Leroi-Gourhan 1984, pp. 81, 84; Guthrie 1984, p. 45.

79. de Sonneville-Bordes 1986, p. 633.

80. Pales 1976/7; Pales & de St Péreuse 1981, pp. 107–40.

81. Pales 1976/7, p. 90/1; Pales & de St Péreuse 1981, pp. 128–31.

82. Pales 1976/7, pp. 91, 101/2; Pales & de St Péreuse 1981, pp. 132/3.

83. Pales & de St Péreuse 1981, pp. 137–9. For an uncritical review of ibex figures in Pyrenean art, see Brielle 1968. On ovcaprids, see also Sacchi in GRAPP 1993, pp. 123–36; Clottes & Courtin 1994, pp. 104–14.

84. Welté 1975/6; 1989.

85. Lister & Bahn 1994; de Spiegeleire 1985; Liubin 1991; Barrière in GRAPP 1993, pp. 151–6. For an earlier, less critical review of mammoths in art, see Berdin 1970.

86. Pales & de St Péreuse 1989.

87. Bosinski 1984; Bosinski & Fischer 1980.

88. Bosinski 1984, p. 301; Bosinski & Fischer 1980, pp. 38–40; Hahn 1984, p. 288; Praslov 1985, p. 186.

89. Ripoll 1984, p. 278; Bandi 1968, p. 15.

90. Barrière 1982.

91. Welté 1975/6, 1989; Leroi-Gourhan 1984, p. 83; for Pech Merle, see Lorblanchet 1977, pp. 54/5.

92. Chauvet et al. 1995, pp. 58, 110.

93. Jordá 1983.

94. Capitan et al. 1924; Breuil, Nougier & Robert 1956; Nougier & Robert 1965, 1966.

95. Rousseau 1967, 1974; Pales 1969; Altuna 1972, pp. 308–15; Barrière in GRAPP 1993, pp. 165–71.

96. Pales 1969, p. 102.

97. Chauvet et al. 1995, pp. 31, 34.

98. Leroi-Gourhan 1984, p. 84; Guthrie 1984, p. 44.

99. Klíma 1984, p. 326.

100. Pales 1969, pp. 53, 84. 102.

101. Mazak 1970; Rousseau 1974.

102. Chauvet et al. 1995, p. 34.

103. Pales 1969, pp. 58–60; Rousseau 1984a, p. 170. For Chauvet, see Chauvet et al. 1995, p. 104.

104. Nougier & Robert 1957a; Millán 1982; Barrière in GRAPP 1993, pp. 157–9. For Gönnersdorf, Bosinski 1996.

105. Chauvet et al. 1995, pp. 72/3, 93. 113.

106. *Ibid.*, pp. 98/9, 113.

107. For general studies of rare species, see Capitan et al. 1910, 1924; Novel 1986; various authors in GRAPP 1993, pp. 119–22 (musk-ox), 131–6 (chamois and saiga), 161–3 (suids), 189–90 (rabbits/hares). For other examples of studies (both critical and uncritical) of individual species, see Barandiarán 1974 (glutton); Nougier & Robert 1958 (saïga), and 1960 (canids).

108. Bosinski & Bosinski 1991.

109. Capitan et al. 1924; de Sonneville-Bordes & Laurent 1983; de Sonneville-Bordes 1986, pp. 639–41; Bahn 1977, pp. 253/4; 1982, p. 255; 1984.

110. Marshack 1970, 1972, 1975. For the whale theory, see Robineau 1984.

111. Fortea et al. 1990, p. 240.

112. Dams 1987, p. 17; 1987a, pp. 219–23; Sanchidrián 1994.

113. Clottes & Courtin 1994, pp. 129–35.

114. Breuil & de St Périer 1927; Lorblanchet in GRAPP 1993, pp. 181–8.

115. Bahn 1984, pp. 153, 157, 243–5, 276–8; Marshack 1970, 1972. For other features of fish depictions, see Barandiarán 1972, pp. 348–51.

116. Dams 1987; 1978, p. 85.

117. Clottes & Courtin 1994, pp. 135–7.

118. On reptiles, see Breuil & de St Périer 1927; Barrière in GRAPP 1993, pp. 191/2.

119. Alcalde, Breuil & Sierra 1911, ch. XVI; Vayson de Pradenne 1934; Breuil & Bégouën 1937; Lorblanchet 1974, pp. 85–96; Crémades 1994, and 1993 in GRAPP, pp. 173–80;

120. Buisson & Pinçon 1986/7, p. 79.

121. Chauvet et al. 1995, p. 49.

122. Clottes & Courtin 1994, pp. 127–9; d'Errico 1994a, 1994b. See also discussion in *Antiquity* 68, 1994, pp. 850–8, and 69, 1995, 1023–5.

123. Eastham 1988. For the more traditional view that these are people, and more specifically men, see Duhard

1991b, p. 139; 1996, pp. 111/12.

124. Bahn & Butlin 1990.

125. Tyldesley & Bahn 1983; Delcourt-Vlaeminck 1975; Marshack 1972, 1975.

126. Leroi-Gourhan 1971.

127. Moure 1988, p. 83.

128. Conkey 1981.

129. González Sainz 1988, pp. 48/9.

130. Lorblanchet 1995, p. 50; Moure 1988, p. 74.

131. Sauvet 1979, pp. 342/3; Leroi-Gourhan 1971, fig. 764; Vialou 1986.

132. Lorblanchet 1995, p. 50.

133. Nougier 1972, p. 265.

134. Chauvet et al. 1995; Clottes 1996a; Hahn 1986.

135. Roussot 1984b. For the statistical problems involved in making cumulative percentage frequency graphs to compare assemblages ranging from hundreds of figures to less than ten, see J. E. Kerrich & D. L. Clarke, 'Notes on the possible misuse and errors of cumulative percentage frequency graphs for the comparison of prehistoric artefact assemblages', in *Proc. Prehist. Soc.* 33, 1967, pp. 91–116.

136. Delporte 1984, p. 132; Capitan & Peyrony 1928, pp. 107–15.

137. Bosinski 1984, p. 320; Bosinski & Fischer 1980, pp. 56–60, 124–6.

138. Ucko & Rosenfeld 1972, p. 156. For an early study of humans in the art, see Capitan et al. 1924, pp. 91–116; for a more recent survey, Gaussen in GRAPP 1993, pp. 87–96.

139. Macintosh 1977, p. 191. See also Moore 1977; Walsh 1979; Wright 1985.

140. Pales & de St Péreuse 1976; 1981, p. 28; Airvaux & Pradel 1984.

141. Brodrick 1963, pp. 126/7.

142. Bahn 1986; Duhard 1993, 1996.

143. Duhard 1992, p. 153; 1996.

144. Pales & de St Péreuse 1976, p. 30; Airvaux & Pradel 1984.

145. Pales & de St Péreuse 1976, pp. 31–4.

146. Delporte 1993, p. 25.

147. Leroi-Gourhan 1971, fig. 777; see Pales & de St Péreuse 1976, p. 49; Ucko & Rosenfeld 1972, p. 194.

148. Duhard 1992a, 1996.

149. de Saint Mathurin 1973; 1975, pp. 25/6.

150. Pales & de St Péreuse 1976, p. 156.

151. Duhard 1993a; Pales & de St Péreuse 1976.

152. Duhard 1989, 1993.

153. Pales & de St Péreuse 1976, pp. 83–5; Ucko & Rosenfeld 1972, pp. 181/2. The abundance of humans with arms raised was noticed by Cartailhac & Breuil 1906, p. 52.

154. Pales & de St Péreuse 1976, p. 52.

155. *Ibid.*, pp. 28/9.

156. Out of a vast literature, see Passemard 1938; Narr 1960; Delporte 1993, 1995; Duhard 1993; Abramova 1995; Gvozdover 1989, 1995.

157. Pales 1972, p. 218.

158. Piette 1895, 1902; Ucko & Rosenfeld 1972, p. 161.

159. Duhard 1988.

160. Pales 1972, p. 220; Pales & de St Péreuse 1976, pp. 32, 91/2.

161. Rice 1981; see also Guthrie 1984, p. 62; the Rice paper is least useful when it compares its results to the supposed age structure of Upper Palaeolithic adult females, since it has lumped together figurines spanning the entire period.

162. Gvozdover 1989, pp. 53, 92; Duhard 1989, 1993.

163. Duhard 1988a; 1993; 1996, p. 162.

164. Duhard 1989, 1991, 1993; for a contrary opinion, Russell 1993.

165. Leroi-Gourhan 1971, fig. 52 bis; Pales 1972; Pales & de St Péreuse 1976, pp. 70–74. See also Delporte 1971,

1993; Duhard 1995.
166. McDermott 1996. In a recent experiment, Russell (pers. comm.) has tried to test this theory by carving figurines in sandstone, using as guides only the views available of her own body from a standing postion. Her experiences and results demolish the idea pretty conclusively.
167. Delporte 1962, p. 58.
168. de Saint Mathurin 1978, 1973.
169. Abramova 1995; 1984, p. 333; Shimkin 1978, p. 278; Berenguer 1986, p. 668.
170. Praslov 1985, pp. 182/3.
171. Gvozdover 1989, 1995.
172. Delporte 1993; Gamble 1982, p. 98.
173. Delluc 1981a, p. 72.
174. Breuil & Lantier 1959, p. 237; Pales & de St Péreuse 1976, p. 101.
175. Carcauzon 1984; Delluc & Delluc 1995.
176. Bosinski & Fischer 1974; Bosinski 1973, 1991.
177. Rosenfeld 1977; Lorblanchet & Welté 1987.
178. Bosinski & Fischer 1974, p. 45 and pl. 59.
179. Ibid., p. 94.
180. Breuil 1905; Breuil & de St Périer 1927; Graziosi 1960, p. 196 and pl. 291; Leroi-Gourhan 1958a, p. 388; 1971, p. 480, fig. 792.
181. Rosenfeld 1977, p. 101.
182. Ucko & Rosenfeld 1972, p. 168; Lorblanchet 1977. On claviforms, see also Stoliar 1977/8a, p. 64.
183. Abramova 1995; Praslov 1985, p. 190; Stoliar 1977/8, pp. 11/12; 1977/8a, pp. 51–9, 73; Shimkin 1978, p. 279. See also Rosenfeld 1977, p. 93.
184. Clottes in GRAPP 1993, pp. 197–9; Tymula 1995.
185. Rousseau 1967, pp. 158/9; Kehoe 1996. On masks, see also chapter in Cartailhac & Breuil 1906.
186. Giedion 1965, p. 374.
187. Ucko & Rosenfeld 1967; 1972, p. 205.
188. Bégouën & Breuil 1958; Rousseau 1967, p. 102.
189. Chauvet et al. 1995, p. 123.
190. Leroi-Gourhan 1983.
191. Pales & de St Péreuse 1976, p. 144.
192. Bosinski 1984, p. 318; Bosinski & Fischer 1974, 1980.
193. Conkey 1981, p. 24.
194. Barandiarán 1994, p. 58.
195. For examples, see Jordá 1979, 1985; on non-figurative markings in general, see Lorblanchet in GRAPP 1993, pp. 211–17, 235–41.
196. Desdemaines-Hugon 1993.
197. Lorblanchet 1977; Bahn 1986, p. 104.
198. See quotations from Forge, Munn and Warner, in Bahn 1986, pp. 104, 109. See also Speck & Schaeffer 1950.
199. And. Leroi-Gourhan 1980; Sauvet & Wlodarczyk 1977, pp. 548–50; Casado 1977.
200. Capdeville 1986; there is also a possible tectiform engraved on a bone tool from Altamira.
201. Clottes, Duport & Féruglio 1990, p. 37.
202. Leroi-Gourhan 1958, 1958a, 1968.
203. Casado 1977; Sauvet & Wlodarczyk 1977; Sauvet in GRAPP 1993, pp. 219–34.
204. Mons 1980/1; Chollot-Varagnac (1980) established twenty-five basic motifs.
205. Conkey 1980, 1981.
206. Sauvet & Wlodarczyk 1977, p. 553.
207. For example, see Piette 1905.
208. Forbes & Crowder 1979.
209. Sauvet & Wlodarczyk 1977, p. 552.
210. See, for example, Clegg 1993, p. 93.
211. Casado 1977, p. 251; Moure 1988, p. 74.
212. Leroi-Gourhan 1958, p. 318; 1971; Sauvet & Wlodarczyk 1977, p. 557.
213. Geoffroy 1974, p. 47.
214. Ibid., p. 57.
215. González García 1985, 1987; Geoffroy 1974, p. 55.

11. READING THE MESSAGES

1. For an account of Piette's view, see Pales 1969, p. 111. On de Mortillet, see Laming-Emperaire 1962, p. 70; Ucko & Rosenfeld 1967, p. 118; Bahn 1992; Richard 1993, p. 61.
2. Quoted in Laming-Emperaire 1962, p. 66; and Pales 1969, p. 112.
3. Halverson 1987; this paper also provides part of the history of the theory. See also Souriau 1971.
4. Chollot-Varagnac 1980, p. 448.
5. Lewis-Williams 1982, p. 429; Bahn 1987a.
6. Reinach 1903; see also Laming-Emperaire 1962, pp. 72–5.
7. Cartailhac & Breuil 1906 (see especially pp. 143 and 225).
8. Richard 1993, p. 66.
9. Breuil 1952; Bégouën 1929.
10. Giedion 1965, p. 57.
11. Lips 1949, pp. 84/5; Lindner 1950, pp. 53–67.
12. Obermaier 1918.
13. Graziosi 1960, p. 152. On the Montespan case and other fallacious evidence for hunting and magic, see Bahn 1991a.
14. Pales 1969, p. 47.
15. Leroi-Gourhan 1958a, p. 390; 1971, p. 30; Delluc & Delluc 1989; Baffier 1990, p. 179.
16. Clottes 1995a; Clottes & Courtin 1994, pp. 149/50.
17. Baffier 1990.
18. Faure 1978, p. 61.
19. Faure 1978.
20. Barandiarán 1984.
21. Clottes & Courtin 1994, pp. 154–61; Rousseau 1996.
22. Guthrie 1984, p. 51, states wrongly that 'portrayals of hunting scenes constitute the bulk of Palaeolithic large animal art'.
23. Guthrie 1984, p. 50.
24. Eaton 1978.
25. Parkington 1969, p. 12. See Rice & Paterson 1986, p. 665, for a similar suggestion.
26. For example, Breuil & Lantier 1959, pp. 237/8; Lindner 1950, p. 69; Marshack 1976b, pp. 71/2.
27. Lindner 1950, pp. 56/7.
28. Kehoe 1989, 1990, 1996.
29. Weissen-Szumlanska 1951, pp. 457/8.
30. Kehoe 1989, 1990, 1996.
31. Pales & de St Péreuse 1976, p. 78.
32. For example, Chollot-Legoux 1961.
33. Delluc 1985, 1981b; Breuil 1925.
34. Bahn 1986; Delluc 1985,
35. Bosinski 1973, p. 45; 1984, p. 315.
36. Klíma 1984, pp. 324–6.
37. Delporte 1984, p. 132.
38. de Sonneville-Bordes & Laurent 1986/7, pp. 69/70.
39. Delluc 1984c; 1981a.
40. Altuna 1983, 1994; Altuna & Apellániz 1978, pp. 106–7. See also Moure 1988, pp. 80–82.
41. Delporte 1984, p. 125.
42. Roussot 1984b, pp. 495/6.
43. Rice & Paterson 1985, 1986.
44. Baffier 1990; Duhard 1991a.
45. Mithen 1988, 1988a.
46. Bahn 1991b; Clark 1992.
47. Kirchner 1952, pp. 254, 260; Davenport & Jochim 1988.
48. See for example Lommel 1967.
49. Glory 1968, pp. 37/8, 57.
50. Smith 1992.
51. Clark 1966, pp. 12/13.
52. Clottes & Lewis-Williams 1996; Lewis-Williams 1991. For a full critique, see Bahn 1997a; Hamayon 1997.
53. Lewis-Williams 1982; Lewis-Williams & Dowson 1988.
54. Bednarik 1986; Davis 1986.
55. Lewis-Williams & Dowson 1988.
56. Comment by Bahn, in Lewis-Williams & Dowson 1988.
57. Lewis-Williams 1982; Lewis-Williams & Dowson 1988; Clottes & Lewis-Williams 1996.
58. Vieira 1996.
59. Bahn 1997, ch. 7; 1997a.
60. Ibid. and, amongst a huge literature, see for example Willcox 1990.
61. Mowaljarlai & Watchman 1989, p. 152.
62. Vialou 1986, p. 139.
63. Laming-Emperaire 1962, pp. 219/20 and pl. 13; Leroi-Gourhan 1958a, p. 396; 1966, p. 41; 1971, p. 387, figs. 491–3; Bouvier 1976.
64. Freeman 1978; 1984, p. 214; but see Breuil & Obermaier 1935, pp. 85/6, who wisely leave the second animal undetermined.
65. Nougier & Robert 1974.
66. For La Vache, Delporte 1993a, p. 135; for Angles, Iakovleva & Pinçon 1996.
67. J.R.B. Speed, Vet.MB, MRCVS, pers. comm.
68. Leroi-Gourhan 1966, p. 39; 1971, pp. 89, 98; Bandi 1968, p. 16; Baffier 1984, pp. 148/9.
69. Glory 1968, p. 57; Rousseau 1984a, p. 195.
70. Pales & de St Péreuse 1976, pp. 114–23. Other researchers deny that any such lines are visible.
71. Bégouën et al. 1982, 1984, 1984/5, pp. 66–70.
72. Cabré 1934.
73. L. Bégouën 1939, pp. 293/4; Bégouën & Clottes 1984; Bégouën et al. 1984/5, pp. 29–33. A further example of such wishful thinking is provided by a small limestone sculpture from Laussel which seems to be the gland of a phallus; Lalanne claimed in 1946 that it came from a carved erection, but in fact this cannot be deduced from a gland alone (see Duhard & Roussot 1988, p. 43).
74. Archambeau 1991, pp. 68/9.
75. St Périer 1936, p. 115.
76. Rice 1981; Gvozdover 1989; Duhard 1993.
77. Praslov 1985, p. 185; Gvozdover 1989; Delporte 1993; Duhard 1993.
78. Leroi-Gourhan 1971, p. 480, fig. 794; Bahn 1986, pp. 109, 119. For the childbirth theory, see Duhard 1993. Coppens (1989) sees the image as a symmetrical stylized female like Lespugue.
79. Delluc 1978, p. 239; Stoliar 1977/8a, p. 42; Bahn 1986, p. 99.
80. Bahn 1986.
81. Collins & Onians 1978; Delluc 1978, p. 353; and see Bahn 1986, p. 102.
82. Stoliar 1977/8a, p. 39.
83. For the phallus theory, see Bégouën & Breuil 1958, p. 98, and Vialou 1986, p. 198; for the horn theory, Ucko & Rosenfeld 1967, pp. 57 and 178; for the sculptor's view, Beasley 1986.
84. de Saint Mathurin 1978, p. 17.
85. Bahn 1986, pp. 107, 117; Delluc 1985, pp. 56/7, 61.
86. Kehoe 1991. For illustrations, see Jelínek 1975, pp. 406–10.
87. Guthrie 1984, pp. 62–6.
88. Collins & Onians 1978.
89. Guthrie 1984, pp. 70/1.
90. Guthrie (ibid.) made the comparison; it was adopted by Kurtén 1986, from whom the quotation is taken.
91. Bahn 1986, 1985a; Gimbutas 1981; Russell 1991, 1993a; Smith 1991.
92. Cartailhac 1902, p. 349.
93. Raphael 1986; see also Chesney 1991, 1994.
94. Laming-Emperaire 1962; for Leroi-Gourhan, see especially 1971, 1982. Two collections of his papers have been published in Spanish: 1984a has his major articles and bibliography from 1935 to 1983; 1984b has his lectures at the Collège de France from 1969 to 1983. A mixture of these

lectures and of published articles has also appeared in French (1992). A critical review of the development of his views on Palaeolithic art can be found in Meylan 1986, and a critique of his early work in Ucko & Rosenfeld 1967, pp. 195–221.

95. Leroi-Gourhan 1982, pp. 45–50.
96. Leroi-Gourhan 1958a, p. 395; 1971.
97. González Morales 1989, p. 275.
98. Leroi-Gourhan 1958, p. 321.
99. Leroi-Gourhan 1982; 1972, p. 285.
100. Leroi-Gourhan 1972, p. 307.
101. Jordá 1979, 1985.
102. Leroi-Gourhan 1971, pp. 86, 322; 1972, p. 308.
103. Parkington 1969, pp. 5/6; Ucko & Rosenfeld 1967; Bandi 1972, p. 315; Lhote 1972, p. 322.
104. Meylan 1986, pp. 45/6.
105. Leroi-Gourhan 1972, p. 291.
106. Leroi-Gourhan 1982, p. 50.
107. For example, Altuna & Apellániz 1976, p. 163; 1978; Jordá 1980; Vialou 1986; González García 1993, 1996.
108. Leroi-Gourhan 1972, p. 306; 1968.
109. Leroi-Gourhan 1972, p. 289.
110. Leroi-Gourhan 1971, p. 481.
111. Lhote 1968a; 1972, pp. 324–8; Lorblanchet 1994.
112. Leroi-Gourhan 1966, p. 44.
113. Leroi-Gourhan 1958a; 1968. On the La Madeleine 'phallus', see Bahn 1986 pp. 103, 114, 118.
114. Breuil & Lantier 1959, p. 237.
115. Leroi-Gourhan 1982, p. 58; see also Bahn 1978b, 1980.
116. Leroi-Gourhan 1971, p. 259; but see Ucko & Rosenfeld 1967, pp. 195–221; 1972, pp. 177, 193–97.
117. Leroi-Gourhan 1971, pp. 95/6; Pales & de St Péreuse 1976, pp. 49, 154.
118. Pales & de St Péreuse 1976, pp. 153–5.
119. *Ibid.*, p. 27; 1981, p. 28; for other examples of associations in Palaeolithic art, see Nougier & Robert 1968, Caralp et al. 1973.
120. Sauvet 1979, p. 347; 1988.
121. See p. 18 in Clottes 1990, vol. 2.
122. Sauvet 1979, p. 348; 1988; Sauvet & Wlodarczyk 1992, 1995.
123. Parkington 1969. For a different statistical test, see Ucko & Rosenfeld 1967, p. 208.
124. Stevens 1975, 1975a.
125. Laming-Emperaire 1962, pp. 115–23. On Raphael, see pp. 118/19; see also Raphael 1986.
126. Laming-Emperaire 1972, pp. 66/7.
127. Laming-Emperaire 1970; 1971; 1972, p. 70.
128. González García 1987; 1985, p. 490.
129. González García 1985, pp. 517–19.
130. González García 1993, 1996.
131. Freeman et al. 1987, pp. 222/3.
132. Clottes & Simonnet 1990, p. 68.
133. Fortea 1995, p. 261.
134. *L'Art des Cavernes* 1984, pp. 544–8.
135. Reznikoff 1987; Reznikoff & Dauvois 1988.
136. Waller 1993, 1993a.
137. Rivenq, pers. comm.; Bégouën 1984, p. 77.
138. Lorblanchet 1995, p. 8 (Pech Merle), 184 (Candamo).
139. González García 1985, pp. 457–63; Eastham 1979, pp. 370/1.
140. Nougier 1975; 1975a, p. 115.
141. Bahn 1978b, 1980.
142. Lorblanchet 1988, p. 277.
143. Jochim 1983. See also Barton, Clark & Cohen 1994 for a similar theory, based on the inevitably skewed database of surviving and discovered parietal art, in which demographic and social change lead to pressures favouring the expression of an assertive and emblemic style.
144. Bahn 1984; Mellars 1985, p. 283.
145. Sieveking 1979a, p. 106, supports the view; against

it, see Bahn 1977, p. 251; 1978b, p. 125; and Conkey 1983, pp. 206, 221.
146. Lorblanchet 1980; and comment in Marshack 1986, p. 64.
147. Nougier, e.g. 1972.
148. Pales & de St Péreuse 1965, p. 230; Pales 1969, pp. 110/11.
149. Apellániz 1982, pp. 38/9; Fritz & Simonnet 1996; Mons 1986/7, p. 91.
150. Moure 1988, p. 78.
151. Almagro 1976, p. 71; Bosinski 1973, p. 43; 1984, p. 274.
152. Breuil 1962, p. 356; for early views on Altamira, see Apellániz 1982, p. 48; 1983, p. 274; Cartailhac & Breuil 1906.
153. Leroi-Gourhan 1971, p. 29.
154. Pales & de St Péreuse 1981, p. 139; see also Baffier 1984, p. 153.
155. Apellániz 1980.
156. Apellániz 1982, pp. 45–63; 1983.
157. Apellániz 1982, pp. 34–8; 1986, pp. 50–52; 1990, 1991, 1992. It will be recalled that, like other authors before him, he attributes the Mas d'Azil disc wrongly to Laugerie-Basse, and that of Laugerie-Basse to the Abri du Souci.
158. Altuna & Apellániz 1976, pp. 148–53; 1978, p. 141; Apellániz 1982.
159. For Chimeneas, see Apellániz 1982, pp. 63–8; 1984. For Lascaux, see 1984a.
160. Altuna & Apellániz 1978, pp. 125–33; Apellániz 1982, pp. 71–92; *Arte Rupestre* 1987, p. 40.
161. Sauvet 1983, pp. 56/7; but see the reply by Apellániz in *Ars Praehistorica* 3/4, 1984/5, pp. 259–60.
162. Apellániz 1984, pp. 534–37; 1987.
163. Bahn 1982, 1984; Conkey 1980; Soffer 1985.
164. Breuil 1952, pp. 194/5; Leroi-Gourhan 1968, p. 69; Casado 1977, p. 16.
165. Leroi-Gourhan 1966, p. 47.
166. Eastham 1979, pp. 378–84; 1991.
167. For Limeuil see Capitan & Bouyssonie 1924; for Mezhirich, see Gladkih et al. 1984, p. 141; Shimkin 1978, pp. 274, 283.
168. Marshack 1979, pp. 287–92.
169. Dewez 1974.
170. Frolov 1977/9; 1981; and in Marshack 1979, pp. 605–7.
171. Marshack 1979, pp. 271, 309, 607; 1972a, p. 329.
172. Marshack 1970a, 1972, 1972a, 1975. See also review by Rosenfeld in *Antiquity* 45, 1971, pp. 317–19, and reply in 46, 1972, pp. 63–5.
173. Marshack 1995; for Taï, see 1991a.
174. d'Errico & Cacho 1994; see also d'Errico 1991, 1995; and *Rock Art Research* 9, 1992, pp. 37–64, and *Cambridge Arch. Journal* 6, 1996, pp. 99–117, for a debate between Marshack and d'Errico about methodology.
175. Couraud 1985; Bahn & Couraud 1984.
176. Frolov 1977/9; 1979; 1981. On the wax, see Marshack in Clottes 1990, vol. 2, p. 161; and in the 1991 edition of Marshack 1972, pp. 336/7.
177. Guthrie 1984, p. 47; Marshack 1970, 1972, 1975; but see Faure 1978, Bahn 1986.
178. Marshack 1977; 1979, p. 305.
179. Marshack 1985, pp. 70, 78–81.
180. Marshack 1969, 1972, 1985.
181. Lorblanchet 1980a, p. 476; and in Marshack 1985, pp. 62/3.
182. Leroi-Gourhan 1981, pp. 25, 32; Lorblanchet 1980a, pp. 474, 476.
183. Chauvet et al. 1995, p. 85.
184. Marshack 1984, 1985.
185. Collot et al. 1982.
186. Pfeiffer 1982.
187. *Ibid.*

188. Yates 1966; Pfeiffer 1982, pp. 210–25.
189. Sauvet 1979.
190. Conkey 1978, 1980, 1985. It should be noted, however, that Conkey (in Clottes 1990, vol. 2, p. 169) has expressed great scepticism about her conclusions concerning Palaeolithic portable art and agglomerations, because they were based on a very patchy corpus of material of uncertain date and from old excavations, and because the social model was not very developed.
191. Conkey 1980; but see note 190.
192. Gamble 1982, p. 104.
193. Duhard 1995, p. 311.
194. Conkey 1983, pp. 218–19; 1984, p. 264.
195. Hammond 1974.
196. Pericot 1962; Bahn 1986, pp. 110, 119.
197. Forbes & Crowder 1979, p. 362 and pl.
198. Atkinson 1992; Hamayon 1995.
199. Layton 1987.

BIBLIOGRAPHY

This is by no means intended to be an exhaustive list of references on the subject of Palaeolithic art: instead it concentrates on recent studies. Extensive bibliographies covering the older literature, and also concerning individual sites, can be found in: Breuil 1952; Zervos 1959; Graziosi 1960; Laming-Emperaire 1962; Leroi-Gourhan 1971; Naber *et al.* 1976; and *L'Art des Cavernes* 1984. The only guidebook available to the decorated caves in France, Spain and Italy which are open to the public is Nougier 1990.

ABRAMOVA, Z. A. 1984. Les corrélations entre l'art et la faune dans le Paléolithique de la Plaine russe (La femme et le mammouth), in Bandi *et al.* 1984, 333–42.
ABRAMOVA, Z. A. 1995. *L'Art Paléolithique d'Europe Orientale et de Sibérie.* Jérôme Millon: Grenoble.
ABSOLON, C. 1937. Les flûtes paléolithiques de l'Aurignacien et du Magdalénien de Moravie (analyse musicale et ethnologique comparative, avec démonstrations). 12e *Congrès Préhist. de France*, Toulouse/Foix 1936, 770–84.
ADAM, K. D. and KURZ, R. 1980. *Eiszeitkunst im süddeutschen Raum.* Konrad Theiss Verlag: Stuttgart.
AIKENS, C. M. and HIGUCHI, T. 1982. *Prehistory of Japan.* Academic Press: London/New York.
AIRVAUX, J. and PRADEL, L. 1984. Gravure d'une tête humaine de face dans le Magdalénien III de La Marche, commune de Lussac-les-Châteaux (Vienne). *Bull. Soc. Préhist. française* 81, 212–15.
AIRVAUX, J. *et al.* 1983. La plaquette gravée du Périgordien supérieur de l'abri Laraux, commune de Lussac-les-Châteaux (Vienne). Nouvelle lecture et comparaisons. *Bull. Soc. Préhist. française* 80, 235–46.
AIRVAUX, J. *et al.* 1991. Les techniques informatiques du traitement de l'image appliquées à l'étude des gravures paléolithiques. *Paléo* 3, 139–42.
ALCALDE DEL RIO, H., BREUIL, H. and SIERRA, L. 1911. *Les Cavernes de la Région Cantabrique.* Monaco.
ALLAIN, J. 1950. Un appeau magdalénien. *Bull. Soc. Préhist. française* 47, 181–92.
ALLAIN, J. and RIGAUD, A. 1986. Décor et fonction. Quelques exemples tirés du Magdalénien. *L'Anthropologie* 90, 713–38.
ALMAGRO BASCH, M. 1969. *Las Pinturas Rupestres de la cueva de Maltravieso en Cáceres.* Min. de Educación y Ciencia: Madrid.
ALMAGRO BASCH, M. 1976. Los omoplatos decorados de la cueva de 'El Castillo', Puente Viesgo (Santander). *Trabajos de Prehistoria* 33, 9–99, 12 pl.
ALMAGRO BASCH, M. 1980. Los grabados de trazo múltiple en el arte cuaternario español, in *Altamira Symposium*, 27–71. Min. de Cultura: Madrid.
ALMAGRO BASCH, M. 1981. La tecnica del grabado de trazos múltiples en el arte cuaternario español, in *Arte Paleolítico*, Comisión XI, Xth Congress UISPP, Mexico City, 1–52.
ALMAGRO BASCH, M., GARCIA GUINEA, M.A. and BERENGUER, M. 1972. La época de las pinturas y esculturas polícromas cuaternarias en relación con los yacimientos de las cuevas: revalorización del Magdaleniense III, in *Santander Symposium* 467–73.
ALTUNA, J. 1972. Fauna de mamíferos de los yacimientos prehistóricos de Guipúzcoa. *Munibe* 24, 464pp.
ALTUNA, J. 1975. *Lehen euskal herria – Guide illustré de préhistoire basque.* Mensajero: Bilbao.
ALTUNA, J. 1983. On the relationship between archaeofaunas and parietal art in the caves of the Cantabrian region, in *Animals and Archaeology, I: Hunters and their Prey* (J. Clutton-Brock and C. Grigson, eds.),

227–38. British Arch. Reports, Int. series 163, Oxford.
ALTUNA, J. 1994. La relación fauna consumida – fauna representada en el Paleolítico Superior Cantábrico. *Complutum* 5, 303–11.
ALTUNA, J. and APELLANIZ, J. M. 1976. Las figuras paleolíticas de la cueva de Altxerri (Guipúzcoa). *Munibe* 28, 3–242.
ALTUNA, J. and APELLANIZ, J. M. 1978. Las figuras rupestres de la cueva de Ekain (Deva). *Munibe* 30, 1–151.
ALTUNA, J., APELLANIZ, J-M. and BARANDIARAN, I. 1992. *Estudio de las Pinturas de Zubialde (Alava). Resumen de los resultados.* Depto. de Cultura, Diputación Foral de Alava: Vitoria.
ANATI, E. 1985. The rock art of Tanzania and the East African sequence. *Bollettino del Centro Camuno di Studi Preistorici* 23, 15–68.
ANATI, E. 1986. Etat de la recherche sur l'art rupestre: rapport mondial. *L'Anthropologie* 90, 783–800.
ANDRIEUX, C. 1974. Premiers résultats sur l'étude du climat de la salle des peintures de la Galerie Clastres (Niaux, Ariège). *Annales de Spéléologie* 29, 3–25.
APELLANIZ, J. M. 1980. El método de determinación de autor en el Cantábrico. Los grabadores de Llonín, in *Altamira Symposium*, 73–84. Min. de Cultura: Madrid.
APELLANIZ, J. M. 1982. *El Arte Prehistórico del Pais Vasco y sus Vecinos.* Desclée de Brouwer: Bilbao.
APELLANIZ, J. M. 1983. El autor de los bisontes tumbados del techo de los polícromos de Altamira, in *Homenaje al Prof. M. Almagro Basch*, vol. 1, 273–80. Min. de Cultura: Madrid.
APELLANIZ, J. M. 1984. La méthode de détermination d'auteur appliquée à l'art pariétal paléolithique. L'auteur des cervidés à silhouette noire de Las Chimeneas (Santander, Espagne). *L'Anthropologie* 88, 531–37.
APELLANIZ, J. M. 1984a. L'auteur des grands taureaux de Lascaux et ses successeurs. *Ibid.* 539–61.
APELLANIZ, J. M. 1986. Análisis de la variación formal y la autoría en la iconografía mueble del Magdaleniense Antiguo de Bolinkoba (Vizcaya). *Munibe* 38, 39–59.
APELLANIZ, J. M. 1987. Aplicación de técnicas estadísticas al análisis iconográfico y al método de determinación de autor. *Munibe* 39, 39–60.
APELLANIZ, J. M. 1990. Modèle d'analyse d'une école dans l'iconographie mobilière paléolithique: l'école des graveurs de chevaux hypertrophiés de La Madeleine, in Clottes 1990, vol. 2, 105–38.
APELLANIZ, J. M. 1991. *Modelo de Análisis de la Autoría en el Arte Figurativo del Paleolítico.* Cuadernos de Arqueología de Deusto 13, Univ. de Deusto: Bilbao.
APELLANIZ, J. M. 1992. Modèle d'analyse d'un auteur de représentations d'animaux de différentes espèces: le tube de Torre (Pays Basque, Espagne). *L'Anthropologie* 96, 453–72.
APELLANIZ, J. M. 1995. *El Análisis de la Autoría y la Autentificación de las Pinturas de Zubialde (Alava).* Cuadernos de Arqueología No. 15, Universidad de Deusto: Bilbao.
ARAMBOUROU, R. 1978. *Le Gisement Préhistorique de Duruthy à Sorde-l'Abbaye (Landes). Bilan des Recherches de 1958 à 1975.* Mémoire 13 de la Soc. Préhist. fr., 158pp.
ARAUJO, A. C. and LEJEUNE, M. 1995. *Gruta do Escoural: Necrópole Neolítica e Arte Rupestre Paleolítica.* Trabalhos de Arqueologia 8, Instituto Português do Património Arquitectónico e Arqueológico.

ARCHAMBEAU, M. and C. 1991. Les figurations humaines pariétales de la Grotte des Combarelles. *Gallia Préhistoire* 33, 53–81.
ARIAS CABAL, P. *et al.* 1996. La Garma. Un nuevo complejo arqueológico con arte rupestre en Cantabria. *Revista de Arqueología* XVII, No. 188, December, 8–17.
L'ART DES CAVERNES 1984. *Atlas des Grottes Ornées Paléolithiques françaises.* Min. de la Culture: Paris.
L'ART PREHISTORIQUE DES PYRENEES 1996. Exhibition catalogue. Réunion des Musées Nationaux: Paris.
ARTE RUPESTRE EN ESPANA 1987. *Revista de Arqueología*, numero extra.
ASCHERO, C. A. and PODESTA, M. M. 1986. El arte rupestre en asentamientos precerámicos de la Puna argentina. *Runa* (Buenos Aires) 16, 29–57.
ASLIN, G. D. and BEDNARIK, R. G. 1984a. Karlie-ngoinpool Cave: A preliminary report. *Rock Art Research* 1 (1), 36–45.
ASLIN, G. D. and BEDNARIK, R. G. 1984b. Koorine Cave, South Australia. *Rock Art Research* 1 (2), 142–4.
ASLIN, G. D., BEDNARIK, E. K. and BEDNARIK, R. G. 1985. The 'Parietal' Markings Project' – a progress report. *Rock Art Research* 2 (1), 71–74.
ASLIN, G. D. and BEDNARIK, R. G. 1985. Mooraa Cave – a preliminary report. *Rock Art Research* 2 (2), 160–65.
ATKINSON, J. M. 1992. Shamanisms today. *Annual Review of Anthropology* 21, 307–30.
ATLAS OF PRIMITIVE MAN IN CHINA 1980. Science Press: Beijing.
AUDOUIN, F. and PLISSON, H. 1982. Les ocres et leurs témoins au Paléolithique en France: enquête et expériences sur leur validité archéologique. *Cahiers du Centre de Recherches Préhistoriques* 8, 33–80.
AUJOULAT, N. 1985. Analyse d'une oeuvre pariétale anamorphosée. *Bull.Soc. Préhist. Ariège* 40, 185–93.
AUJOULAT, N. 1987. *Le Relevé des Oeuvres Pariétales Paléolithiques. Enregistrement et Traitement des Données.* Documents d'Archéologie française No. 9. Maison des Sciences de l'Homme: Paris.
AUJOULAT, N., ROUSSOT, A. and RIGAUD, J-P. 1984. Traces peu explicites et attributions douteuses ou erronées, in *L'Art des Cavernes*, 72–85. Min. de Culture: Paris.
AVELEYRA ARROYO DE ANDA, L. 1965. The Pleistocene carved bone from Tequixquiac, Mexico: A reappraisal. *American Antiquity* 30, 261–77.
AZEMA, M. 1992. La représentation du mouvement dans l'art animalier paléolithique des Pyrénées. *Bull. Soc. Préhistorique Ariège-Pyrénées* 47, 19–76.

BACHECHI, L., FABBRI, P-F. and MALLEGNI, F. 1997. An arrow-caused lesion in a Late Upper Palaeolithic human pelvis. *Current Anthropology* 38, 135–40.
BADER, O. N. 1965. *La Caverne Kapovaïa, Peinture Paléolithique.* Nauka: Moscow.
BADER, O. N. 1967. Die paläolithischen Höhlenmalereien in Osteuropa. *Quartär* 18, 127–38.
BADER, O. N. 1976. Upper Palaeolithic Burials and the Grave at the Sungir Site. *Sovetskaia Arkheologiia* 3, 1967, 142–59, translated for the US Geological Survey.
BAFFIER, D. 1984. Les caractères sexuels secondaires des mammifères dans l'art pariétal paléolithique franco-cantabrique, in Bandi *et al.* 1984, 143–54.
BAFFIER, D. 1990. Lecture technologique des représentations paléolithiques liées à la chasse et au gibier. *Paléo* 2, 177–90.

BAHN, P. G. 1977. Seasonal migration in S.W. France during the late glacial period. *Journal of Arch. Science* 4, 245–57.

BAHN, P. G. 1978. Palaeolithic pottery, the history of an anomaly. *Anthropos* (Athens), 5, 98–110.

BAHN, P. G. 1978a. The 'unacceptable face' of the West European Upper Palaeolithic. *Antiquity* 52, 183–92, 1 pl.

BAHN, P. G. 1978b. Water mythology and the distribution of Palaeolithic parietal art. *Proc. Prehist. Society* 44, 125–34.

BAHN, P. G. 1980. 'Histoire d'Eau': L'art pariétal préhistorique des Pyrénées. *Travaux de l'Inst. d'Art Préhist. Toulouse* 22, 129–35.

BAHN, P. G. 1982. Inter-site and inter-regional links during the Upper Palaeolithic: the Pyrenean evidence. *The Oxford Journal of Archaeology* 1, 247–68.

BAHN, P. G. 1982a. Homme et cheval dans le Quaternaire des Pays de l'Adour, in *Les Pays de l'Adour, Royaume du Cheval*, Guide-Catalogue, Musée Pyrénéen, Lourdes, 21–6.

BAHN, P. G. 1983. A Palaeolithic treasure-house in the Pyrenees. *Nature* 302, 571–72.

BAHN, P. G. 1983a. New finds at Pincevent. *Nature* 304, 682–3.

BAHN, P. G. 1984. *Pyrenean Prehistory*. Aris & Phillips: Warminster.

BAHN, P. G. 1984a. How to spot a fake azilian pebble. *Nature* 308, p. 229.

BAHN, P. G. 1985. Ice Age drawings on open rock faces in the Pyrenees. *Nature* 313, 530–31.

BAHN, P. G. 1985a. Review of Bandi *et al*. 1984. *Antiquity* 59, 57–8.

BAHN, P. G. 1986. No sex, please, we're Aurignacians. *Rock Art Research* 3, 99–120.

BAHN, P. G. 1987. A la recherche de l'iconographie paléolithique hors de l'Europe. *Travaux de l'Institut d'Art Préhist. de Toulouse* 29, 7–18.

BAHN, P. G. 1987a. Comment on article by J. Halverson. *Current Anthropology* 28, 72–3.

BAHN, P. G. 1987b. Excavation of a Palaeolithic plank from Japan. *Nature* 329, p. 110.

BAHN, P. G. 1990. Familiar fakes and unfamiliar faces. *European Cultural Heritage Newsletter on Research* 4 (1), Feb., 51–3.

BAHN, P. G. 1991. Pleistocene images outside Europe. *Proceedings of the Prehistoric Society* 57, 91–102.

BAHN, P. G. 1991a. Where's the beef? The myth of hunting magic in Palaeolithic art, in *Rock Art and Prehistory* (P. G. Bahn and A. Rosenfeld, eds), 1–13. Oxbow Monograph 10: Oxford.

BAHN, P. G. 1991b. Review of 'Thoughtful Foragers' by S. J. Mithen. *Antiquity* 65, 158–62.

BAHN, P. G. 1992. Expecting the Spanish Inquisition: Altamira's rejection in its 19th century context, in *Ancient Images, Ancient Thought: The Archaeology of Ideology*, (S. Goldsmith, S. Garvie, D. Selin and J. Smith, eds.), Proceedings of the 23rd Annual Chacmool Conference, Calgary 1990, 339–46. Archaeological Association, University of Calgary.

BAHN, P. G. 1992a. Open air rock art in the Palaeolithic, in *Rock Art in the Old World* (M. Lorblanchet, ed.), 395–400. Indira Gandhi National Centre for the Arts, New Delhi.

BAHN, P. G. 1993. The 'dead wood stage' of prehistoric art studies: style is not enough, in *Rock Art Studies: The Post-Stylistic Era or Where do we go from here?* (M. Lorblanchet and P. G. Bahn, eds), 51–9. Oxbow Monograph 35: Oxford.

BAHN, P. G. 1994. Lascaux: Composition or accumulation? *Zephyrus* 47, 3–13.

BAHN, P. G. 1995. Cave art without the caves. *Antiquity* 69, 231–7.

BAHN, P. G. 1997. *The Cambridge Illustrated History of Prehistoric Art*. Cambridge University Press: Cambridge.

BAHN, P. G. 1997a. Membrane and numb brain: a close look at a recent claim for shamanism in Palaeolithic art. *Rock Art Research* 14 (1).

BAHN, P. G. and BUTLIN, R. K. 1990. Les insectes dans l'art paléolithique. Quelques observations nouvelles sur la sauterelle d'Enlène (Ariège), in *L'Art des Objets au Paleolithique (Actes du Colloque International d'Art Mobilier Paléolithique), Foix/Le Mas d'Azil*, November 1987, Tome 1, 247–53.

BAHN, P. G. and COLE, G. 1986. La préhistoire pyrénéenne aux Etats-Unis. *Bull. Soc. Préh. Ariège-Pyrénées* 41, 95–149.

BAHN, P. G. and COURAUD, C. 1984. Azilian pebbles – an unsolved mystery. *Endeavour* 8, 156–8.

BAHN, P. G. and OTTE, M. 1985. La poterie 'paléolithique' de Belgique: analyses récentes. *Helinium* 25, 238–41.

de BALBIN BEHRMANN, R. and ALCOLEA GONZALEZ, J. J. 1994. Arte paleolítico de la meseta española. *Complutum* 5, 97–138.

de BALBIN BEHRMANN, R., ALCOLEA GONZALEZ, J. J. and CRUZ NAIMI, L. A. 1995. El yacimiento rupestre paleolítico al aire libre de Siega Verde (Salamanca, España): una visión de conjunto. *Actas VII, 11 Congresso de Arqueol. Peninsular. Trabalhos de Antropologia e Etnologia* (Porto) 35, 73–102.

de BALBIN BEHRMANN, R., ALCOLEA GONZALEZ, J. J. and SANTONJA GOMEZ, M. 1995. Siega Verde. Un art rupestre paléolithique à l'air libre dans la vallée du Douro. *Dossier d'Archéologie* 209 (Dec. 1995/Jan 1996), 98–105.

de BALBIN BEHRMANN, R. and MOURE ROMANILLO, J. A. 1982. El panel principal de la cueva de Tito Bustillo (Ribadesella, Asturias). *Ars Praehistorica* 1, 47–97.

de BALBIN BEHRMANN, R. and MOURE ROMANILLO, J. A. 1988. El arte rupestre en Domingo García (Segovia). *Revista de Arqueología* 87, 16–24.

de BALBIN BEHRMANN, R. *et al* 1991. Siega Verde (Salamanca). Yacimiento artístico paleolítico al aire libre, 33–48 in *Del Paleolítico a la Historia*. Museo de Salamanca: Salamanca.

BALDEON, A. 1993. El yacimiento de Lezetxiki (Gipuzkoa, País Vasco). Los niveles musterienses. *Munibe* 45, 3–97.

BANDI, H.-G. 1968. Art quaternaire et zoologie, in *Simposio de Arte Rupestre*, Barcelona 1966, 13–19.

BANDI, H.-G. 1972. Quelques réflexions sur la nouvelle hypothèse de A. Leroi-Gourhan concernant la signification de l'art quaternaire, in *Santander Symposium*, 309–19.

BANDI, H.-G. 1988. Mise bas et non défécation. Nouvelle interprétation de trois propulseurs magdaléniens sur des bases zoologiques, éthologiques et symboliques. *Espacio, Tiempo y Forma* 1, 133–47.

BANDI, H.-G. and DELPORTE, H. 1984. Propulseurs magdaléniens décorés en France et en Suisse, in *Eléments de Pré et Protohistoire européenne, Hommages à J-P. Millotte*, Annales Litt. de l'Univ. de Besançon, 203–17.

BANDI, H.-G. *et al.* (eds.) 1984. *La Contribution de la Zoologie et de l'Ethologie à l'interprétation de l'art des peuples chasseurs préhistoriques*. 3e colloque de la Soc. suisse des Sciences Humaines, Sigriswil 1979. Editions Universitaires: Fribourg.

BARANDIARAN, I. 1968. Rodetes paleolíticos de hueso. *Ampurias* 30, 1–37.

BARANDIARAN, I. 1969. Representaciones de reno en el arte paleolítico español. *Pyrenae* 5, 1–33.

BARANDIARAN, I. 1971. Hueso con grabados pale-

olíticos en Torre (Oyarzun, Guipúzcoa). *Munibe* 23, 37–69.

BARANDIARAN, I. 1971a. 'Bramaderas' en el Paleolítico superior peninsular. *Pyrenae* 7, 7–18.

BARANDIARAN, I. 1972. Algunas convenciones de representación en las figuras animales del arte paleolítico, in *Santander Symposium* 345–84.

BARANDIARAN, I. 1973. *Arte Mueble del Paleolítico Cantábrico*. Monog. Arq. 14: Zaragoza.

BARANDIARAN, I. 1974. El Glotón (*Gulo gulo* L.) en el arte paleolítico. *Zephyrus* 25, 177–96.

BARANDIARAN, I. 1984. Signos asociados a hocicos de animales en el arte paleolítico. *Veleia* 1, 7–24.

BARANDIARAN, I. 1988. Datation C14 de l'art mobilier magdalénien cantabrique. *Bull. Soc. Préhist. Ariège-Pyrénées* 43, 63–84.

BARANDIARAN, I. 1993. El lobo feroz: la vacuidad de un cuento magdaleniense. *Veleia* 10, 7–37.

BARANDIARAN, I. 1994. Arte mueble paleolítico cantábrico: una visión de síntesis. *Complutum* 5, 45–79.

BARANDIARAN, I. 1995 La datación de la gráfica rupestre de apariencia paleolítica: un siglo de conjeturas y datos. *Veleia* 12, 7–48.

de BARANDIARAN, J-M. 1980. Excavaciones en Axlor (Campaña de 1969), in *Obras Completas* XVII, 129–384.

BARRIERE, C. 1976. *L'Art Pariétal de la Grotte de Gargas*. Mémoire III de l'Inst. Art Préhist. Toulouse; British Arch. Reports (Oxford), Int. series No. 14, 2 vols.

BARRIERE, C. 1982. *L'Art Pariétal de Rouffignac*. Picard: Paris.

BARRIERE, C. and SAHLY, A. 1964. Les empreintes humaines de Lascaux, in *Miscelánea en Homenaje al Abate H. Breuil*, vol. 1, 173–80. Barcelona.

BARRIERE, C. and SUERES, C. 1993. Les mains de Gargas, in *La Main dans la Préhistoire, Les Dossiers d'Arch.*, Jan., 178, 46–54.

BARTON, C. M., CLARK, G. A. and COHEN, A. E. 1994. Art as information: explaining Upper Palaeolithic art in western Europe. *World Arch.* 26, 185–207.

BASLER, D. 1979. Le Paléolithique final en Herzégovine. in *La Fin des Temps Glaciaires en Europe* (D. de Sonneville-Bordes, ed.), vol. 1, 345–55. C. N. R. S.: Paris.

BAULOIS, A. 1980. Les sagaies décorées du Paléolithique supérieur dans la zone franco-cantabrique. *Bull. Soc. Préhist. Ariège* 35, 125–28.

BEASLEY, B. 1986. Les bisons d'argile de la grotte du Tuc d'Audoubert. *Bull. Soc. Préhist. Ariège-Pyrénées* 41, 23–30.

BEAUMONT, P. B. 1973. Border Cave – a progress report. *South African Journal of Science* 69, 41–6.

de BEAUNE, S. A. 1984. Comment s'éclairaient les hommes préhistoriques? *La Recherche* 15, No. 152, 247–9.

de BEAUNE, S. A. 1987. *Lampes et Godets au Paléolithique*. XXIIIe Supplément à Gallia Préhistoire.

de BEAUNE, S. A. 1987a. Palaeolithic lamps and their specialization: a hypothesis. *Current Anth.* 28, 569–77.

BEDNARIK, R. G. 1986. Parietal finger markings in Europe and Australia. *Rock Art Research* 3, 30–61 and 159–70.

BEDNARIK, R. G. 1989. On the Pleistocene settlement of South America. *Antiquity* 63, 101–11.

BEDNARIK, R. G. 1991. On natural cave markings. *Helictite* 29 (2), 27–41.

BEDNARIK, R. G. 1992. Natural line markings on palaeolithic objects. *Anthropologie* 30 (3), 233–40.

BEDNARIK, R. G. 1992a. Palaeoart and archaeological myths. *Cambridge Arch. Journal* 2, 27–57.

BEDNARIK, R. G. 1993. Palaeolithic art in India. *Man*

and Environment 18 (2), 33–40.

BEDNARIK, R. G. 1994. The discrimination of rock markings. *Rock Art Research* 11, 23– 44.

BEDNARIK, R. G. 1994a. The Pleistocene art of Asia. *Journal of World Prehistory* 8 (4), 351–75.

BEDNARIK, R. G. 1994b. Conceptual pitfalls in Palaeolithic rock art dating. *Préhistoire Anthropologie Méditerranéennes* 3, 95–102.

BEDNARIK, R. G. 1995. Concept-mediated marking in the Lower Palaeolithic. *Current Anthropology* 36, 605–34.

BEDNARIK, R. G. 1995a. The age of the Côa Valley petroglyphs in Portugal. *Rock Art Research* 12, 86–103.

BEDNARIK, R. G. 1995b. The Côa petroglyphs: an obituary to the stylistic dating of palaeolithic rock-art. *Antiquity* 69, 877–83.

BEDNARIK, R. G. 1995c. Towards a better understanding of the origins of body decoration. *Anthropologie* (Brno) 33 (3), 201–12.

BEDNARIK, R. G. 1996. Palaeolithic rock art in the Alps? *The Artefact* 19, 97–8.

BEDNARIK, R. G. and YOU, Y. 1991. Palaeolithic art from China. *Rock Art Research* 8, 119–23.

BEGOUEN, H. 1925/6. Observations nouvelles dans les grottes des Pyrénées, in *Mélanges Gorjanovitch-Kramberger*, 501–9. Zagreb.

BEGOUEN, H. 1929. The magic origin of prehistoric art. *Antiquity* 3, 5–19.

BEGOUEN, H. 1931. La technique des gravures pariétales de quelques cavernes pyrénéennes. *XVe Congrès Int. d'Anth. et d'Arch. Préhistorique*, Portugal 1930, 8pp., 13 pl.

BEGOUEN, H. 1932. La Préhistoire à la Société Archéologique du Midi de la France. *Mémoires de la Soc. Arch. du Midi de la France*, 1–11.

BEGOUEN, H. 1942. *De la Lecture des Gravures Préhistoriques. Conseils à mes Etudiants*. Editions du Museum: Toulouse. 10pp.

BEGOUEN, H. 1947. Eloge de M. Emile Cartailhac. *Bull. Soc. Arch. du Midi de la France* 3e série, 5 (1942–5), 438–51.

BEGOUEN, H. 1952. Séance du 24 juillet. *Bull. Soc. Préh. fr.* 49, 289–91.

BEGOUEN, H. and L. 1934. Quelques plaquettes de pierre gravées ou peintes des cavernes pyrénéennes. *11e Congrès Préhist. de France*, Périgueux, 3pp.

BEGOUEN, H., BEGOUEN, L. and VALLOIS, H. 1937. Une pendeloque faite d'un fragment de mandibule humaine (Epoque magdalénienne). *12e Congrès Préhist. de France*, Toulouse-Foix 1936, 559–64.

BEGOUEN, H. and BREUIL, H. 1958. *Les Cavernes du Volp: Trois-Frères —Tuc d'Audoubert*. Arts et Métiers Graphiques: Paris.

BEGOUEN, H., CASTERET, N. and CAPITAN, L. 1923. La caverne de Montespan (Haute-Garonne). *Revue Anth.* 33, 1–18, 1 pl.

BEGOUEN, J. 1939. De quelques signes gravés et peints des grottes de Montesquieu-Avantès, in *Mélanges Bégouën*, 281–7. Toulouse.

BEGOUEN, L. 1939. Pierres gravées et peintes de l'époque magdalénienne, in *Mélanges Bégouën*, 289–305. Toulouse.

BEGOUEN, R. 1980. La conservation des cavernes du Volp. Son histoire, son bilan. in *Altamira Symposium*, 681–93. Min. de Cultura: Madrid.

BEGOUEN, R. 1984. Les bisons d'argile du Tuc d'Audoubert. *Dossiers de l'Archéologie* 87, octobre, 'Les Premiers Artistes', 77–9.

BEGOUEN, R. and CLOTTES, J. 1979. Le bâton au saumon d'Enlène. *Bull. Soc. Préhist. Ariège* 34, 17–25.

BEGOUEN, R. and CLOTTES, J. 1979a. Galet gravé de la caverne d'Enlène à Montesquieu-Avantès (Ariège). *Caesaraugusta* 49/50, 57–64.

BEGOUEN, R. and CLOTTES, J. 1980. Apports mobiliers dans les cavernes du Volp (Enlène, Les Trois-Frères, Le Tuc d'Audoubert), in *Altamira Symposium*, 157–88. Min. de Cultura: Madrid.

BEGOUEN, R. and CLOTTES, J. 1981. Nouvelles fouilles dans la Salle des Morts de la Caverne d'Enlène à Montesquieu-Avantès (Ariège), in *21e Congrès Préhist. de France*, Montauban/Cahors 1979, vol. 1, 33–56.

BEGOUEN, R. and CLOTTES, J. 1983. El arte mobiliar de las cavernas del Volp (en Montesquieu-Avantès/Ariège). *Revista de Arqueologia* Año IV, No. 27, 6–17.

BEGOUEN, R. and CLOTTES, J. 1984. Un cas d'érotisme préhistorique. *La Recherche* 15, 992–5.

BEGOUEN, R. and CLOTTES, J. 1985. L'art mobilier des Magdaléniens. *Archéologia* 207, novembre, 40–49.

BEGOUEN, R. and CLOTTES, J. 1986/7. Le grand félin des Trois-Frères. *Antiquités Nationales* 18/19, 109–13.

BEGOUEN, R. and CLOTTES, J. 1987. Les Trois-Frères after Breuil. *Antiquity* 61, 180–87.

BEGOUEN, R., CLOTTES, J. and DELPORTE, H. 1977. Le retour du petit bison au Tuc d'Audoubert. *Bull. Soc. Préhist. fr.* 74, 112–20.

BEGOUEN, R. *et al.* 1982. Plaquette gravée d'Enlène, Montesquieu-Avantès (Ariège). *Ibid.* 79, 103–12.

BEGOUEN, R. *et al.* 1984. Compléments à la grande plaquette gravée d'Enlène. *Ibid.* 81, 142–8.

BEGOUEN, R. *et al.* 1984/5. Art mobilier sur support lithique d'Enlène (Montesquieu- Avantès, Ariège), Collection Bégouën du Musée de l'Homme. *Ars Praehistorica* III/IV, 25–80.

BEGOUEN, R. *et al* 1987. Un bison sculpté en grès à Enlène. *Bull. Soc. Mérid. de Préhist. Spél.* 27, 23–7.

BELFER-COHEN, A. and BAR-YOSEF, O. 1981. The Aurignacian at Hayonim Cave. *Paléorient* 7 (2), 19–42.

BELLIER, C. 1981/2. *Contribution à l'étude de l'art paléolithique en Europe occidentale, le contour découpé en os*. Mémoire de licence en histoire de l'art et arch., Univ. libre de Bruxelles, 2 vols.

BELLIER, C. 1984. Contribution à l'étude de l'industrie préhistorique: les contours découpés du type 'têtes d'herbivores'. *Bull. Soc. royale belge Anth. Préhist.* 95, 21– 34.

BELTRAN, A., GAILLI, R. and ROBERT, R. 1973. *La Cueva de Niaux*. Monograf. Arq. No. 16, Zaragoza.

BELTRAN, A., ROBERT, R. and GAILLI, R. 1967. *La Cueva de Bédeilhac*. Monograf. Arq. No. 2, Zaragoza.

BERDIN, M.O. 1970. La répartition des mammouths dans l'art pariétal quaternaire. *Travaux Inst. Art Préhist. Toulouse* 12, 181–367.

BERENGUER ALONSO, M. 1986. Art pariétal paléolithique occidental. Techniques d'expression et identification chronologique. *L'Anthropologie* 90, 665–77.

BIBIKOV, S. N. 1981. *The Oldest Musical Complex made of mammoth bones*. Naukova Dumka: Kiev. (In Russian).(see also 1975, A Stone Age orchestra. *UNESCO Courier*, June, 8–15).

BISSON, M. S. and BOLDUC, P. 1994. Previously undescribed figurines from the Grimaldi Caves. *Current Anthropology* 35, 458–68.

BLANCHARD, J. 1964. Informations recherchées d'après les équidés européens figurés, in *Prehistoric Art of the Western Mediterranean and the Sahara* (L. Pericot and E. Ripoll, eds.), 3–34. Viking Fund Publications in Anth. No. 39, New York.

BLANCHARD, J. 1964a. Sélection intentionnelle des belles têtes de cerfs gravées et peintes, in *Miscelánea en Homenaje al abate H. Breuil*, vol. 1, 249–58. Barcelona.

BOHIGAS ROLDAN, R. and SARABIA ROGINA, P. 1988. Nouvelles découvertes d'art paléolithique dans la région cantabrique. La 'Fuente del Salín' à

Muñorrodero. *L'Anthropologie* 92, 133–7.

BORDES, F. 1952. Sur l'usage probable de la peinture corporelle dans certaines tribus moustériennes. *Bull. Soc. Préhist. française* 49, 169–71.

BORDES, F. 1969. Os percé moustérien et os gravé acheuléen du Pech de l'Azé II. *Quaternaria* 11, 1–6.

BOSINSKI, G. 1973. Le site Magdalénien de Gönnersdorf (Commune de Neuwied, Vallée du Rhin Moyen, RFA). *Bull. Soc. Préhist. Ariège* 28, 25–48.

BOSINSKI, G. 1982. *Die Kunst der Eiszeit in Deutschland und in der Schweiz*. Habelt: Bonn.

BOSINSKI, G. 1984. The mammoth engravings of the Magdalenian site Gönnersdorf (Rhineland, Germany), in Bandi *et al.* 1984, 295–322.

BOSINSKI, G. 1991. The representation of female figures in the Rhineland Magdalenian. *Proceedings of the Prehistoric Society* 57, 51–64.

BOSINSKI, G. 1996 Die Nashorndarstellungen von Gönnersdorf, in *Spuren der Jagd – Die Jagd nach Spuren* (I. Campen *et al.*, eds), 177–89. Tübinger Monographien zur Urgeschichte 11.

BOSINSKI, G. and BOSINSKI, H. 1991. Robbendarstellungen von Gönnersdorf. *Sonderveröffentlichungen, Geologisches Institut der Universität zu Köln* 82 (Festschrift Karl Brunnacker), 81–87.

BOSINSKI, G. and FISCHER, G. 1974. *Die Menschendarstellungen von Gönnersdorf der Ausgrabungen von 1968*. Steiner: Wiesbaden.

BOSINSKI, G. and FISCHER, G. 1980. *Mammut-und Pferdedarstellungen von Gönnersdorf*. Steiner: Wiesbaden.

BOUCHUD, J. 1966. *Essai sur le Renne et la Climatologie du Paléolithique Moyen et Supérieur*. Magne: Périgueux.

BOUET, B. *et al.* 1986/7. Le Centre d'Information et de Documentation (C.I.D.) H. Breuil. *Antiquités Nationales* 18/19, 9–15.

BOURDELLE, E. 1956. Les parentés morphologiques des équidés caballins d'après les gravures rupestres du Sud-Ouest de la France. *Mammalia* 20, 22–3.

BOUVIER, J-M. 1976. La Chaire à Calvin, Mouthiers (Charente). Données et problèmes, in *Sud-Ouest (Aquitaine et Charente)*, Livret-Guide de l'Excursion A4, U.I.S.P.P. Nice, 133–6.

BOUVIER, J-M. 1977. *Un Gisement Préhistorique: La Madeleine*. Fanlac: Périgueux.

BOYLE, M. *et al.* 1963. Recollections of the Abbé Breuil. *Antiquity* 37, 12–18.

BREUIL, H. 1905. La dégénérescence des figures d'animaux en motifs ornementaux à l'époque du Renne. *C.r. Acad. Inscr. Belles-Lettres*, 105–20.

BREUIL, H. 1909. L'évolution de l'art quaternaire et les travaux d'Edouard Piette. *Revue Arch.* 4e série, 13, 378–411.

BREUIL, H. 1912. L'âge des cavernes et roches ornées de France et d'Espagne. *Revue Arch.* 19, 193–234.

BREUIL, H. 1925. Les origines de l'art. *Journal de Psychologie* 22, 289–96.

BREUIL. H. 1926. Les origines de l'art décoratif. *Ibid.* 23, 364–75.

BREUIL, H. 1936/7. De quelques oeuvres d'art magdaléniennes inédites ou peu connues. *IPEK* 11, 1–16.

BREUIL, H. 1949. Les fresques de la Galerie Vidal à la Caverne de Bédeilhac (Ariège). *Bull. Soc. Préhist. Ariège* 4, 11–16, 11 pl.

BREUIL, H. 1952. *Four Hundred Centuries of Cave Art*. Centre d'Etudes et de Documentation Préhistoriques: Montignac.

BREUIL, H. 1952a. La Caverne de Niaux. Compléments inédits sur sa décoration. *Bull. Soc. Préhist. Ariège* 7, 11–35.

BREUIL, H. 1957. En lisant 'L'Arte dell'Antica Età della Pietra' de P. Graziosi. *Quaternaria* 4, 231–42.

BREUIL, H. 1959. Notre art de l'Epoque du Renne: sa

répartition, sa succession, ses musées, in Zervos 1959, 13–20 and 109.

BREUIL, H. 1962. Théories et faits cantabriques relatifs au Paléolithique supérieur et à son art des cavernes. *Munibe* 14, 353–8.

BREUIL, H. 1964. Préface, in *Musée des Antiquités Nationales. Collection Piette*, by M. Chollot. Musées Nationaux, Paris, 11–13.

BREUIL, H. 1977. La perspective dans les dessins paléolithiques antérieurs au Solutréen, in *Cougnac, grotte peinte*, by L. Méroc and J. Mazet, 53–61. Editions des grottes de Cougnac: Gourdon.

BREUIL, H. and BEGOUEN, H. 1937. Quelques oiseaux inédits ou méconnus de l'art préhistorique. *12e Congrès Préhist. de France*, Toulouse-Foix 1936, 475–88.

BREUIL, H. and LANTIER, R. 1959. *Les Hommes de la Pierre Ancienne*. Payot: Paris. 2nd edition.

BREUIL, H., NOUGIER, L-R and ROBERT, R. 1956. Le 'lissoir aux ours' de la grotte de la Vache, à Alliat, et l'ours dans l'art franco-cantabrique occidental. *Bull. Soc. Préhist. Ariège* 11, 15–78, 7 pl. (and 12, 1957, 53–4).

BREUIL, H. and OBERMAIER, H. 1935. *The Cave of Altamira at Santillana del Mar, Spain*. Tipografía de Archivos: Madrid.

BREUIL, H. and de SAINT PERIER, R. 1927. *Les Poissons, les Batraciens et les Reptiles dans l'Art Quaternaire*. Archives de l'Inst. Paléont. Humaine 2, Paris.

BREWSTER, R. 1986. Aboriginal cave paintings among oldest in world. *Popular Archaeology* 7 (6), 44.

BRIELLE, G. 1968. Les bouquetins dans l'art pariétal quaternaire pyrénéen. *Travaux Inst. Art Préhist. Toulouse* 10, 31–96.

BRODRICK, A.H. 1963. *The Abbé Breuil, Prehistorian; a Biography*. Hutchinson: London.

BROGLIO, A. 1992. Le pietre dipinte dell'Epigravettiano recente del Riparo Villabruna-A (Dolomiti Venete), in *Atti della XXVIII Riunione Scientifica, L'Arte in Italia dal Paleolitico all'Età del Bronzo*, Firenze Nov. 1989, 223–37. Istituto Italiano di Preistoria e Protostoria.

BROWN, S. 1987. The 1986 AURA Franco-Cantabrian field trip: summary report. *Rock Art Research* 4 (1), 56–60.

BRUNET, J. 1981. Niaux: Du charbon de bois dans les peintures. *Archéologia* 156, juillet, 72.

BRUNET, J. 1982. La Grotte de Niaux. *Archéologia* 162, janvier, 74.

BUISSON, D. 1990. Les flûtes paléolithiques d'Isturitz. *Bull. Soc. Préhist. française* 87, 420–33.

BUISSON, D. and PINÇON, G. 1986/7. Nouvelle lecture d'un galet gravé de Gourdan et essai d'analyse des figurations d'oiseaux dans l'art paléolithique français. *Antiquités Nationales* 18/19, 75–90.

BUISSON, D. *et al.* 1989. Les objets colorés du Paléolithique Supérieur. Cas de la grotte de la Vache (Ariège). *Bull. Soc. Préhist. française* 86, 183–91.

BUISSON, D. *et al* 1996. Analyse formelle des contours découpés de têtes de chevaux: implications archéologiques, in *Pyrénées Préhistoriques, Arts et Sociétés*. 118e Congrès Nat. des Sociétés Historiques et Scientifiques, Pau 1993, 327–40. Editions du Comité des Travaux Historiques et Scientifiques: Paris.

BUTZER, K.W. *et al.* 1979. Dating and context of rock engravings in southern Africa. *Science* 203, 1201–14.

CABRE AGUILO, J. 1934. *Las Cuevas de Los Casares y de la Hoz*. Archivo Esp. de Arte y Arq. 30, Madrid.

CABRERA GARRIDO, J. M. 1978. Les matériaux des peintures de la grotte d'Altamira. *Actes de la 5e réunion triennale de l'ICOM*, Zagreb, 1–9.

CAMPBELL, J.B. 1977. *The Upper Palaeolithic of Britain*. 2 vols, Clarendon Press: Oxford.

CANBY, T.Y. 1979. The search for the first Americans. *National Geographic* 156 (3), Sept., 330–63.

CAPDEVILLE, E. 1986. Aperçus sur le problème des signes tectiformes dans l'art pariétal paléolithique supérieur d'Europe. *Travaux de l'Institut d'Art Préhistorique de Toulouse* 28, 59–104.

CAPITAN, L. and BOUYSSONIE, J. 1924. *Un atelier d'art préhistorique. Limeuil. Son gisement à gravures sur pierres de l'âge du Renne*. Nourry: Paris.

CAPITAN, L., BREUIL, H. and PEYRONY, D. 1910. *La Caverne de Font-de-Gaume aux Eyzies (Dordogne)*. Monaco.

CAPITAN, L., BREUIL, H. and PEYRONY, D. 1924. *Les Combarelles aux Eyzies (Dordogne)*. Masson: Paris.

CAPITAN, L. and PEYRONY, D. 1928. *La Madeleine. Son gisement, son industrie, ses oeuvres d'art*. Nourry: Paris.

CARALP, E., NOUGIER, L-R. and ROBERT, R. 1973. Le thème du 'mammifère aux poissons' dans l'art magdalénien. *Bull. Soc. Préhist. Ariège* 28, 11–3.

CARCAUZON, C. 1984. Une nouvelle découverte en Dordogne: la grotte préhistorique de Fronsac. *Revue Archéologique Sites* 22, août, 7–15.

CARCAUZON, C. 1986. La grotte de Font-Bargeix. *Bull. Soc. Hist. et Arch. du Périgord* 113, 191–8.

CARCIUMARU, M. and BITIRI, M. 1983. Peintures rupestres de la grotte Cuciulat (Roumanie). *Bull. Soc. Préhist. française* 80, 94–6.

CARDICH, A. 1987. Arqueología de Los Toldos y El Ceibo (Provincia de Santa Cruz, Argentina). *Estudios Atacameños* (San Pedro de Atacama, Chile) 8 (No. especial), 98– 117.

CARTAILHAC, E. 1902. Les cavernes ornées de dessins. La grotte d'Altamira, Espagne. 'Mea culpa d'un sceptique'. *L'Anthropologie* 13, 348–54.

CARTAILHAC, E. 1908. Les plus anciens artistes de l'Humanité. *Revue des Pyrénées* 20, 509–27.

CARTAILHAC, E. and BREUIL, H. 1905. Les peintures et gravures murales des cavernes pyrénéennes: Marsoulas. *L'Anthropologie* 16, 431–43.

CARTAILHAC, E. and BREUIL, H. 1906. *La Caverne d'Altamira à Santillane, près Santander (Espagne)*. Imp. de Monaco.

CARTAILHAC, E. and BREUIL, H. 1908. Les peintures et gravures murales des cavernes pyrénéennes. III, Niaux. *L'Anthropologie* 19, 15–46.

CASADO LOPEZ, M.P. 1977. *Los Signos en el Arte Paleolítico de la Peninsula Ibérica*. Monografías Arq. 20, Zaragoza.

CATTELAIN, P. 1977/8. *Les propulseurs au Paléolithique Supérieur. Essai d'un inventaire descriptif et critique*. Mémoire de licence en histoire de l'art et arch., Univ. libre de Bruxelles, 2 vols.

CHABREDIER, L. 1975. Les gravures paléolithiques de la grotte d'Ebbou (Ardèche). *Archéocivilisation*, nlle série, numéro spécial, 14/15.

CHAFFEE, S. D., HYMAN, M. and ROWE, M. 1993. AMS 14C dating of rock paintings, in *Time and Space. Dating and Spatial Considerations in Rock Art Research* (J. Steinbring *et al.*, eds), 67–73. AURA Publication 8: Melbourne.

CHALOUPKA, G. 1984. *From Palaeoart to Casual Paintings*. Monograph 1, Northern Territory Museum of Arts and Sciences: Darwin.

CHALOUPKA, G. 1993. *Journey in Time. The World's Longest Continuing Art Tradition. The 50,000–Year Story of the Australian Aboriginal Rock Art of Arnhem Land*. Reed: Chatswood, NSW.

CHAPA BRUNET, T. and MENENDEZ FERNANDEZ, M. (eds) 1994. *Arte Paleolítico*. Complutum 5.

CHAUVET, J-M., BRUNEL DESCHAMPS, E. and HILLAIRE, C. 1995. *Chauvet Cave. The Discovery of the World's Oldest Paintings*. Thames & Hudson: London (US edition: *Dawn of Art, The Chauvet Cave: The Oldest Known Paintings in the World*. Abrams: New York).

CHESNEY, S. 1991. Max Raphael's contributions to the study of prehistoric symbol systems, in *Rock Art and Prehistory* (P. G. Bahn and A. Rosenfeld, eds), 14–22. Oxbow Monograph 10: Oxford.

CHESNEY, S. 1994. Max Raphael (1889–1952): a pioneer of the semiotic approach to Palaeolithic art. *Semiotica* 100 (2/4), 109–24.

CHOLLOT-LEGOUX, M. 1961. Une figuration magdalénienne de dépouille animale. *Antiquités Nationales et Internationales* IIe année, mars–juin, p. 13 and 1 pl.

CHOLLOT-LEGOUX, M. 1964. *Musée des Antiquités Nationales: Collection Piette (Art Mobilier Préhistorique)*. Musées Nationaux: Paris.

CHOLLOT-VARAGNAC, M. 1980. *Les Origines du Graphisme Symbolique*. Singer- Polignac: Paris.

CHRISTENSEN, M. 1996. Un exemple de travail de l'ivoire du paléolithique allemand: apport de l'étude fonctionnelle des outils aurignaciens de Geissenklösterle. *Techne, Arts Préhistoriques* 3, 39–53.

CLARK, G. A. 1992. A comment on Mithen's ecological interpretation of Palaeolithic art. *Proc. Prehist. Soc.* 58, 107–9.

CLARK, L.H. 1966. *They Sang For Horses: The Impact of the Horse on Navajo and Apache Folklore*. Univ. of Arizona Press.

CLEGG, J. 1993. Pictures, jargon and theory —our own ethnography and roadside rock art. *Records of the Australian Museum*, Supplement 17, 91–103.

CLOT, A., MENU, M. and WALTER, P. 1995. Manières de peindre des mains à Gargas et Tibiran (H-P). *L'Anthropologie* 99, 221–35.

CLOTTES, J. 1989. The identification of human and animal figures in European Palaeolithic art, in *Animals into Art* (H. Morphy, ed.), 21–56. Unwin Hyman: London.

CLOTTES, J. (ed.) 1990. *L'Art des Objets au Paléolithique*. 2 vols. Ministère de la Culture: Paris.

CLOTTES, J. 1990a. The parietal art of the late Magdalenian. *Antiquity* 64, 527–48.

CLOTTES, J. 1993. Paint analyses from several Magdalenian caves in the Ariège region of France. *Journal of Arch. Science* 20, 223–35.

CLOTTES, J. 1995. Paleolithic petroglyphs at Foz Côa, Portugal. *International Newsletter on Rock Art* 10, p. 2.

CLOTTES, J. 1995a. *Les Cavernes de Niaux. Art Préhistorique en Ariège*. Le Seuil: Paris.

CLOTTES, J. 1996. Le Parc Pyrénéen d'Art Préhistorique, in *Revivre le Passé grâce à l'Archéologie, Dossiers d'Archéologie* 216, Septembre, 66– 67.

CLOTTES, J. 1996a. Thematic changes in Upper Palaeolithic art: a view from the Grotte Chauvet. *Antiquity* 1970, 276–88.

CLOTTES, J. and COURTIN, J. 1994. *La Grotte Cosquer. Peintures et Gravures de la Caverne Engloutie*. Le Seuil: Paris. (US edition, *The Cave Beneath the Sea: Paleolithic Images at Cosquer*. Abrams: New York).

CLOTTES, J., COURTIN, J. and VALLADAS, H. 1992. A well-dated Palaeolithic cave: the Cosquer Cave at Marseille. *Rock Art Research* 9, 122–9.

CLOTTES, J., COURTIN, J. and VALLADAS, H. 1996. New direct dates for the Cosquer Cave. *International Newsletter on Rock Art* 15, 2–4.

CLOTTES, J., DUPORT, L. and FERUGLIO, V. 1990. Les signes du Placard. *Bull. Soc. Préhist. Ariège-Pyrénées* 45, 15–49.

CLOTTES, J., GARNER, M. and MAURY, G. 1994. Magdalenian bison in the caves of the Ariège. *Rock Art Research* 11, 58–70.

CLOTTES, J. and LEWIS-WILLIAMS, D. 1996. *Les Chamanes de la Préhistoire. Transe et Magie dans les Grottes Ornées*. Le Seuil: Paris.

CLOTTES, J., MENU, M. and WALTER, P. 1990. La préparation des peintures magdaléniennes des cavernes ariégeoises. *Bull. Soc. Préhist. française* 87, 170–92.

CLOTTES, J., MENU, M. and WALTER, P. 1990a. New light on the Niaux paintings. *Rock Art Research* 7, 21–6.

CLOTTES, J. and SIMONNET, R. 1990. Retour au Réseau Clastres (Niaux, Ariège). *Bull. Soc. Préhist. Ariège-Pyrénées* 45, 51–139.

CLOTTES, J., VALLADAS, H., CACHIER, H. and ARNOLD, M. 1992. Des dates pour Niaux et Gargas. *Bull. Soc. Préhist. française* 89, 270–74.

CLOTTES, J. *et al* 1995. Les peintures paléolithiques de la Grotte Chauvet-Pont d'Arc, à Vallon-Pont d'Arc (Ardèche, France): datations directes et indirectes par la méthode du radiocarbone. *C. r. Acad. Sciences de Paris* 320, série II a, 1133–40.

COLE, N. A., WATCHMAN, A. and MORWOOD, M. J. 1995. Chronology of Laura rock art, in *Quinkan Prehistory. The Archaeology of Aboriginal Art in S. E. Cape York Peninsula, Australia* (M. J. Morwood and D. R. Hobbs, eds), Tempus vol. 3, 147–60. Univ. of Queensland.

COLLINS, D. and ONIANS, J. 1978. The origins of art. *Art History* 1, 1–25.

COLLOT, F. *et al.* La théorie de l'information en Préhistoire. *Bio-Mathématique* 77, 32pp.

COMBIER, J. 1984. Grottes ornées de l'Ardèche, in *Les Premiers Artistes, Dossier de l'Arch.* 87, 80–86.

COMBIER, J., DROUOT, E. and HUCHARD, P. 1958. Les grottes solutréennes à gravures pariétales du canyon inférieur de l'Ardèche. *Mém. Soc. Préhist. fr.* 5, 61–117.

CONKEY, M. W. 1978. Style and information in cultural evolution: toward a predictive model for the Paleolithic, in *Social Archaeology, Beyond Subsistence and Dating* (C. L. Redman *et al.*, eds), 61–85. Academic Press: New York/London.

CONKEY, M. W. 1980. The identification of prehistoric hunter-gatherer aggregation sites; the case of Altamira. *Current Anth.* 21, 609–30.

CONKEY, M. W. 1981. A century of Palaeolithic cave art. *Archaeology* 34, 20–28.

CONKEY, M. W. 1983. On the origins of Paleolithic art: a review and some critical thoughts, in *The Mousterian Legacy* (E. Trinkaus, ed.), 201–27. British Arch. Reports, Int series 164: Oxford.

CONKEY, M. W. 1984. To find ourselves: art and social geography of prehistoric hunter-gatherers, in *Past and Present in Hunter-Gatherer Studies* (C. Schrire, ed.), 253–76. Academic Press: New York/London.

CONKEY, M. W. 1985. Ritual communication, social elaboration and the variable trajectories of Paleolithic material culture, in *Prehistoric Hunter-Gatherers: the Emergence of Social and Cultural Complexity* (T. D. Price and J. A. Brown, eds.), 299–323. Academic Press: New York/London.

COOK, J. and WELTE, A-C. 1995. La Grotte du Courbet (Tarn): sa contribution dans l'histoire de l'homme fossile et de l'art paléolithique. *Bull. Soc. Préhist. Ariège-Pyrénées* 50, 85–96.

COPPENS, Y. 1989. L'ambiguité des doubles Vénus du Gravettien de France. *C. r. Acad. des Inscriptions et Belles-Lettres*, July–Dec., 566–71.

CORCHON, M. S. 1986. *El Arte Mueble Paleolítico Cantábrico: Contexto y Análisis Interno.* Monografía 16 del Centro de Investigación y Museo de Altamira.

COSGROVE, R. and JONES, R. 1989. Judds Cavern: a subterranean aboriginal dating site, Southern Tasmania. *Rock Art Research* 6, 96–104.

COURAUD, C. 1982. Techniques de peintures préhistoriques: Expériences. *Information Couleur* 19, 3–6.

COURAUD, C. 1983. Pour une étude méthodologique des colorants préhistoriques. *Bull. Soc.*

COURAUD, C. *Préhist. française* 80, 104–10.

COURAUD, C. 1983a. La couleur dans l'art paléolithique. *Information Couleur* 22, 5–9.

COURAUD, C. 1985. *L'Art Azilien. Origine – Survivance.* XXe Supplément à Gallia Préhistoire. CNRS: Paris.

COURAUD, C. 1988. Pigments utilisés en préhistoire. Provenance, préparation, mode d'utilisation. *L'Anthropologie* 92, 17–28.

COURAUD, C. 1991. Les pigments des grottes d'Arcy-sur-Cure (Yonne). *Gallia Préhistoire* 33, 17–52.

CREMADES, M. 1989. *Contribution à l'étude de l'art mobilier du Paléolithique Supérieur du Bassin Aquitain: Techniques de gravure sur os et matériaux organiques.* 2 vols. Thèse de Doctorat de l'Univ. de Bordeaux I.

CREMADES, M. 1994. Sédentarité et migrations animales à travers les figurations d'oiseaux de l'art paléolithique français. *Bull. Soc. Préhist. Ariège-Pyrénées* 49, 191–210.

CREMADES, M. 1996. L'expression graphique au Paléolithique inférieur et moyen: L'exemple de l'Abri Suard (La Chaise de Vouthon, Charente). *Bull. Soc. Préhist. française* 93, 494–501.

CREMADES, M. *et al.* 1995. Une pierre gravée de 50,000 ans B.P. dans les Balkans. *Paléo* 7, 201–9

CRIVELLI MONTERO, E. A. and FERNANDEZ, M. M. 1996. Palaeoindian bedrock petroglyphs at Epullán Grande Cave, northern Patagonia, Argentina. *Rock Art Research* 13 (2), 124–28.

DAMS, L. 1978. *L'Art Paléolithique de la Caverne de la Pileta.* Akademische Druck- u. Verlagsanstalt: Graz.

DAMS, L. 1984. Preliminary findings at the 'organ' sanctuary in the cave of Nerja, Málaga, Spain. *Oxford Journal of Arch.* 3, 1–14.

DAMS, L. 1985. Palaeolithic lithophones: descriptions and comparisons. *Oxford Journal of Arch.* 4, 31–46.

DAMS, L. 1987. Fish images in palaeolithic cave art. *Archaeology Today* 8 (2), 16–20.

DAMS, L. 1987a. *L'Art Paléolithique de la Grotte de Nerja (Málaga, Espagne).* British Arch. Reports, Int. Series No. 385, Oxford.

DART, R. A. 1974. The waterworn australopithecine pebble of many faces from Makapansgat. *South African Journal of Science* 70, 167–9.

DASTUGUE, J. and DE LUMLEY, M.A. 1976. Les maladies des hommes préhistoriques du Paléolithique supérieur et du Mésolithique, in *La Préhistoire française*, vol. I:1, 612–22 (H. de Lumley, ed.). CNRS: Paris.

DAUVOIS, M. 1977. Travail expérimental de l'ivoire: sculpture d'une statuette féminine, in *Méthodologie appliquée à l'industrie de l'os préhistorique* (H. Camps-Fabrer, ed.), 270–73. CNRS: Paris.

DAUVOIS, M. 1989. Son et musique paléolithiques, in *La Musique dans l'Antiquité, Les Dossiers d'Arch.* 142, Nov., 2–11.

DAUVOIS, M. 1994. Les témoins sonores paléolithiques extérieur et souterrain, in *'Sons Originels', Préhistoire de la Musique* (M. Otte, ed,), Actes du Colloque de Musicologie, Dec. 1992, 11–31. ERAUL 61: Liège.

DAUVOIS, M. and BOUTILLON, X. 1990. Etudes acoustiques au Réseau Clastres: Salle des Peintures et lithophones Naturels. *Bull. Soc. Préhist. Ariège-Pyrénées* 45, 175–86.

DAVENPORT, D. and JOCHIM, M. A. 1988. The scene in the shaft at Lascaux. *Antiquity* 62, 558–62.

DAVIS, W. 1986. The origins of image making. *Current Anthropology* 27, 193–215, 371, 515–16.

DAVIS, W. 1987. Replication and depiction in Paleolithic art. *Representations* 19, 111–47.

DECOUVERTES DE L'ART DES GROTTES ET DES ABRIS. 1984. Catalogue de l'Exposition, 59pp. Réjou:

Périgueux.

DELCOURT-VLAEMINCK, M. 1975. Les représentations végétales dans l'art du Paléolithique supérieur. *Paléontologie et Préhistoire: Rev. Soc. Tournaisienne de Géol., Préh. et Arch.* 31, 70–96.

DELLUC, B. and G. 1973. Quelques figurations paléolithiques inédites des environs des Eyzies (Dordogne): Grottes Archambeau, du Roc et de la Mouthe. *Gallia Préhistoire* 16, 201–9.

DELLUC, B. and G. 1978. Les manifestations graphiques aurignaciens sur support rocheux des environs des Eyzies (Dordogne). *Gallia Préhistoire* 21, 213–438.

DELLUC, B. and G. 1981. Une visite à la grotte de Rouffignac en 1759. *Bull. Soc. Hist. et Arch. du Périgord* 108, 3–11.

DELLUC, B. and G. 1981a. La grotte ornée de Comarque à Sireuil (Dordogne). *Gallia Préhistoire* 24, 1–97.

DELLUC. B. and G. 1981b. Les signes en 'empreinte' du début du Paléolithique supérieur, in *21e Congrès Préhist. de France*, vol. 2, 111–16.

DELLUC, B. and G. 1983. Les grottes ornées de Domme (Dordogne): La Martine, Le Mammouth et le Pigeonnier. *Gallia Préhistoire* 26, 7–80.

DELLUC, B. and G. 1983a. La Croze à Gontran, grotte ornée aux Eyzies-de-Tayac (Dordogne). *Ars Praehistorica* 2, 13–48.

DELLUC, B. and G. 1984. L'art pariétal avant Lascaux. *Dossiers de l'Archéologie* 87, octobre, 'Les Premiers Artistes', 52–60.

DELLUC, B. and G. 1984a. Lascaux II, a faithful copy. *Antiquity* 58, 194–6, 2 pl.

DELLUC, B. and G. 1984b. Lecture analytique des supports rocheux gravés et relevé synthétique. *L'Anthropologie* 88, 519–29.

DELLUC, B. and G. 1984c. Faune figurée et faune consommée: une magie de la chasse? *Les Premiers Artistes, Dossiers de l'Arch.* 87, oct., 28–9.

DELLUC, B. and G. 1985. De l'empreinte au signe, in *Traces et Messages de la Préhistoire, Dossiers de l'Arch.* 90, 56–62.

DELLUC, B. and DELLUC, G. 1989. Le sang, la souffrance et la mort dans l'art paléolithique. *L'Anthropologie* 93, 389–406.

DELLUC, B. and DELLUC, G. 1991. *L'Art Pariétal Archaïque en Aquitaine.* 28e Supplément à Gallia Préhistoire. CNRS: Paris.

DELLUC, B. and DELLUC, G. 1995. Les figures féminines schématiques du Périgord. *L'Anthropologie* 99, 236–57.

DELPORTE, H. 1962. Le problème des statuettes féminines au leptolithique occidental. *Mitteilungen der Anth. Gesellschaft in Wien* 92, 53–60.

DELPORTE, H. 1971. A propos du style des figurations féminines gravettiennes. *Antiquités Nationales* 3, 5–20.

DELPORTE, H. 1973. Les techniques de la gravure paléolithique, in *Estudios dedicados al Prof. Dr. Luis Pericot*, 119–29. Inst. de Arq. y Preh., Univ. de Barcelona.

DELPORTE, H. 1984. L'art mobilier et ses rapports avec la faune paléolithique, in Bandi *et al.* 1984, 111–42.

DELPORTE, H. 1987. *Piette, pionnier de la Préhistoire.* Picard: Paris.

DELPORTE, H. 1990. *L'Image des Animaux dans l'Art Préhistorique.* Picard: Paris.

DELPORTE, H. 1993. *L'Image de la Femme dans l'Art Préhistorique.* (2nd ed.). Picard: Paris.

DELPORTE, H. 1993a. L'art mobilier de la grotte de La Vache: premier essai de vue générale. *Bull. Soc. Préhist. française* 90, 131–6.

DELPORTE, H. (ed.) 1995. *La Dame de Brassempouy.*

Actes du Colloque de Brassempouy (July 1994). ERAUL 74, Liège.

DELPORTE, H., KANDEL, D. and PINÇON, G. 1986. Le C.I.D. Breuil, domaine A.P.M.: Un système documentaire sur l'Art Paléolithique Mobilier. *Bull. Soc. Préhist. française* 83, 299–303.

DELPORTE, H. and MONS, L. 1975. Omoplate décorée du Mas d'Azil (Ariège). *Antiquités Nationales* 7, 14–23.

DELPORTE, H. and MONS, L. 1980. Gravure sur os et argile, in *Revivre la Préhistoire, Dossier de l'Archéologie* 46, 40–45.

DEMAILLY, S. 1990. Premiers résultats de l'étude des peintures rupestres concernant la Grotte de Lascaux, in *50 Ans Après la Découverte de Lascaux. Journées Int. d'Etude sur la Conservation de l'Art Rupestre*, 101–16. ICOM.

DEMARS, P-Y. 1992. Les colorants dans le Moustérien du Périgord. L'apport des fouilles de F. Bordes. *Bull. Soc. Préhist. Ariège-Pyrénées* 47, 185–94.

DESBROSSE, R., FERRIER, J. and TABORIN, Y. 1976. La parure. in *La Préhistoire française*, I:1 (H. de Lumley, ed.), 710–13. CNRS: Paris.

DESDEMAINES-HUGON, C. 1993. Les décors non-figuratifs du mobilier osseux des niveaux du Magdalénien IV et V de la Madeleine (Dordogne). Observations et comparaisons. *Bull. Soc. Préhist. Ariège-Pyrénées* 48, 123–203.

DEWEZ, M. 1974. New hypotheses concerning two engraved bones from La Grotte de Remouchamps, Belgium. *World Arch.* 5, 337–45, 1 pl.

DJINDJIAN, F. and PINÇON, G. 1986. Un exemple de banque de données sur microserveur Vidéotex: L'Art Pariétal Paléolithique. *Bull. Soc. Préhist. française* 83, 332–4.

DORN, R. I. 1997. Constraining the age of the Côa valley (Portugal) engravings with radiocarbon dating. *Antiquity* 71, 105–15.

DORN, R. I., NOBBS, M. and CAHILL, T. A. 1988. Cation-ratio dating of rock engravings from the Olary Province of arid South Australia. *Antiquity* 62, 681–89.

DORTCH, C. 1979. Australia's oldest known ornaments. *Antiquity* 53, 39–43.

DORTCH, C. 1984. *Devil's Lair: A study in prehistory.* Western Australia Museum: Perth.

DRAGOVICH, D. 1986. Minimum age of some desert varnish near Broken Hill, New South Wales. *Search* 17, 149–51.

DUBOURG, C. 1994. Les expressions de la saisonnalité dans les arts paléolithiques – les arts sur support lithique du Bassin d'Aquitaine. *Bull. Soc. Préhistorique Ariège-Pyrénées* 49, 145–89.

DUHARD, J-P. 1987. Le soi-disant 'hermaphrodite' de Grimaldi: une nouvelle interprétation obstétricale. *Bull. Soc. Anth. Sud-Ouest* 22, 139–44.

DUHARD, J-P. 1987a. La statuette de Monpazier représente-elle une parturiente? *Bull. Soc. Préhist. Ariège-Pyrénées* 42, 155–63.

DUHARD, J-P. 1988. Peut-on parler d'obésité chez les femmes figurées dans les oeuvres pariétales et mobilières paléolithiques? *Bull. Soc. Préhist. Ariège-Pyrénées* 43, 85–103.

DUHARD, J-P. 1988a. Le calendrier obstétrical de la femme à la corne de Laussel. *Bull. Soc. Historique et Archéologique du Périgord* 115, 23–39.

DUHARD, J-P. 1989. La gestuelle du membre supérieur dans les figurations féminines sculptées paléolithiques. *Rock Art Research* 6, 105–17.

DUHARD, J-P. 1991. The shape of Pleistocene women. *Antiquity* 65, 552–61.

DUHARD, J-P. 1991a. Images de la chasse au Paléolithique. *Oxford Journal of Arch.* 10, 127–57.

DUHARD, J-P. 1992. Les humains ithyphalliques dans l'art paléolithique. *Bull. Soc. Préhist. Ariège-Pyrénées* 47, 133–59.

DUHARD, J-P. 1992a. La dichotomie sociale sexuelle dans les figurations humaines magdaléniennes. *Rock Art Research* 9, 111–18.

DUHARD, J-P. 1993. *Réalisme de l'Image Féminine Paléolithique.* Cahiers du Quaternaire 19. CNRS: Paris.

DUHARD, J-P. 1993a. Upper Palaeolithic figures as a reflection of human morphology and social organization. *Antiquity* 67, 83–91.

DUHARD, J-P. 1995. De la confusion entre morphologie et géométrie dans les figurations féminines gravettiennes et du supposé style gravettien. *Bull. Soc. Préhist. française* 92, 302–12.

DUHARD, J-P. 1996. *Réalisme de l'Image Masculine Paléolithique.* Jérôme Millon: Grenoble.

DUHARD, J-P. and ROUSSOT, A. 1988. Le gland pénien sculpté de Laussel (Dordogne). *Bull. Soc. Préhist. française* 85, 41–4.

EASTHAM, A. 1988. The season or the symbol: the evidence of swallows in the Paleolithic of Western Europe. *Archaeozoologica* 2 (1–2), 243–52.

EASTHAM, A. and M. 1979. The wall art of the Franco-Cantabrian deep caves. *Art History* 2, 365–87.

EASTHAM, M. and A. 1991. Palaeolithic parietal art and its topographical context. *Proc. Prehist. Soc.* 57, 115–28.

EATON, R. L. 1978. The evolution of trophy hunting. *Carnivore* 1, 110–21.

EPPEL, F. 1972. Les objets d'art paléolithique en Autriche. *Bull. Soc. Préh. Ariège* 27, 73–81.

d'ERRICO, F. 1987. Nouveaux indices et nouvelles techniques microscopiques pour la lecture de l'art gravé mobilier. *C.R. Acad. Sc. Paris* 304, Série II, No. 13, 761–4, 1 pl.

d'ERRICO, F. 1991. Microscopic and statistical criteria for the identification of prehistoric systems of notation. *Rock Art Research* 8, 83–93.

d'ERRICO, F. 1994. *L'Art Gravé Azilien. De la technique à la signification.* 31e Suppl. à Gallia Préhistoire. C. N. R. S.: Paris.

d'ERRICO, F. 1994a. Birds of the Grotte Cosquer: the Great Auk and Palaeolithic prehistory. *Antiquity* 68, 39–47.

d'ERRICO, F. 1994b. Birds of Cosquer Cave. *Rock Art Research* 11, 45–57.

d'ERRICO, F. 1995. A new model and its implications for the origin of writing: the La Marche antler revisited. *Cambridge Arch. Journal* 5, 163–206.

d'ERRICO, F. and CACHO, C. 1994. Notation versus decoration in the Upper Palaeolithic: a case-study from Tossal de la Roca, Alicante, Spain. *Journal of Arch. Science* 21, 185–200.

d'ERRICO, F., WILLIAMS, C. T. and STRINGER, C. B. 1997. AMS dating and microscopic analysis of the Sherborne bone. *Journal of Arch. Science* (in press).

FAGES, G. and MOURER-CHAUVIRE, C. 1983. La flûte en os d'oiseau de la grotte sépulcrale de Veyreau (Aveyron) et inventaire des flûtes préhistoriques d'Europe, in *La Faune et l'Homme Préhistoriques*, Mém. 16 de la Soc. Préhist. française, 95–103.

FAURE, M. 1978. Révision critique d'une collection de gravures mobilières paléolithiques: les galets et les os gravés de la Colombière (Neuville-sur-Ain, Ain, France). *Nouv. Arch. Mus. Hist. Nat. Lyon fasc.* 16, 41–99, 6 pl.

FORBES, A. and CROWDER, T. R. 1979. The problem of Franco-Cantabrian abstract signs: agenda for a new approach. *World Arch.* 10, 350–66, 2 pl.

FORMOZOV, A. A. 1983. On the problem of 'centres of primitive art'. *Sovetskaya Arkheologia* 3, 5–13. (in Russian).

FORTEA, F. J. 1978. Arte paleolítico del Mediterraneo español. *Trabajos de Prehistoria* 35, 99–149.

FORTEA, J. 1981. Investigaciones en la cuenca media del Nalón, Asturias (España). Noticia y primeros resultados. *Zephyrus* 32/33, 5–16.

FORTEA PEREZ, J. 1995. Covaciella, in *Excavaciones Arqueológicas en Asturias 1991–1994*, 3, 258–70. Oviedo.

FORTEA, J. *et al.* 1990. Travaux récents dans les Vallées du Nalón et du Sella (Asturies), in *L'Art des Objets au Paléolithique* (J. Clottes, ed.), vol. 1, 219–44.

FREEMAN, L. G. 1978. Mamut, jabalí y bisonte en Altamira: reinterpretaciones sugeridas por la historia natural, in *Curso de Arte Rupestre Paleolítico* (A. Beltrán, ed.), 157–79. Universidad Int. 'Menéndez Pelayo': Santander.

FREEMAN, L. G. 1984. Techniques of figure enhancement in Paleolithic cave art, in *Scripta Praehistorica. Francisco Jordá Oblata*, 209–32. Univ. de Salamanca.

FREEMAN, L. G. 1987. Altamira revisited: First steps in a new investigation, in *Altamira Revisited and other essays on early art* (L. G. Freeman *et al.*, eds), 67–97. Inst. for Preh. Investigations: Chicago.

FREEMAN, L. G., BERNALDO DE QUIROS, F. and OGDEN, J. 1987. Animals, faces and space at Altamira: a restudy of the Final Gallery ('Cola de Caballo'), in *ibid.*, 179–247.

FREEMAN, L. G. and GONZALEZ ECHEGARAY, J. 1983. Tally-marked bone from Mousterian levels at Cueva Morin (Santander, Spain), in *Homenaje al Prof. M. Almagro Basch*, I, 143–7. Min. de Cultura: Madrid.

FREUND, G. 1957. L'art aurignacien en Europe centrale. *Bull. Soc. Préh. Ariège* 12, 55– 78, 2 pl.

FRITZ, C. 1990. *Les Plaquettes Gravées de Gourdan.* 2 vols. Maîtrise de Préhistoire, Univ. de Paris I.

FRITZ, C. and SIMONNET, R. 1996. Du geste à l'objet: les contours découpés de Labastide. Résultats préliminaires. *Techne. Arts Préhistoriques* 3, 63–77.

FROLOV, B. A. 1977/9. Numbers in Paleolithic graphic art and the initial stages in the development of mathematics. *Soviet Anth. and Arch.* 16, 1977/8, 142–66; and 17, 1978/9, 73–93, 41–74 and 61–113.

FROLOV, B. A. 1979. Les bases cognitives de l'art paléolithique, in *Valcamonica Symposium III. The Intellectual Expressions of Prehistoric Man: Art and Religion*, 295–98.

FROLOV, B. A. 1981. L'art paléolithique: Préhistoire de la science? in *Arte Paleolítico*, Comisión XI, Xth Congress UISPP, Mexico City, 60–81.

FULLAGAR, R. L. K., PRICE, D. M. and HEAD, L. M. 1996. Early human occupation of northern Australia: archaeology and thermoluminescence dating of Jinmium rock-shelter, Northern Territory. *Antiquity* 70, 751–73.

GAMBLE, C. 1982. Interaction and alliance in Palaeolithic society. *Man* 17, 92–107.

GARCIA, M. 1979. Les silicones élastomères R.T.V. appliqués aux relevés de vestiges préhistoriques (Arts, Empreintes Humaines et Animales). *L'Anthropologie* 83, 5–42 and 189–222.

GARCIA GUINEA, M. A. 1979. *Altamira y otras cuevas de Cantabria.* Silex: Madrid.

GAUSSEN, J. 1964. *La Grotte Ornée de Gabillou (près Mussidan, Dordogne).* Mémoire 4, Inst. de Préhist. de l'Univ. de Bordeaux.

GEOFFROY, C. 1974. *La couleur dans l'art pariétal paléolithique.* Cahiers du Centre de Recherches Préhist. 3, 45–64.

GIEDION, S. 1965. *L'Eternel Présent. La Naissance de l'Art.* Editions de la Connaissance: Bruxelles.

GIMBUTAS, M. 1981. Vulvas, breasts and buttocks of

the Goddess Creatress: Commentary on the origins of art, in *The Shape of the Past: Studies in Honor of Franklin D. Murphy* (G. Buccellati and C. Speroni, eds.), 15–42. Inst. of Arch.: Los Angeles.

GIRARD, M., BAFFIER, D. and CHEVALIER, J. 1995. Témoins d'éclairages à Arcy-sur-Cure. *Archéologia* 314, July/Aug., 74–81.

GIRARD, M., BAFFIER, D. and CHEVALIER, J. 1996. Découverte d'une lampe paléolithique à usage vertical. *Archéologia* 319, Jan., 8–9.

GIRARD, M., STRICH, J-D., BAFFIER, D. and MOLEZ, D. 1992. Premiers enregistrements photographiques Infra-Rouge dans la Grande Grotte d'Arcy-sur-Cure (Yonne). *Bull. Soc. Préhist. française* 89, 163–4.

GLADKIH, M. I., KORNIETZ, N. L. and SOFFER, O. 1984. Mammoth-bone dwellings on the Russian Plain. *Scientific American* 251, 136–42.

GLORY, A. 1964. La stratigraphie des peintures à Lascaux, in *Miscelánea en Homenaje al Abate H. Breuil*, vol. 1, 449–55. Barcelona.

GLORY, A. 1968. L'énigme de l'art quaternaire peut-elle être résolue par la théorie du culte des ongones? in *Simposio de Arte Rupestre*, Barcelona 1966, 25–60.

GONZALEZ ECHEGARAY, J. 1968. Sobre la datación de los santuarios paleolíticos, in Simposio de Arte Rupestre, Barcelona 1966, 61–65.

GONZALEZ ECHEGARAY, J. 1972. Notas para el estudio cronológico del arte rupestre de la cueva del Castillo, in *Santander Symposium* 409–22.

GONZALEZ ECHEGARAY, J. 1974. *Pinturas y grabados de la cueva de las Chimeneas (Puente Viesgo, Santander)*. Monografias de Arte Rupestre, Arte Paleolítico No. 2, Barcelona.

GONZALEZ GARCIA, R. 1985. *Aproximació al Desenvolupament i Situació de les Manifestacions Artistiques Quaternàries: a les cavitats del Monte del Castillo*. Tesis de Llicenciatura, Univ. de Barcelona, Fac. de Geografia i Historia, Dept. d'Historia de l'Art. 547pp.

GONZALEZ GARCIA, R. 1987. Organisation, distribution and typology of the cave art of Monte del Castillo, Spain. *Rock Art Research* 4, 127–36.

GONZALEZ GARCIA, R. 1993. The validity of generalised stylistic comparisons in Palaeolithic parietal figures, in *Rock Art Studies: The Post-Stylistic Era* (M. Lorblanchet and P. G. Bahn, eds), 37–50. Oxbow Monograph 35: Oxford.

GONZALEZ GARCIA, R. 1996. *Arte Rupestre Paleolítico: Organización Espacial y Programa Decorativo en las Cavidades de la Región Cantábrica*. Tesi Doctoral, Universitat de Barcelona. 4 vols.

GONZALEZ MORALES, M. R. 1989. Problemas de documentación del arte rupestre cantábrico: algunos resultados. *Almansor, Revista de Cultura* 7 (Colóquio Int. de Arte Pré-histórica), 271–88.

GONZALEZ SAINZ, C. 1988. Le fait artistique à la fin du Paléolithique: quelques réflexions. *Bull. Soc. Préhist. Ariège-Pyrénées* 43, 35–62.

GOREN-INBAR, N. 1986. A figurine from the Acheulian site of Berekhat Ram. *Mi'Tekufat Ha'Even* 19, 7–12.

GRAINDOR, M. J. and MARTIN, Y. 1972. *L'Art Préhistorique de Gouy*. Presses de la Cité: Paris.

GRAPP: GROUPE DE REFLEXION SUR L'ART PARIETAL PALEOLITHIQUE 1993. *L'Art Pariétal Paléolithique. Techniques et Méthodes d'Etude*. Editions du Comité des Travaux Historiques et Scientifiques: Paris.

GRAZIOSI, P. 1960. *Palaeolithic Art*. Faber: London.

GRAZIOSI, P. 1973. *L'Arte Preistorica in Italia*. Sanzoni: Florence.

GRIFFIN, J. B. et al 1988. A mammoth fraud in science. *American Antiquity* 53, 578–82.

GROENEN, M. 1988. Les représentations de mains

négatives dans les grottes de Gargas et de Tibiran (H-P. Approche méthodologique. *Bull. Soc. roy. belge Anthrop. Préhist*. 99, 81–113.

GUIDON, N. and DELIBRIAS, G. 1986. Carbon-14 dates point to man in the Americas 32,000 years ago. *Nature* 321, 769–71.

GUTHRIE, R. D. 1984. Ethological observations from Palaeolithic art, in Bandi *et al*. 1984, 35–74.

GVOZDOVER, M. D.1989. The typology of female figurines of the Kostenki Paleolithic culture. *Soviet Anth. and Archeology* 27 (4), 32–94.

GVOZDOVER, M. D. 1995. *Art of the Mammoth Hunters. The Finds from Avdeevo*. Oxbow Monograph 49: Oxford.

HACHI, S. 1985. Figurines en terre cuite du gisement ibéro-maurusien d'Afalou Bou Rhummel. *Travaux du Lab. de Préhistoire et d'Ethnographie des Pays de la Méditerranée Occidentale*, étude 9, 6pp. Aix-en-Provence.

HACHID, M. 1992. *Les Pierres Ecrites de l'Atlas Saharien. El-Hadjra, El-Mektouba*. Ed. Enag: Algiers. 2 vols.

HAHN, J. 1971. La statuette masculine de la grotte du Hohlenstein-Stadel (Württemberg). *L'Anthropologie* 75, 233–44.

HAHN, J. 1972. Aurignacian signs, pendants and art objects in central and eastern Europe. *World Arch*. 3, 252–66, 2 pl.

HAHN, J. 1979. Elfenbeinplastiken des Aurignacien aus dem Geissenklösterle. *Archäologisches Korrespondenzblatt* 9, 135–42, 2 pl.

HAHN, J. 1982. Demi-relief aurignacien en ivoire de la grotte Geissenklösterle, près d'Ulm (Allemagne fédérale). *Bull. Soc. Préhist. fr*. 79, 73–7.

HAHN, J. 1984. L'art mobilier aurignacien en Allemagne du Sud-Ouest. Essai d'analyse zoologique et éthologique, in Bandi *et al*. 1984, 283–93.

HAHN, J. 1986. *Kraft und Aggression. Die Botschaft der Eiszeitkunst im Aurignacien Süddeutschlands?* Verlag Archaeologica Venatoria, Institut für Urgeschichte der Universität Tübingen.

HAHN, J. 1991. Höhlenkunst aus dem Hohlen Fels bei Schelklingen, Alb-Donau-Kreis. *Archäol. Ausgrabungen Baden-Württemberg* 10, 19–22.

HAHN, J. *et al*. (eds) 1995. *Le Travail et l'Usage de l'Ivoire au Paléolithique Supérieur*. Istituto Poligrafico e Zecca della Stato, Libreria dello Stato: Rome.

HALLAM, S. J. 1971. Roof marking in the 'Orchestra Shell' Cave, Wanneroo, near Perth, Western Australia. *Mankind* 8, 90–103.

HALVERSON, J. 1987. Art for art's sake in the Paleolithic. *Current Anth*. 28, 63–89, 203–5.

HAMAYON, R. N. 1995. Pour en finir avec la 'transe' et l''extase' dans l'étude du chamanisme. *Etudes mongoles et sibériennes* 26, 155–90.

HAMAYON, R. N. 1997. Review of *Les Chamanes de la Préhistoire* by J. Clottes and D. Lewis-Williams. *Les Nouvelles de L'Archéologie* (in press).

HAMMOND, N. 1974. Paleolithic mammalian faunas and parietal art in Cantabria: a comment on Freeman. *American Antiquity* 39, 618–19.

HARLE, E. 1881. La grotte d'Altamira, près de Santander (Espagne). *Matériaux pour l'Histoire Primitive et Naturelle de l'Homme* 2e série, XII, 275–83, 1 pl.

HARRISON, R. A. 1978. A pierced reindeer phalanx from Banwell Bone Cave and some experimental work on phalangeal whistles. *Proc. Univ. Bristol Spel. Soc*. 15, 7–22, 1 pl.

HAYDEN, B. 1993. The cultural capacities of Neandertals: a review and re-evaluation. *Journal of Human Evolution* 24, 113–46.

HENNIG, H. 1960. L'art magdalénien en Europe centrale. *Bull. Soc. Préh. Ariège* 15, 24–45.

HORNBLOWER, G. D. 1951. The origin of pictorial art. *Man* 51, 2–3.

HOVERS, E. 1990. Art in the Levantine Epi-Palaeolithic: an engraved pebble from a Kebaran site in the lower Jordan Valley. *Current Anthropology* 31, 317–22.

HUBLIN, J-J. *et al*. 1996. A late Neanderthal associated with Upper Palaeolithic artefacts. *Nature* 382, 224–6.

HUYGE, D. 1990. Mousterian skiffle? Note on a Middle Palaeolithic engraved bone from Schulen, Belgium. *Rock Art Research* 7, 125–32.

HUYGE, D. 1991. The 'Venus' of Laussel in the light of Ethnomusicology. *Archeologie in Vlaanderen* 1, 11–18.

IAKOVLEVA, L. and PINÇON, G. 1996. Une composition de deux bisons sculptés de la frise de l'abri Bourdois à Angles-sur-l'Anglin (Vienne). *Bull. Soc. Préhist. française* 93, 195–200.

IGLER, W. *et al*.1994. Datation radiocarbone de deux figures pariétales de la grotte du Portel (Commune de Loubens, Ariège). *Bull. Soc. Préhist. Ariège-Pyrénées* 49, 231–6.

IMAZ, M. 1990. Estratigrafía de los moluscos marinos en los yacimientos prehistóricos vascos. *Munibe* 42, 269–74.

JAGOR, F. 1882. Cueva de Altamira. *Verhandlungen der Berliner Gesellschaft für Anthropologie, Ethnologie und Urgeschichte*, 170–71.

JANSSENS, P. A. 1957. Medical views on prehistoric representations of human hands. *Medical History* 1, 318–22, 2 pl.

JELINEK, J. 1975. *Encyclopédie illustrée de l'Homme Préhistorique*. Gründ: Paris.

JELINEK, J. 1988. Considérations sur l'art paléolithique mobilier de l'Europe centrale. *L'Anthropologie* 92, 203–38.

JIA LANPO 1980. *Early Man in China*. Foreign Languages Press: Beijing.

JOCHIM, M. 1983. Palaeolithic cave art in ecological perspective, in *Hunter-Gatherer Economy in Prehistory* (G. N. Bailey, ed.), 212–19. Cambridge Univ. Press.

JORDA CERDA, F. 1964. El arte rupestre paleolítico de la región cantabrica: nueva secuencia cronológico-cultural, in *Prehistoric Art of the Western Mediterranean and the Sahara* (L. Pericot and E. Ripoll, eds), 47–81. Viking Fund Publications in Anth., No. 39, New York.

JORDA CERDA, F. 1964a. Sobre técnicas, temas y etapas del Arte Paleolítico de la Región Cantábrica. *Zephyrus* 15, 5–25.

JORDA CERDA, F. 1970. Cueva de Lledias, tesoro artístico o falsificación? Respuesta de Francisco Jordá. *Asturias Semanal*, 27 Nov., 34–5.

JORDA CERDA, F. 1972. Las superposiciones en el gran techo de Altamira, in *Santander Symposium*, 423–56.

JORDA CERDA, F. 1979. Sur des sanctuaires monothématiques dans l'art rupestre cantabrique, in *Valcamonica Symposium III. The Intellectual Expressions of Prehistoric Man: Art and Religion*, 331–48.

JORDA CERDA, F. 1980. El gran techo de Altamira y sus santuarios superpuestos, in *Altamira Symposium*, 277–87. Min. de Cultura: Madrid.

JORDA CERDA, F. 1983. El mamut en el arte paleolítico peninsular y la hierogamia de Los Casares, in *Homenaje al Prof. M. Almagro Basch*, vol. 1, 265–72. Min. de Cultura: Madrid.

JORDA CERDA, F. 1985. Los grabados de Mazouco, los santuarios monotémicos y los animales dominantes en el arte paleolítico peninsular. *Revista de Guimarães* 94, 23pp.

JORDA CERDA, F. 1987. Sobre figuras rupestres de

posibles caballos domesticados. *Archivo de Prehistoria Levantina* 17, 49–58.

JORGE, S. O. *et al.* 1981. Gravuras rupestres de Mazouco (Freixo de Espada à Cinta). *Arqueologia* (Porto) 3, 3–12.

JORGE, S. O. *et al.* 1982. Descoberta de gravuras rupestres em Mazouco, Freixo de Espada-à-Cinta (Portugal). *Zephyrus* 34/35, 65–70.

JORGE, V. O. 1987. Arte rupestre en Portugal. *Revista de Arqueología* 76, 10–19.

JORGE, V. O. 1981/2. Mazouco (Freixo de Espada à Cinta). Nótula arqueológica. *Portugalia*, nova serie II/III, 143–145.

JUDSON, S. 1959. Paleolithic paint. *Science* 130, 708.

JULIEN, M. and LAVALLEE, D. 1987. Les chasseurs de la préhistoire, in *Ancien Pérou: Vie, Pouvoir et Mort.* Exhibition catalogue, Musée de l'Homme, 45–60. Nathan: Paris.

KEHOE, A. B. 1991. No possible, probable shadow of doubt. *Antiquity* 65, 129–31.

KEHOE, T. F. 1989. Corralling: evidence from Upper Paleolithic cave art, in *Hunters of the Recent Past* (L. B. Davis and B. O. K. Shreeve, eds), 34–46. Unwin Hyman: London.

KEHOE, T. F. 1990. Corralling life, in *The Life of Symbols* (M. L. Foster and L. J. Botscharow, eds), 175–93. Westview Press: Boulder.

KEHOE, T. F. 1996. The case for an ethnographic approach to understanding the European Paleolithic, in *Spuren der Jagd – Die Jagd nach Spuren* (I. Campen *et al.,* eds), 193–202. Tübinger Monographien zur Urgeschichte 11.

KIRCHNER, H. 1952. Ein archäologischer Beitrag zur Urgeschichte des Schamanismus. *Anthropos* 47, 244–86.

KLIMA, B. 1982. Neue Forschungsergebnisse in Dolní Vestonice, in *Aurignacien et Gravettien en Europe,* Fascicule II, Cracovie/Nitra 1980. Etudes et Recherches Arch. de l'Univ. de Liège, 13, 171–80.

KLIMA, B. 1983. Une nouvelle statuette paléolithique à Dolní Vestonice. *Bull. Soc. Préhist. fr.* 80, 176–8.

KLIMA, B. 1984. Les représentations animales du Paléolithique Supérieur de Dolní Vestonice (Tchécoslovaquie), in Bandi *et al.* 1984, 323–32.

KNIGHT, C., POWER, C. and WATTS, I. 1995. The human symbolic revolution: a Darwinian account. *Cambridge Arch. Journal* 5, 75–114.

KOBY, F. E. 1954. Y a-t-il eu, à Lascaux, 'un *Bos longifrons'? Bull. Soc. Préhist. fr.* 51, 434–41.

KOBY, F. E. 1968. Les 'rennes' de Tursac paraissent être plutôt des daims. *Bull. Soc. Préhist. Ariège* 23, 123–30.

KOHL, H. and BURGSTALLER, E. 1992. *Eiszeit in Oberösterreich. Paläolithikum Felsbilder.* Österreichisches Felsbildermuseum: Spital am Pyhrn.

KÖKTEN, K. 1955. Ein allgemeiner Überblick über die prähistorischen Forschungen im Karain-Höhle bei Anatolya. *Belleten* 19, 271–93.

KOZLOWSKI, J. K. (ed.) 1982. *Excavation in the Bacho Kiro Cave (Bulgaria): Final Report.* Panstwowe Wydawnictwo Naukowe: Warsaw.

KOZLOWSKI, J. K. 1992. *L'Art de la Préhistoire en Europe Orientale.* CNRS: Paris.

KRAFT, J. C. and THOMAS, R. A. 1976. Early man at Holly Oak, Delaware. *Science* 192, 756–61.

KÜHN, H. 1955. *On the Track of Prehistoric Man.* Hutchinson: London.

KÜHN, H. 1971. *Die Felsbilder Europas.* Kohlhammer: Stuttgart.

KUJUNDZIC, Z. 1989. Gravure na stijeni i gravirani ukrasi na upotrebnim predmetima – Badanj i Pecina pod lipom. *GZM(A),* NS sv. 44, 21–38.

KUMAR, G. 1996. Daraki-Chattan: a Palaeolithic cupule site in India. *Rock Art Research* 13, 38–46.

KUMAR, G., NARVARE, G. and PANCHOLI, R. 1988. Engraved ostrich eggshell objects: new evidence of Upper Palaeolithic art in India. *Rock Art Research* 5 (1), 43–53.

KURTEN, B. 1986. *How to Deep-Freeze a Mammoth.* Columbia Univ. Press: New York.

LADIER, E. and WELTE, A-C. 1994. *Bijoux de la Préhistoire. La Parure Magdalénienne dans la Vallée de l'Aveyron.* Muséum d'Hist. Naturelle: Montauban.

LAMING-EMPERAIRE, A. 1959. *Lascaux. Paintings and Engravings.* Pelican: Harmondsworth.

LAMING-EMPERAIRE, A. 1962. *La Signification de l'Art Rupestre Paléolithique.* Picard: Paris.

LAMING-EMPERAIRE, A. 1970. Système de penser et organisation sociale dans l'art rupestre paléolithique, in *L'Homme de Cro-Magnon,* 197–211. Arts et Métiers Graphiques: Paris.

LAMING-EMPERAIRE, A. 1971. Une hypothèse de travail pour une nouvelle approche des sociétés préhistoriques, in *Mélanges Varagnac,* 541–51. Ecole Pratique des Hautes Etudes: Paris.

LAMING-EMPERAIRE, A. 1972. Art rupestre et organisation sociale, in *Santander Symposium,* 65–82.

LARRIBAU, J-D. and PRUDHOMME, S. 1989. Etude préliminaire de la Grotte d'Erberua (P-A). *L'Anthropologie* 93, 475–93.

LARTET, E. 1861. Nouvelles recherches sur la coexistence de l'Homme et des grands mammifères fossiles. *Annales des Sciences Naturelles (Zoologie),* 4e série, XV, 177–253, 3 pl.

LARTET, E. and CHRISTY, H. 1864. Sur des figures d'animaux gravées ou sculptées et autres produits d'art et d'industrie rapportables aux temps primordiaux de la période humaine. *Revue Arch.* IX, 233–67, 2 pl.

LARTET, E. and CHRISTY, H. 1875. *Reliquiae Aquitanicae.* Jones: London (1865–1875).

LASHERAS, J. A., MUZQUIZ, M. and SAURA, P. 1995. Altamira en Japón, Proceso de una reproducción facsimilar. *Revista de Arqueología* 171, July, Año XVI, 12–27.

LAURENT, P. 1963. La tête humaine gravée sur bois de renne de la grotte du Placard (Charente). *L'Anthropologie* 67, 563–9.

LAURENT, P. 1971. Iconographie et copies successives: la gravure anthropomorphe du Placard. *Mém. Soc. Arch. Hist. Charente,* 215–28.

LAYTON, R. 1987. The use of ethnographic parallels in interpreting Upper Palaeolithic rock art, in *Comparative Anthropology* (L. Holy, ed.), 210–39. Blackwell: Oxford.

LAYTON, R. 1991. Figure, motif and symbol in the hunter-gatherer rock art of Europe and Australia, in *Rock Art and Prehistory* (P. Bahn and A. Rosenfeld, eds), 23–38. Oxbow Monograph 10: Oxford.

LEAKEY, M. 1983. *Africa's Vanishing Art.* Doubleday: New York.

LEASON, P. A. 1939. A new view of the Western European group of Quaternary cave art. *Proc. Prehist. Society* 5, 51–60.

LEASON, P. A. 1956. Obvious facts of Quaternary cave art. *Medical and Biological Illustration* 6 (4), 209–14.

LEGUAY, L. 1877. Les procédés employés pour la gravure et la sculpture des os avec le silex. *Bull. Soc. Anth.* Paris 2e série, 12, 280–96.

LEGUAY, L. 1882. Sur la gravure des os au moyen du silex. *C.r. Assoc. française pour l'Avancement des Sciences* 11, La Rochelle, 677–9.

LEJEUNE, M. 1981. *L'Utilisation des Accidents Naturels dans le tracé des figurations pariétales du Paléolithique Supérieur franco-cantabrique.* Mémoire de fin d'études,

Univ. de Liège, 274 pp.

LEJEUNE, M. 1987. *L'Art Mobilier Paléolithique et Mésolithique en Belgique.* Artefacts, Editions du Centre d'Etudes et de Documentation Archéologiques: Treignes-Viroinval, Belgium.

LE MORT, F. 1985. Un exemple de modification intentionnelle: la dent humaine perforée de Saint-Germain-la-Rivière (Pal. Sup.). *Bull. Soc. Préhist. fr.* 82, 190–92.

LEMOZI, A. 1929. *La Grotte-Temple du Pech-Merle. Un nouveau sanctuaire préhistorique.* Picard: Paris.

LEONARDI, P. 1983. Incisioni musteriane del Riparo Tagliente in Valpantena nei Monte Lessini presso Verona (Italia), in *Homenaje al Prof. M. Almagro Basch,* vol. 1, 149–54. Min. de Cultura: Madrid.

LEONARDI, P. 1988. Art paléolithique mobilier et pariétal en Italie. *L'Anthropologie* 92, 139–202.

LEONARDI, P. 1989. *Sacralità Arte e Grafia Paleolitiche – Splendori e Problemi.* Museo Civico di Storia Naturale di Trieste.

LEROI-GOURHAN, And. 1958. La fonction des signes dans les sanctuaires paléolithiques. *Bull. Soc. Préhist. fr.* 55, 307–21.

LEROI-GOURHAN, And. 1958a. Le symbolisme des grands signes dans l'art pariétal paléolithique. *Ibid.* 55, 384–98.

LEROI-GOURHAN, And. 1958b. Répartition et groupement des animaux dans l'art pariétal paléolithique. *Ibid.* 55, 515–28.

LEROI-GOURHAN, And. 1962. Chronologie de l'art paléolithique, in *Atti del VI Congresso UISPP,* Rome, vol. 3, 341–55.

LEROI-GOURHAN, And. 1966. Réflexions de méthode sur l'art paléolithique. *Bull. Soc. Préhist. fr.* 63, 35–49.

LEROI-GOURHAN, And. 1967. Les mains de Gargas. Essai pour une étude d'ensemble. *Ibid.* 64, 107–22.

LEROI-GOURHAN, And. 1968. Les signes pariétaux du Paléolithique supérieur franco-cantabrique, in *Simposio de Arte Rupestre,* Barcelona 1966, 67–77.

LEROI-GOURHAN, And. 1971. *Préhistoire de l'Art Occidental.* Mazenod: Paris. 2nd edition. (3rd edition 1995, Citadelles - Mazenod: Paris).

LEROI-GOURHAN, And. 1972. Considérations sur l'organisation spatiale des figures animales dans l'art pariétal paléolithique, in *Santander Symposium* 281–308.

LEROI-GOURHAN, And. 1978. Le cheval sur galet de la galerie Breuil au Mas d'Azil (Ariège). *Gallia Préhistoire* 21, 439–45.

LEROI-GOURHAN, And. 1980. Les signes pariétaux comme 'marqueurs' ethniques, in *Altamira Symposium* 289–94. Min. de Cultura: Madrid.

LEROI-GOURHAN, And. 1981. Le cas de Lascaux. *Monuments Historiques* 118, 24–32.

LEROI-GOURHAN, And. 1982. *The Dawn of European Art.* Cambridge Univ. Press.

LEROI-GOURHAN, And. 1983. Les entités imaginaires. Esquisse d'une recherche sur les monstres pariétaux paléolithiques, in *Homenaje al Prof. M. Almagro Basch,* vol. 1, 251–63. Min. de Cultura: Madrid.

LEROI-GOURHAN, And. 1984. Le réalisme du comportement dans l'art paléolithique d'Europe de l'Ouest, in Bandi *et al.* 1984, 75–90.

LEROI-GOURHAN, And. 1984a. *Simbolos, Artes y Creencias de la Prehistoria.* Istmo: Madrid.

LEROI-GOURHAN, And. 1984b. *Arte y Grafismo en la Europa Prehistorica.* Istmo: Madrid.

LEROI-GOURHAN, And. 1992. *L'Art Pariétal. Langage de la Préhistoire.* Jérôme Millon: Grenoble.

LEROI-GOURHAN, Arl. 1983. Du fond des grottes aux terrasses ensoleillées, in *Homenaje al Prof. M. Almagro Basch,* vol. 1, 239–49. Min. de Cultura: Madrid.

LEROI-GOURHAN, Arl. and ALLAIN, J. (eds) 1979. *Lascaux Inconnu*. XIIe Suppl. à Gallia Préhistoire. CNRS: Paris.

LEROY-PROST, C. 1984. L'art mobilier, in *Les Premiers Artistes, Dossier de l'Archéologie* 87, 45–51.

LEWIS, D. 1988. *The Rock Paintings of Arnhem Land, Australia. Social, Ecological and Material Culture Change in the Post-glacial Period*. British Archaeological Reports, Int. series No. 415: Oxford.

LEWIS-WILLIAMS, J. D. 1982. The economic and social context of Southern San rock art. *Current Anth.* 23, 429–49.

LEWIS-WILLIAMS, J. D. 1991. Wrestling with analogy: a methodological dilemma in Upper Palaeolithic art research. *Proc. Prehist. Soc.* 57, 149–62.

LEWIS-WILLIAMS, J. D. and DOWSON, T. A. 1988. The signs of all times: entoptic phenomena in Upper Palaeolithic art. *Current Anth.* 29, 201–45.

LHOTE, H. 1968. La plaquette dite de 'la femme au Renne', de Laugerie-Basse, et son interprétation zoologique, in *Simposio de Arte Rupestre*, Barcelona 1966, 79–97.

LHOTE, H. 1968a. A propos de l'identité de la femme et du bison selon les théories récentes de l'art pariétal préhistorique, in *ibid.*, 99–108.

LHOTE, H. 1972. Observations sur la technique et la lecture des gravures et peintures quaternaires du Sud-ouest de la France, in *Santander Symposium*, 321–30.

LIGER, J-C. 1995. Concrétionnement et Archéologie aux Grottes d'Arcy-sur-Cure (Yonne). *Bull. Soc. Préhist. française* 92, 445–9.

LINARES MALAGA, E. 1988. Arte mobiliar con tradición rupestre en el Sur del Perú. *Rock Art Research* 5 (1), 54–66.

LINDNER, K. 1950. *La Chasse Préhistorique*. Payot: Paris.

LION VALDERRABANO, R. 1971. *El Caballo en el Arte Cantábro-Aquitano*. Public. del Patronato de las cuevas prehist. de la prov. de Santander VIII: Santander.

LIPS, J. E. 1949. *The Origin of Things*. Harrap: London.

LISTER, A. and BAHN, P. G. 1994. *Mammoths*. Macmillan: New York / Boxtree: London.

LIUBIN, V. P. 1991. The images of mammoths in Palaeolithic art. *Sovietskaya Arkheologya* 20–42 (in Russian).

LOMMEL, A. 1967. *The World of the Early Hunters. Medicine-men, Shamans and Artists*. Evelyn, Adams and Mackay: London.

LORBLANCHET, M. 1974. *L'Art Préhistorique en Quercy. La Grotte des Escabasses (Thémines – Lot)*. PGP: Morlaas.

LORBLANCHET, M. 1977. From naturalism to abstraction in European prehistoric rock art, in *Form in Indigenous Art* (P. J. Ucko, ed.), 44–56. Duckworth: London.

LORBLANCHET, M. 1980. Peindre sur les parois des grottes, in *Revivre la Préhistoire, Dossier de l'Arch.* 46, 33–9.

LORBLANCHET, M. 1980a. Les gravures de l'ouest australien. Leur rénovation au cours des âges. *Bull. Soc. Préhist. fr.* 77, 463–77.

LORBLANCHET, M. 1981. Les dessins noirs du Pech-Merle, in *XXIe Congrès Préhist. de France*, Montauban/Cahors 1979, vol. 1, 178–207.

LORBLANCHET, M. 1984. Les relevés d'art préhistorique, in *L'Art des Cavernes. Atlas des grottes ornées paléolithiques françaises*, 41–51. Min. de la Culture: Paris.

LORBLANCHET, M. 1988. De l'art pariétal des chasseurs de rennes à l'art rupestre des chasseurs de kangourous. *L'Anthropologie* 92, 271–316.

LORBLANCHET, M. 1989. From man to animal and sign in Palaeolithic art, in *Animals into Art* (H. Morphy, ed.), 109–43. Unwin Hyman: London.

LORBLANCHET, M. 1990. Etudes des pigments des grottes ornées paléolithiques du Quercy. *Bull. Soc. des Etudes Lit., Sci. Artistiques du Lot* 111, 93–143.

LORBLANCHET, M. 1991. Spitting images. Replicating the spotted horses of Pech Merle. *Archaeology*, 44 (6), Nov/Dec., 24–31.

LORBLANCHET, M. 1992. Finger markings in Pech Merle and their place in prehistoric art, in *Rock Art in the Old World* (M. Lorblanchet, ed.), 451–90. Indira Gandhi Centre for the Arts: New Delhi.

LORBLANCHET, M. 1993. Etude et conservation de l'art pariétal du Quercy, in *La Protección y Conservación del Arte Rupestre Paleolítico* (J. Fortea, ed.), 115–30. Servicio Publicaciones, Principado de Asturias: Oviedo.

LORBLANCHET, M. 1993a. From styles to dates, in *Rock Art Studies: The Post-Stylistic Era* (M. Lorblanchet and P. G. Bahn, eds), 61–72. Oxbow Monograph 35: Oxford.

LORBLANCHET, M. 1994. Le mythe des Femmes-Bisons, in *Psychanalyse et Préhistoire* (A. Fine, R. Perron and F. Sacco, eds), 41–54. Presses Universitaires de France: Paris.

LORBLANCHET, M. 1995. *Les Grottes Ornées de la Préhistoire. Nouveaux Regards*. Editions Errance: Paris.

LORBLANCHET, M. 1995a. Les gravures rupestres de la vallée de Côa (Portugal). *L'Archéologue, Archéologie Nouvelle* No. 17, Dec. 1995/Jan. 1996, 38–41.

LORBLANCHET, M. and WELTE, A-C. 1987. Les figurations féminines stylisées du Magdalénien Supérieur du Quercy. *Bull. Soc. Etudes du Lot* 108, f. 3, 3–58.

LORBLANCHET, M. *et al.* 1973. La grotte de Sainte Eulalie à Espagnac (Lot). *Gallia Préhistoire* 16, 4–62 and 233–325.

LORBLANCHET, M. *et al.* 1990. Palaeolithic pigments in the Quercy, France. *Rock Art Research* 7, 4–20.

LORBLANCHET, M. *et al.* 1995. Direct date for one of the Pech-Merle spotted horses. *Int. Newsletter on Rock Art* 12, 2–3.

LORENZO, J. L. 1968. Sur les pièces d'art mobilier de la préhistoire mexicaine, in *La Préhistoire: Problèmes et Tendances*, 283–9, 5 pl. CNRS.: Paris.

LOY, T. H. 1994. Direct dating of rock art at Laurie Creek (NT), Australia: a reply to Nelson. *Antiquity* 68, 147–8.

LOY, T. H., *et al.* 1990. Accelerator radiocarbon dating of human blood protein pigments from Late Pleistocene art sites in Australia. *Antiquity* 64, 110–16.

de LUMLEY, H. 1968. Le bison gravé de Ségriès, Moustiers-Ste-Marie, Bassin du Verdon (Basses-Alpes), in *Simposio de Arte Rupestre*, Barcelona 1966, 109–21.

de LUMLEY, H. 1968a. Proportions et constructions dans l'art paléolithique: le bison, in *ibid.*, 123–45.

MACINTOSH, N. W. G. 1977. Beswick Creek cave two decades later, in *Form in Indigenous Art* (P. J. Ucko, ed.), 191–7. Duckworth: London.

MADARIAGA DE LA CAMPA, B. 1972. *Hermilio Alcalde del Rio. Una escuela de prehistoria en Santander*. Patronato de las Cuevas Prehistóricas de Santander.

MADARIAGA DE LA CAMPA, B. 1981. Historia del descubrimento y valoración del arte rupestre español, in *Altamira Symposium*, 299–310. Min. de Cultura: Madrid.

MALLERY, G. 1893. *Picture-Writing of the American Indians*. Government Printing Office: Washington DC. (Reprint in 2 vols, 1972. Dover Publications Inc.: New York).

MANIA, D. and MANIA, U. 1988. Deliberate engravings on bone artefacts of *Homo erectus*. *Rock Art Research* 5 (2), 91–106.

MARINGER, J. and BANDI, H-G. 1953. *Art in the Ice Age*. Allen & Unwin: London.

MARSHACK, A. 1969. Polesini: a reexamination of the engraved Upper Palaeolithic mobiliary materials of Italy by a new methodology. *Rivista di Scienze Preistoriche* 24, 219–81.

MARSHACK, A. 1970. Le bâton de commandement de Montgaudier (Charente): Réexamen au microscope et interprétation nouvelle. *L'Anthropologie* 74, 321–52.

MARSHACK, A. 1970a. *Notation dans les Gravures du Paléolithique Supérieur*. Mémoire 8, Inst. Préhist. Univ. Bordeaux. Delmas: Bordeaux.

MARSHACK, A. 1972. *The Roots of Civilization*. Weidenfeld & Nicolson: London. (2nd edition, 1991, Moyer Bell: New York).

MARSHACK, A. 1972a. Cognitive aspects of Upper Paleolithic engraving. *Current Anth.* 13, 445–77; also 15, 1974, 327–32 and 16, 1975, 297–8.

MARSHACK, A. 1972b. Aq Kupruq: art and symbol, in *Prehistoric Research in Afghanistan* (1959–66) (L. Dupree *et al.*, eds), Transactions of the American Philosophical Society 62 (4), 66–72.

MARSHACK, A. 1975. Exploring the mind of Ice Age man. *National Geographic* 147 (1), 64–89.

MARSHACK, A. 1976. Implications of the Paleolithic symbolic evidence for the origin of language. *American Scientist* 64, 136–45.

MARSHACK, A. 1976a. Some implications of the Paleolithic symbolic evidence for the origin of language. *Current Anth.* 17, 274–81.

MARSHACK, A. 1976b. The message in the markings. *Horizon* 18, 64–73.

MARSHACK, A. 1977. The meander as a system: the analysis and recognition of iconographic units in Upper Paleolithic compositions, in *Form in Indigenous Art* (P. J. Ucko, ed.), 286–317. Duckworth: London.

MARSHACK, A. 1979. Upper Paleolithic symbol systems of the Russian Plain: cognitive and comparative analysis. *Current Anth.* 20, 271–311, 604–8.

MARSHACK, A. 1981. On Paleolithic ochre and the early uses of color and symbol. *Current Anth.* 22, 188–91.

MARSHACK, A. 1984. Concepts théoriques conduisant à de nouvelles méthodes analytiques, de nouveaux procédés de recherche et catégories de données. *L'Anthropologie* 88, 575–86.

MARSHACK, A. 1985. Theoretical concepts that lead to new analytic methods, modes of inquiry and classes of data. *Rock Art Research* 2, 95–111; 3 (1986), 62–82 and 175–7.

MARSHACK, A. 1988. La pensée symbolique et l'art, in *L'Homme de Néandertal, Dossiers de l'Archéologie* 124, 80–90.

MARSHACK, A. 1988a. The Neanderthals and the human capacity for symbolic thought: cognitive and problem-solving aspects of Mousterian symbol, in *L'Homme de Néandertal 5, La Pensée*, 57–91. ERAUL 32, Liège.

MARSHACK, A. 1988b. An Ice Age ancestor? *National Geographic* 174 (4), 478–81.

MARSHACK, A. 1990. Early hominid symbol and evolution of the human capacity, in *The Emergence of Modern Humans* (P. Mellars, ed.), 457–98. Edinburgh Univ. Press.

MARSHACK, A. 1991. A reply to Davidson on Mania and Mania. *Rock Art Research* 8, 47–58.

MARSHACK, A. 1991a. The Taï plaque and calendrical notation in the Upper Paleolithic. *Cambridge Archaeological Journal* 1, 25–61.

MARSHACK, A. 1995. Methodology and the search for notation among engraved pebbles of the European Late Palaeolithic. *Antiquity* 69, 1049–51.

MARSHACK, A. 1996. A Middle Palaeolithic symbolic composition from the Golan Heights: the earliest known depictive image. *Current Anthropology* 37, 357–65.

MARSHACK, A. 1997. The Berekhat Ram figurine: a late Acheulian carving from the Middle East. *Antiquity* 71, 327–37.

MARTI, J. 1977. *Informe sobre los estudios realizados en las cuevas de Altamira.* Instituto de Catalisis CSIC: Madrid.

MARTIN, Y. 1973. *L'Art Paléolithique de Gouy.* Martin: Gouy.

MARTIN, Y. 1974. Technique de moulage de gravures rupestres. *Bull. Soc. Préhist. fr.* 71, 146–8.

MARTIN, Y. 1989. Nouvelles découvertes de gravures à Gouy. *L'Anthropologie* 93, 513–46.

MARTIN SANTAMARIA, E. and MOURE ROMANILLO, J. A. 1981. El grabado de estilo paleolítico de Domingo García (Segovia). *Trabajos de Prehistoria* 38, 97–105, 2 pl.

MARTINEZ GARCIA, J. 1986/7. Un grabado paleolítico al aire libre en Piedras Blancas (Escúllar, Almería). *Ars Praehistorica* V/VI, 49–58.

MARTINEZ GARCIA, J. 1992. Arte paleolítico en Almería. Los primeros documentos. *Revista de Arqueología* 130, 24–33.

MAZAK, V. 1970. On a supposed prehistoric representation of the Pleistocene scimitar cat, *Homotherium* FABRINI 1890 (*Mammalia. Machairodontidae*). *Zeitschrift f. Säugetierkunde* 35, 359–62.

McDERMOTT, L. 1996. Self-representation in Upper Paleolithic female figurines. *Current Anth.* 37, 227–75.

McGOWAN, A., SHREEVE, B., BROLSMA, H. and HUGHES, C. 1993. Photogrammetric recording of Pleistocene cave paintings in Southwest Tasmania, in *Sahul in Review* (M. A. Smith, M. Spriggs and B. Fankhauser, eds), 225–32. Occasional papers in prehistory 24, ANU: Canberra.

MELLARS, P. A. 1985. The ecological basis of social complexity in the Upper Palaeolithic of Southwestern France, in *Prehistoric Hunter-Gatherers: the Emergence of Social and Cultural Complexity* (T. D. Price and J. A. Brown, eds), 271–97. Academic Press: New York/London.

MELTZER, D. J. and STURTEVANT, W. C. 1983. The Holly Oak shell game: an historic archaeological fraud, in *Lulu Linear Punctuated: Essays in Honor of G.I. Quimby*, 325–52. Anth. Papers 72, Museum of Anth., Univ. of Michigan: Ann Arbor.

MENENDEZ FERNANDEZ, M. 1994. Arte rupestre y arte mueble paleolítico: relaciones. *Complutum* 5, 343–55.

MENU, M. *et al.* 1993. Façons de peindre au Magdalénien. *Bull. Soc. Préhist. française* 90, 426–32.

MESSMACHER, M. 1981. El arte paleolítico en México, in *Arte Paleolítico*, Comisión XI, Xth Congress, UISPP, Mexico City, 82–110.

MESZAROS, G. and VERTES, L. 1955. A paint mine from the early Upper Palaeolithic age near Lovas (Hungary, County Veszprém). *Acta Arch. Hungarica* 5, 1–34.

MEYLAN, C. 1986. Leroi-Gourhan au bestiaire de la préhistoire. *Bull. Soc. Préhist. Ariège-Pyrénées* 41, 31–62.

MILLAN CASCALLO, M. 1982. El rhinoceros en el arte pleistocénico. *Helike* 1, 31–68.

MINELLONO, F. 1985/6. L'incisione di testa maschile dal Riparo di Vado all'Arancio (Grosseto). *Rivista di Scienze Preistoriche* 40 (1–2), 115–35.

MINVIELLE, P. 1972. *Sur les Chemins de la Préhistoire.* Denoël: Paris.

MITHEN, S. J. 1988. Looking and learning: Upper Palaeolithic art and information gathering. *World Arch.* 19, 297–327.

MITHEN, S. J. 1988a. To hunt or to paint: animals and art in the Upper Palaeolithic. *Man* 23, 671–95.

MOLARD, Cdt. 1908. Les grottes de Sabart (Ariège):

Niaux et les dessins préhistoriques. *Spelunca* VII, No. 53, 177–91.

MONS, L. 1980/1. Les baguettes demi-rondes du Paléolithique supérieur occidental: analyse et réflexions. *Antiquités Nationales* 12/13, 7–19.

MONS, L. 1986. Les statuettes animalières en grès de la grotte d'Isturitz (Pyr-Atl). Observations et hypothèses de fragmentation volontaire. *L'Anthropologie* 90, 701–11.

MONS, L. 1986/7. Les figurations de bisons dans l'art mobilier de la grotte d'Isturitz (Pyrénées-Atlantiques). *Antiquités Nationales* 18/19, 91–9.

MOORE, D. R. 1977. The hand stencil as symbol, in *Form in Indigenous Art* (P. J. Ucko, ed.), 318–24. Duckworth: London.

MORI, F. 1974. The earliest Saharan rock-engravings. *Antiquity* 48, 87–92.

MORSE, K. 1993. Shell beads from Mandu Mandu Creek rockshelter, Cape Range peninsula, Western Australia, dated before 30,000 bp. *Antiquity* 67, 877–83.

de MORTILLET, G. 1898. Statuette fausse des Baoussé-Roussé. *Bull. Soc. Anth. Paris* IX, 4e série, 146–53 (with comment by E. Rivière).

MOURE ROMANILLO, J. A. 1985. Nouveautés dans l'art mobilier figuratif du Paléolithique cantabrique. *Bull. Soc. Préhist. Ariège* 40, 99–129.

MOURE ROMANILLO, J. A. 1986. New data on the chronology and context of Cantabrian Paleolithic cave art. *Current Anth.* 27, 65.

MOURE ROMANILLO, J. A. 1988. Composition et variabilité dans l'art pariétal paléolithique cantabrique. *L'Anthropologie* 92, 73–86.

MOURE ROMANILLO, J. A. 1993. Investigación y conservación, in *La Protección y Conservación del Arte Rupestre Paleolítico* (J. Fortea, ed.), 175–80. Servicio Publicaciones, Principado de Asturias: Oviedo.

MOURE ROMANILLO, J. A. *et al.* 1984/5. Las pinturas paleolíticas de la cueva de la Fuente del Salín (Muñorrodero, Cantabria). *Ars Praehistorica* III/IV, 13–23.

MOURE ROMANILLO, J. A. *et al.* 1996. Dataciones absolutas de pigmentos en cuevas cantábricas: Altamira, El Castillo, Chimeneas y Las Monedas, in *'El Hombre Fósil' 80 Años Después. Homenaje a Hugo Obermaier* (A. Moure, ed.), 295–324. Servicio de Publicaciones, Univ. de Cantabria: Santander.

MOWALJARLAI, D. and WATCHMAN, A. 1989. An Aboriginal view of rock art management. *Rock Art Research* 6, 151–3.

MÜLLER-BECK, H. and ALBRECHT, G. (eds) 1987. *Die Anfänge der Kunst vor 30,000 Jahren.* Theiss: Stuttgart.

MURRAY, P. and CHALOUPKA, G. 1983/4. The Dreamtime animals: extinct megafauna in Arnhem Land rock art. *Archaeology in Oceania* 18/19, 105–16.

MUZZOLINI, A. 1986. *L'Art Rupestre Préhistorique des Massifs Centraux Sahariens.* British Arch. Reports, Int. Series 318, Oxford.

NABER, F. B. *et al.* 1976. L'Art Pariétal Paléolithique en Europe Romane. 3 vols. *Bonner Hefte für Vorgeschichte* Nos. 14–16.

NARR, K. J. 1960. Weibliche symbol-plastik der älteren Steinzeit. *Antaios* II (2), 132–57.

NELSON, D. E. 1993. Second thoughts on a rock-art date. *Antiquity* 67, 893–5.

NEUGEBAUER-MARESCH, C. 1995. La statuette du Galgenberg (entre Stratzing et Krems- Rehberg) et les figurines féminines d'Autriche, in *La Dame de Brassempouy* (H. Delporte, ed.), 187–94. ERAUL 74: Liège.

NEUGEBAUER-MARESCH, C. 1996. Die Arbeiten

zum Paläolithikum im Raum Krems (1990–1995), in *Le Paléolithique Supérieur Européen, Bilan Quinquennal 1991–1996* (M. Otte, ed.), 97–103. Etudes et Recherches Archéologiques de l'Université de Liège No. 76.

NIEDHORN, U. 1990. *The Lady from Brassempouy. A Fake – a Hoax?* Isernhägener Studien zur frühen Skulptur 3, Haag & Herchen: Frankfurt.

NOBBS, M. and DORN, R. I. 1988. Age determination for rock varnish formation within petroglyphs. *Rock Art Research* 5 (2), 108–46.

NOBBS, M. and DORN, R. 1993. New surface exposure ages for petroglyphs from the Olary province, South Australia. *Archaeology in Oceania* 28, 18–39.

NOIRET, P. 1991. Le décor des bâtons percés. *Bull. Soc. Préhist. Ariège-Pyrénées* 46, 145–59.

NOUGIER, L-R. 1972. Nouvelles approches de l'art préhistorique animalier, in *Santander Symposium*, 263–78.

NOUGIER, L-R. 1975. L'importance du choix dans l'explication religieuse de l'art quaternaire, in *Valcamonica Symposium 1972: Les Religions de la Préhistoire*, 57–64.

NOUGIER, L-R. 1975a. Art préhistorique et topographie. *Bull. Soc. Préhist. Ariège* 30, 115–17.

NOUGIER, L-R. 1990. *Les Grottes Préhistoriques Ornées de France, d'Espagne et d'Italie.* Editions Balland: Paris.

NOUGIER, L-R. and ROBERT, R. 1957. *Rouffignac, ou la Guerre des Mammouths.* La Table Ronde: Paris.

NOUGIER, L-R. and ROBERT, R. 1957a. Le rhinocéros dans l'art franco-cantabrique occidental. *Bull. Soc. Préhist. Ariège* 12, 15–52, 2 pl.

NOUGIER, L-R. and ROBERT, R. 1958. Le 'lissoir aux Saïgas' de la grotte de la Vache à Alliat et l'antilope Saïga dans l'art franco-cantabrique. *Ibid.* 13, 13–28, 2 pl.

NOUGIER, L-R. and ROBERT, R. 1960. Les 'loups affrontés' de la grotte de la Vache (Ariège) et les canidés dans l'art franco-cantabrique, in *Festschrift für Lothar Zotz*, 399–420, 1 pl. Röhrscheid Verlag: Bonn.

NOUGIER, L-R. and ROBERT, R. 1965. Les félins dans l'art quaternaire. *Bull. Soc. Préhist. Ariège* 20, 17–84; and 21, 1966, 35–46, 2 pl.

NOUGIER, L-R. and ROBERT, R. 1968. Les processions, les associations et les juxtapositions d'anthropomorphes et d'animaux dans l'art quaternaire. *Ibid.* 23, 33–98.

NOUGIER, L-R. and ROBERT, R. 1974. De l'accouplement dans l'art préhistorique. *Ibid.* 29, 15–63.

NOVEL, P. 1986. Les animaux rares dans l'art pariétal aquitain. *Bull. Soc. Préhist. Ariège-Pyrénées* 41, 63–93 (and in 42, 1987, 83–118).

NOVGORODOVA, E. 1980. *Alte Kunst der Mongolia.* E. A. Seemann Verlag: Leipzig.

OAKLEY, K. P. 1981. Emergence of higher thought 3.0–0.2 Ma B.P., in *'The Emergence of Man'*, *Phil. Trans. Royal Society London* B 292, 205–11.

OBERMAIER, H. 1918. Trampas cuaternarias para espiritus malignos. *Bol. Real. Soc. esp. de Hist. Nat.* 18, 162–9.

O'CONNOR, S. 1995. Carpenter's Gap Rockshelter 1: 40,000 years of Aboriginal occupation in the Napier Ranges, Kimberley, WA. *Australian Archaeology* 40, 58–9.

OKLADNIKOV, A. P. 1972. *Tsentral'noaziatskiy ochag pervobîtnogo iskusstva* (Centre of primitive art in Central Asia). Novosibirsk.

OKLADNIKOV, A. P. and MYEDVYEDYEV, V.Y. 1983. Isslyedovaniye mnogosloinogo posyelyeniya Gasya na niznyem Amurye (Investigation of the stratified settlement of Gasja on the lower Amur). *Izwestiya Sibirskogo otdyelyeniya Akad. Nauk USSR, Novosibirsk*, 1 (1), 93–7.

OLIVA, M. 1996. Le Paléolithique Supérieur de la République Tchèque: (1991–95), in *Le Paléolithique Supérieur Européen, Bilan Quinquennal 1991–1996* (M. Otte, ed.), 115–29. Etudes et Recherches Archéologiques de l'Université de Liège No. 76.

OLIVIE MARTINEZ-PEÑALVER, A. 1984/5. Representaciones de cérvidos en el arte parietal paleolítico de Cantabria. *Ars Praehistorica* 3/4, 95–103.

OLLIER DE MARICHARD, P. 1973. Un pionnier de la préhistoire ardéchoise: Jules Ollier de Marichard (1824–1901). *Etudes Préhistoriques* 4, mars, 25–9.

OMNES, J. 1982. *La Grotte Ornée de Labastide (Hautes-Pyrénées)*. Omnès: Lourdes.

OTTE, M. 1974. Observations sur le débitage et le façonnage de l'ivoire dans l'Aurignacien en Belgique, in *1er Colloque Int. sur l'Industrie de l'Os dans la Préhistoire* (H. Camps- Fabrer, ed.), 93–6. Univ. de Provence.

OTTE, M. *et al.* 1995. The Epi-Palaeolithic of Öküzini Cave (SW Anatolia) and its mobiliary art. *Antiquity* 69, 931–44.

PAILLET, P. 1993. *Les traitements magdaléniens de l'image du bison dans l'art pariétal et mobilier du Périgord. Nouvelle approche d'un theme du bestiaire paléolithique*. Thèse de Doctorat du Muséum d'Hist. Nat., Paris. 2 vols.

PALES, L. 1964. Préface, in Gaussen 1964, 11–16.

PALES, L. 1969. *Les Gravures de La Marche: I, Félins et Ours*. Mémoire 7, Publications de l'Inst. de Préhistoire de Bordeaux. Delmas: Bordeaux.

PALES, L. 1970. Le 'Coco des Roseaux' ou la fin d'une erreur. *Bull. Soc. Préhist. française* 67, 85–8.

PALES, L. 1972. Les ci-devant Vénus stéatopyges aurignaciennes, in *Santander Symposium*, 217–61.

PALES, L. 1976/7. Les ovicapridés préhistoriques franco-ibériques au naturel et figurés. *Sautuola* II, 67–105, 1 pl. Publicaciones del patronato de las cuevas prehist. de la Prov. de Santander XV.

PALES, L. and de ST PEREUSE, M. T. 1965. En compagnie de l'abbé Breuil devant les bisons gravés magdaléniens de la grotte de la Marche, in *Miscelánea en Homenaje al Abate H. Breuil*, vol. II, 217–50. Barcelona.

PALES, L. and de St PEREUSE, M.T. 1966. Un cheval-prétexte: retour au chevêtre. *Objets et Mondes* 6, 187–206.

PALES, L. and de ST PEREUSE, M. T. 1976. *Les Gravures de La Marche: II, Les Humains*. Ophrys: Paris.

PALES, L. and de ST PEREUSE, M. T. 1979. L'abri Durif à Enval (Vic-le-Comte, Puy-de-Dôme). Gravures et sculptures sur pierre. *Gallia Préhistoire* 22, 113–42.

PALES, L. and de ST PEREUSE, M. T. 1981. *Les Gravures de La Marche: III, Equidés et Bovidés*. Ophrys: Paris.

PALES, L. and de ST PEREUSE, M. T. 1989. *Les Gravures de La Marche: IV, Cervidés, Eléphants et Divers*. Ophrys: Paris.

PALES, L. *et al.* 1976. Les Empreintes de Pieds Humains dans les Cavernes. *Archives de l'Inst. de Paléontologie Humaine* 36, Paris.

PARKINGTON, J. 1969. Symbolism in cave art. *South African Arch. Bull.* 24, 3–13.

PASSEMARD, E. and BREUIL, H. 1928. La plus grande gravure magdalénienne à contours découpés. *Rev. Arch.* 27, 4

PASSEMARD, L. 1938. *Les Statuettes Féminines Paléolithiques dites Vénus stéatopyges*. Teissier: Nîmes.

PEPE, C. *et al.* 1991. Le liant des peintures préhistoriques ariégeoises. *C.r. Acad. Sc. Paris* 312, série II, 929–34.

PEQUART, M. and S-J. 1939/40. Fouilles archéologiques et nouvelles découvertes au Mas d'Azil. *L'Anthropologie* 49, 450–53.

PEQUART, M. and S-J. 1960/3. Grotte du Mas d'Azil

(Ariège) – Une nouvelle galerie magdalénienne. *Annales de Paléontologie*, collected papers, 351 pp.

PERE, P. 1988. Les contours découpés dans l'art mobilier du Paléolithique occidental. *Travaux de l'Inst. d'Art Préhist. Toulouse* 30, 155–206.

PERICOT, L. 1942. *La Cueva del Parpalló (Gandia)*. Consejo Sup. de Investigaciones Científicas, Inst. Diego Velázquez: Madrid.

PERICOT, L. 1962. Un curioso paralelo. *Munibe* 14, 456–8.

PERPERE, M. 1984. Les instruments de l'artiste, in *Les Premiers Artistes, Dossier de l'Arch.* 87, 41–4.

PETRIN, V. 1992. *The Palaeolithic Sanctuary of Ignatiev Cave in the Southern Urals*. NAUKA: Novosibirsk. (In Russian.)

PETRIN, V. T. and SHIROKOV, V. N. 1991. Die Ignatieva-Höhle (Ural). *Jahrbuch Römisch-Germanischen Zentralmuseums Mainz* 38, 17–31.

PFEIFFER, J. E. 1982. *The Creative Explosion. An inquiry into the origins of art and religion*. Harper & Row: New York.

PHILIBERT, S. 1994. L'ocre et le traitement des peaux: révision d'une conception traditionnelle par l'analyse fonctionnelle des grattoirs ocrés de la Balma Margineda (Andorre). *L'Anthropologie* 98, 447–53.

PHILLIPS. F. M., FLINSCH, M., ELMORE, D. and SHARMA, P. 1997. Maximum ages of the Côa valley (Portugal) engravings measured with Chlorine-36. *Antiquity* 71, 100–104.

PIETTE, E. 1887. Equidés de la période quaternaire d'après les gravures de ce temps. *Matériaux* 3e série, 4, 359–66.

PIETTE, E. 1889. L'époque de transition intermédiaire entre l'âge du Renne et l'époque de la pierre polie. *Comptes rendus du 10e Congr. Int. d'Anth. et d'Arch. Préhist.*, Paris (publ. 1891), 203–13.

PIETTE, E. 1894. Notes pour servir à l'histoire de l'art primitif. *L'Anthropologie* 5, 129–46.

PIETTE, E. 1895. La station de Brassempouy et les statuettes humaines de la période glyptique. *L'Anthropologie* 6, 129–51, 7 pl.

PIETTE, E. 1902. Gravure du Mas d'Azil et statuettes de Menton. *Bull. Soc. Anth. Paris*, 5e série, 3, 771–9.

PIETTE, E. 1905. Les écritures de l'Age glyptique. *L'Anthropologie* 16, 1–11.

PIETTE, E. 1906. Le chevêtre et la semi-domestication des animaux aux temps pléistocènes. *L'Anthropologie* 17, 27–53.

PIETTE, E. 1907. *L'Art pendant l'Age du Renne*. Masson: Paris.

PITTARD, E. 1929. La première découverte d'art préhistorique (gravure et sculpture) a été faite dans la station de Veyrier (Hte-Savoie) par le Genevois François Mayor. *Revue Anth.* 39e année, No. 7–9, 296–304.

PLENIER, A. 1971. *L'Art de la Grotte de Marsoulas*. Mémoire 1, Inst. Art Préhist. Toulouse.

POND, A. W. 1925. The oldest jewelry in the world. *Art and Archaeology* 19, 131–4, 1 pl.

POPLIN, F. 1983. Incisives de renne sciées du Magdalénien d'Europe occidentale, in *La Faune et l'Homme Préhistoriques*, Mém. 16, Soc. Préhist. française, 55–67.

POWERS, R. and STRINGER, C. B. 1975. Palaeolithic cave art fauna. *Studies in Speleology* 2 (7/8), 265–98.

PRADEL, L. 1975. Les mains incomplètes de Gargas, Tibiran et Maltravieso. *Quartär* 26, 159–66.

PRASLOV, N. D. 1985. L'art du Paléolithique Supérieur à l'est de l'Europe. *L'Anthropologie* 89, 181–92.

PRASLOV, N. D. 1993. Eine neue Frauenstatuette aus Kalkstein von Kostenki I (Don, Russland). *Archäologisches Korrespondenzblatt* 23, 165–73.

PRAT, F. 1986. Le cheval dans l'art paléolithique et les données de la paléontologie. *Arqueologia* (Porto) 14, Dec., 27–33.

PROUS, A. 1994. L'art rupestre du Brésil. *Bull. Soc. Préhist. Ariège-Pyrénées* 49, 77–144.

PURDY, B. A. 1996. *Indian Art of Ancient Florida*. University Press of Florida: Gainesville.

PURDY, B. A. 1996a. *How to do Archaeology the right way*. University Press of Florida; Gainesville.

RAPHAEL, M. 1986. *L'Art Pariétal Paléolithique*. Kronos: Paris.

REGNAULT, F. 1906. Empreintes de mains humaines dans la grotte de Gargas (Hautes-Pyrénées). *Bull. Soc. Anth.* 5e série, 7, 331–2.

REINACH, S. 1899. Gabriel de Mortillet. *Revue Historique* 24e année, 69, 67–95.

REINACH, S. 1903. L'art et la magie. A propos des peintures et des gravures de l'Age du Renne. *L'Anthropologie* 14, 257–66.

REZNIKOFF, I. 1987. Sur la discussion sonore des grottes à peintures du Paléolithique. *C. r. Acad. Sc. Paris* 304, série II, 153–6; and 305, 307–10.

REZNIKOFF, I. and DAUVOIS, M. 1988. La dimension sonore des grottes ornées. *Bull. Soc. Préhist. française* 85, 238–46.

RICE, P. C. 1981. Prehistoric venuses: symbols of motherhood or womanhood? *Journal of Anth. Research* 37, 402–14.

RICE, P. C. and PATERSON, A. L. 1985. Cave art and bones: exploring the interrelationships. *American Anthropologist* 87, 94–100.

RICE, P. C. and PATERSON, A. L. 1986. Validating the cave art – archeofaunal relationship in Cantabrian Spain. *Ibid.* 88, 658–67.

RICHARD, N. 1993. De l'art ludique à l'art magique. Interprétations de l'art pariétal au XIXe siècle. *Bull. Soc. Préhist. française* 90, 60–68.

RIDDELL, W. H. 1940. Dead or alive? *Antiquity* 14, 154–62.

RIPOLL LOPEZ, S. and MUNICIO GONZALEZ, L. 1992, Las representaciones de estilo paleolítico en el conjunto de Domingo García (Segovia). *Espacio, Tiempo y Forma*, Serie 1, Prehistoria y Arqueología, V, Madrid, 107–38.

RIPOLL LOPEZ, S. and MUNICIO GONZALEZ, L. J. 1994. A large open air grouping of Paleolithic rock art in the Spanish Meseta. *International Newsletter on Rock Art* 7: 2–5.

RIPOLL LOPEZ, S. *et al.* 1994. Un conjunto excepcional del arte paleolítico. El Cerro de San Isidro en Domingo García. Nuevos descubrimientos. *Revista de Arqueología* 157, 12–21.

RIPOLL LOPEZ, S. *et al.* 1997. Cueva de Maltravieso. *Revista de Arqueología* 193, 6–15

RIPOLL PERELLO, E. 1964. Problemas cronológicos del arte paleolítico, in *Prehistoric Art of the Western Mediterranean and the Sahara* (L. Pericot and E. Ripoll, eds), 83–100. Viking Fund Publications in Anth. 39, New York.

RIPOLL PERELLO, E. 1972. *La Cueva de las Monedas en Puente Viesgo (Santander)*. Monografías de Arte Rupestre, Arte Paleolítico No. 1, Barcelona.

RIPOLL PERELLO, E. 1984. Notes sur certaines représentations d'animaux dans l'art paléolithique de la péninsule ibérique, in Bandi *et al.* 1984, 263–82.

RIPOLL PERELLO, E. 1994. *El Abate Henri Breuil (1877–1961)*. Univ. Nacional de Educación a Distancia: Madrid.

RIVENQ, C. 1976. *La 'Scene de Chasse' de Ganties-Montespan*. Rivenq: Toulouse.

RIVENQ, C. 1984. Grotte de Ganties-Montespan, in *L'Art des Cavernes. Atlas des Grottes Ornées Paléolithiques françaises*, 438–45. Min. de la Culture: Paris.

ROBERT, R. *et al.* 1953. Sur l'existence possible d'une école d'art dans le Magdalénien pyrénéen. *Bull. Arch.* 187–93.

ROBINEAU, D. 1984. Sur les mammifères marins du bâton gravé préhistorique de Montgaudier. *L'Anthropologie* 88, 661–4.

RONEN, A. and BARTON, G. M. 1981. Rock engravings on western Mount Carmel, Israel. *Quartär* 31/32, 121–37.

ROOSEVELT, A. C. *et al.* 1996. Paleoindian cave dwellers in the Amazon: the peopling of the Americas. *Science* 272, 373–84.

ROPER, M. K. 1969. A survey of the evidence for intrahuman killing in the Pleistocene. *Current Anth.* 10, 427–59.

ROSENFELD, A. 1977. Profile figures: schematisation of the human figure in the Magdalenian culture of Europe, in *Form in Indigenous Art* (P. J. Ucko, ed.), 90–109. Duckworth: London.

ROSENFELD, A. 1984. The identification of animal representations in the art of the Laura region, North Queensland (Australia), in Bandi *et al.* 1984, 399–422.

ROSENFELD, A. 1993. A review of the evidence for the emergence of rock art in Australia, in *Sahul in Review* (M. A. Smith, M. Spriggs and B. Fankhauser, eds), 71–80. Occasional Papers in Prehistory 24, ANU: Canberra.

ROSENFELD, A., HORTON, D. and WINTER, J. 1981. *Early Man in North Queensland.* Terra Australis 6, Australian National University: Canberra.

ROTTLÄNDER, R. 1965. Zur Frage des Pigmentbinders der Franko-Kantabrischen Höhlenmalereien. *Fundamenta* Reihe A, Band 2, 340–44.

ROUSSEAU, M. 1967. *Les Grands Félins dans l'Art de notre Préhistoire.* Picard: Paris.

ROUSSEAU, M. 1973. Darwin et les chevaux peints paléolithiques d'Ekain. *Bull. Soc. Préhist. Ariège* 28, 49–55.

ROUSSEAU, M. 1974. Récentes déterminations et réfutations de 'Félin' dans l'art paléolithique. *Säugetierkundliche Mitteilungen* 2, 97–103.

ROUSSEAU, M. 1984. Torsions conventionnelles et flexions naturelles dans l'art animalier paléolithique et au delà, in Bandi *et al.* 1984, 243–9.

ROUSSEAU, M. 1984a. Les pelages dans l'iconographie paléolithique, in *ibid*, 161–97.

ROUSSEAU, M. 1996. Dans l'art paléolithique: "L'homme tué" de la grotte Cosquer et d'ailleurs, les hommes blessés. *Bull. Soc. Préhist. française* 93, 204–7.

ROUSSOT, A. 1965. Les découvertes d'art pariétal en Périgord, in *Centenaire de la Préhistoire en Périgord (1864–1964)*, 99–125. Fanlac: Périgueux.

ROUSSOT, A. 1970. Flûtes et sifflets paléolithiques en Gironde. *Rev. Historique de Bordeaux et du Dept. de la Gironde*, 5–12.

ROUSSOT, A. 1972/3. La découverte des gravures de Pair-non-Pair d'après les notes de François Daleau. *Cahiers du Vitrezais* 1, July, 5–7, and Oct. 15–17; 2, April, 22–24.

ROUSSOT, A. 1981. Observations sur le coloriage de sculptures paléolithiques. *Bull. Soc. Préhist. française* 78, 200.

ROUSSOT, A. 1984. Les premiers bas-reliefs de l'humanité, in *Les Premiers Artistes, Dossier de l'Arch.* 87, 73–6.

ROUSSOT, A. 1984a. Peintures, gravures et sculptures de l'abri du Poisson aux Eyzies. Quelques nouvelles observations. *Bull. Soc. Préhist. Ariège* 39, 11–26.

ROUSSOT, A. 1984b. Approche statistique du bestiaire figuré dans l'art pariétal. *L'Anthropologie* 88, 485–98.

ROUSSOT, A. 1984c. La rondelle 'aux chamois' de Laugerie-Basse, in *Eléments de Pré et Protohistoire européenne, Hommages à J-P. Millotte*, Annales Litt. Univ. de Besançon, 219–31.

ROUSSOT, A. 1995. Review of "Les Cavernes de Niaux" by J. Clottes. *Bull. Soc. Préhist. Ariège-Pyrénées* 50, 315–23.

ROUZAUD, F. 1978. *La Paléospéléologie. L'Homme et le Milieu Souterrain pyrénéen au Paléolithique supérieur.* Archives d'Ecologie Préhist. 3, Toulouse.

ROZOY, J. G. 1985. Peut-on identifier les animaux de Roc-la-Tour I? *Anthropozoologica* 2, 5–7.

ROZOY, J. G. 1988. Le Magdalénien Supérieur du Roc-La-Tour I. *Helinium* 28, 157–91.

ROZOY, J. G. 1990. Les plaquettes gravées magdaléniennes de Roc-La-Tour I, in *L'Art des Objets au Paléolithique* (J. Clottes, ed.), vol. 1, 261–77.

RUSPOLI, M. 1987. *The Cave of Lascaux: The Final Photographic Record.* Thames & Hudson: London.

RUSSELL, P. M. 1987. *Women in Upper Palaeolithic Europe.* M. A. Dissertation, University of Auckland.

RUSSELL, P. M. 1989. Plaques as palaeolithic slates: an experiment to reproduce them. *Rock Art Research* 6, 68–9 and 40–1.

RUSSELL, P. M. 1989a. Who and why in Palaeolithic art. *Oxford Journal of Archaeology* 8, 237–49.

RUSSELL, P. M. 1991. Men only? The myths about European Palaeolithic artists, in *The Archaeology of Gender* (D. Walde and N. D. Willows, eds), 346–51. Proc. of 22nd Chacmool Conference, Arch. Assoc. of the Univ. of Calgary.

RUSSELL, P. M. 1993. Forme et imagination: L'image féminine dans l'Europe paléolithique. *Paléo* 5, 375–88.

RUSSELL, P. M. 1993a. The Palaeolithic mother-goddess: fact or fiction?, in *Women in Archaeology* (H. du Cros and L. Smith, eds), 93–7. Occasional Papers in Prehistory 23, A. N. U.: Canberra.

SACCHI, D. 1987. L'art paléolithique des Pyrénées roussillonnaises, in *Etudes Roussillonnaises offertes à Pierre Ponsich*, Perpignan, 47–52.

SACCHI, D. 1990. Bases objectives de la chronologie de l'art mobilier paléolithique dans les Pyrénées septentrionales, in Clottes 1990, vol. 1, 13–28.

SACCHI, D., ABELANET, J. and BRULE, J-L. 1987. Le rocher gravé de Fornols-Haut. *Archéologia* 225, juin, 52–7.

SACCHI, D. *et al.* 1988. Les gravures rupestres de Fornols-Haut, Pyrénées-Orientales. *L'Anthropologie* 92, 87–100.

SAHLY, A. 1963. Nouvelles découvertes dans la grotte de Gargas. *Bull. Soc. Préhist. Ariège* 18, 65–74, 2 pl.

de SAINT MATHURIN, S. 1958. Rouffignac, ses textes, ses plans. *Bull. Soc. Préhist. fr.* 55, 588–92.

de SAINT MATHURIN, S. 1971. Les biches du Chaffaud (Vienne). Vicissitudes d'une découverte. *Antiquités Nationales* 3, 22–8.

de SAINT MATHURIN, S. 1973. Bas-relief et plaquette d'homme magdalénien d'Angles-sur- l'Anglin. *Antiquités Nationales* 5, 12–19.

de SAINT MATHURIN, S. 1975. Reliefs magdaléniens d'Angles-sur-l'Anglin (Vienne). *Antiquités Nationales* 7, 24–31.

de SAINT MATHURIN, S. 1978. Les 'Vénus' pariétales et mobilières du Magdalénien d'Angles-sur-l'Anglin. *Antiquités Nationales* 10, 15–22.

de SAINT PERIER, R. 1930. La Grotte d'Isturitz, I: Le Magdalénien de la Salle de St-Martin. *Archives de l'Inst. Pal. Humaine* 7, Paris.

de SAINT PERIER, R. 1936. La Grotte d'Isturitz, II: Le Magdalénien de la Grande Salle. *Archives de l'Inst. Pal. Humaine* 17, Paris.

de SAINT PERIER, R-S. 1965. Inventaire de l'art mobilier paléolithique du Périgord, in *Centenaire de la Préhistoire en Périgord (1864–1964)*, 139–59. Fanlac: Périgueux.

SANCHIDRIAN TORTI, J. L. 1994. *Arte Rupestre de la Cueva de Nerja.* Trabajos sobre la Cueva de Nerja No. 4, Patronato de la Cueva de Nerja.

SANZ de SAUTUOLA, M. 1880. *Breves apuntos sobre algunos objetos prehistóricos de la provincia de Santander.* Santander, 27pp, 4 pl.

SAUVET, G. 1983. Les représentations d'équidés paléolithiques de la grotte de la Griega (Pedraza, Segovia). A propos d'une nouvelle découverte. *Ars Praehistorica* 2, 49–59.

SAUVET, G. 1988. La communication graphique paléolithique. *L'Anthropologie* 92, 3–16.

SAUVET, G. and S. 1979. Fonction sémiologique de l'art pariétal animalier franco-cantabrique. *Bull. Soc. Préhist. fr.* 76, 340–54.

SAUVET, G. and S., and WLODARCZYK, A. 1977. Essai de sémiologie préhistorique (pour une théorie des premiers signes graphiques de l'homme). *Ibid.* 74, 545–58.

SAUVET, G. and WLODARCZYK, A. 1992. Structural interpretation of statistical data from European Palaeolithic cave art, in *Ancient Images, Ancient Thought. The Archaeology of Ideology* (A. S. Goldsmith *et al.*, eds), 223–34. Proc. of 23rd Chacmool Conference, Arch. Assoc. of Univ. of Calgary.

SAUVET, G. and WLODARCZYK, A. 1995. Eléments d'une grammaire formelle de l'art pariétal paléolithique. *L'Anthropologie* 99, 193–211

SAXON, E. C. 1976. Pre-Neolithic pottery: new evidence from North Africa. *Proceedings of the Prehistoric Society* 42, 327–9.

SCHMID, E. 1984. Some anatomical observations on Palaeolithic depictions of horses, in Bandi *et al.* 1984, 155–60.

SCHUSTER, C./ CARPENTER, E. 1988. *Materials for the Study of Social Symbolism in Ancient and Tribal Art.* Vol. 2 (5). Rock Foundation: New York.

SHCHELINSKY, V. 1989. Some results of new investigations at Kapova Cave in the southern Urals. *Proceedings of the Prehistoric Society* 55, 181–91.

SHIMKIN, E. M. 1978. The Upper Paleolithic in North-Central Eurasia: evidence and problems, in *Views of the Past* (L. G. Freeman, ed.), 193–315. Mouton: The Hague.

SIEVEKING, A. 1971. Palaeolithic decorated bone discs. *British Museum Quarterly* 35, 206–29, 11 pl.

SIEVEKING, A. 1978. La significación de las distribuciónes en el arte paleolítico. *Trabajos de Prehistoria* 35, 61–80.

SIEVEKING, A. 1979. *The Cave Artists.* Thames & Hudson: London.

SIEVEKING, A. 1979a. Style and regional grouping in Magdalenian cave art. *Bull. Inst. of Arch.* 16, 95–109.

SIEVEKING, A. 1983. Decorated scapulae from the Upper Palaeolithic of France and Cantabrian Spain, in *Homenaje al Prof. M. Almagro Basch*, vol. 1, 313–17. Min. de Cultura: Madrid.

SIEVEKING, A. 1987. *Engraved Magdalenian Plaquettes.* British Arch. Reports, Int. series No. 369. Oxford.

SIEVEKING, G. 1972. Art mobilier in Britain, in *Santander Symposium*, 385–8.

SIMEK, J. 1986. A Paleolithic sculpture from the abri Labattut in the American Museum of Natural History collection. *Current Anth.* 27, 402–7.

SIMONNET, G. 1952. Une belle parure magdalénienne, in *12e Congrès Préhist. de France*, Paris 1950, 564–8.

SIMONNET, G., L. and R. 1984. Quelques beaux objets d'art venant de nos recherches dans la grotte ornée de Labastide (H-P). Approche naturaliste. *Bull. Soc. Méridionale de Spéléol. et Préhist.* 24, 25–36.

SIMONNET, G., L. and R. 1991. Le propulseur au faon de Labastide (H-P). *Bull. Soc. Préhist. Ariège-Pyrénées* 46, 133–43.

SIMONNET, R. 1980. Emergence de la préhistoire en pays ariégeois. Aperçu critique d'un siècle de recherches. *Bull. Soc. ariégeoise Sciences, Lettres et Arts* 35, 5–88.

SINGER, R. and WYMER, J. 1982. *The Middle Stone Age at Klasies River Mouth in South Africa*. Chicago University Press: Chicago.

SMITH, C. E. 1991. Female artists: the unrecognized factor in sacred rock art production, in *Rock Art and Prehistory* (P. G. Bahn and A. Rosenfeld, eds), 45–52. Oxbow Monograph 10: Oxford.

SMITH, N. W. 1992. *An Analysis of Ice Age Art. Its Psychology and Belief System*. Peter Lang: New York.

SOFFER, O. 1985. *The Upper Paleolithic of the Central Russian Plain*. Academic Press: Orlando.

SOFFER, O. 1985a. Patterns of intensification as seen from the Upper Paleolithic of the Central Russian Plain, in *Prehistoric Hunter-Gatherers, The Emergence of Cultural Complexity* (T. D. Price and J. A. Brown, eds), 235–70. Academic Press: New York/London.

SOFFER, O. 1987. Upper Paleolithic connubia, refugia, and the archaeological record from Eastern Europe, in *The Pleistocene Old World, Regional Perspectives* (O. Soffer, ed.), 333–48. Plenum Press: New York.

SOFFER, O. *et al.* 1993. The pyrotechnology of performance art: Moravian venuses and wolverines, in *Before Lascaux, The Complex Record of the Early Upper Paleolithic* (H. Knecht *et al.*, eds), 259–75. CRC Press: Boca Raton.

SOHN POW-KEY 1974. Palaeolithic culture of Korea. *Korea Journal*, April, 4–11.

SOHN POW-KEY 1981. Inception of art mobilier in the Middle Palaeolithic period at Chommal Cave, Korea, in *Resumenes de Comunicaciones, Paleolítico Medio*, Xth Congress UISPP, Mexico City, 31–2.

de SONNEVILLE BORDES, D. 1986. Le bestiaire paléolithique en Périgord. Chronologie et signification. *L'Anthropologie* 90, 613–56.

de SONNEVILLE BORDES, D. and LAURENT, P. 1983. Le phoque à la fin des temps glaciaires, in *La Faune et l'Homme Préhistoriques*, Mém. Soc. Préhist. fr. 16, 69–80.

de SONNEVILLE BORDES, D. and LAURENT, P. 1986/7. Figurations de chevaux à l'abri Morin. Observations complémentaires. *Antiquités Nationales* 18/19, 69–74.

SOUBEYRAN, F. 1991. Nouveau regard sur la pathologie des figures pariétales. *Bull. Soc. Historique et Archéologique du Périgord* 118, 523–60.

SOUBEYRAN, F. 1995. Lascaux: Proposition de nouvelle lecture de la scène du puits. *Paléo* 7, 275–88

SOURIAU, E. 1971. Art préhistorique et esthétique du mouvement, in *Mélanges A. Varagnac*, 697–705. Sevpen: Paris.

SPECK, F. G. and SCHAEFFER, C. E. 1950. The deer and rabbit hunting drive in Virginia. *Southern Indian Studies* 2, 3–20.

de SPIEGELEIRE, M. A. 1985. Figurations paléolithiques et réalité anatomique du mammouth (*Mammuthus primigenius*): essai d'interprétation. *Bull. Soc. roy. belge Anthrop. Préhist.* 96, 93–116.

STEPANCHUK, V. N. 1993. Prolom II, a Middle Palaeolithic cave site in the eastern Crimea with non-utilitarian bone artefacts. *Proc. Prehist. Society* 59, 17–37.

STEVENS, A. 1975. Animals in Palaeolithic cave art: Leroi-Gourhan's hypothesis. *Antiquity* 49, 54–57.

STEVENS, A. 1975a. Association of animals in Palaeolithic cave art: the second hypothesis. *Science and Archaeology* 16, 3–9.

STOLIAR, A. D. 1977/8. On the genesis of depictive

activity and its role in the formation of consciousness (Toward a formulation of the problem). *Soviet Anth. and Arch*. 16, 3–42; and 17 (1978/9), 3–33.

STOLIAR, A. D. 1977/8a. On the sociohistorical decoding of Upper Palaeolithic female signs. *Ibid*. 16, 36–77.

STOLIAR, A. D. 1981. On the archeological aspect of the problem of the genesis of animalistic art in the Eurasian Paleolithic. *Ibid*. 20, 72–108.

STOLIAR, A. D. 1985. *Proisxozdenie izobrazitel'nogo iskusstva* (The origin of figurative art). Iskusstvo: Moscow.

STOLIAR, A. D. 1995. Origins of art as an archaeological reality. (in Russian). *Archaeological News* 4, 217–33. Inst. of History of Material Culture: St Petersburg.

STORDEUR, D. and JAMMOUS, B 1995. Pierre à rainure à décor animal trouvée dans l'horizon PPNA de Jerf el-Ahmar (Syric). *Paléorient* 21, 129–30.

STORDEUR, D. *et al.* 1996. Jerf el-Ahmer: a new Mureybetian site (PPNA) on the Middle Euphrates. *Neo-Lithics, A Newsletter of Southwest Asian Lithics Research* 2, 1–2.

STRAUS, L. G. 1982. Observations on Upper Paleolithic art: old problems and new directions. *Zephyrus* 34/5, 71–80.

STRAUS, L. G. 1987. The Paleolithic cave art of Vasco-Cantabrian Spain. *Oxford Journal of Arch*. 6, 149–63.

STRINGER, C. B. *et al.* 1995. Solution for the Sherborne problem. *Nature* 378, 452.

SUERES, M. 1991. Les mains de Gargas. Approche expérimental et statistique du problème des mutilations. *Travaux de l'Inst. d'Art Préhist. Toulouse* 33, 9–200.

TABORIN, Y. 1977. Quelques objets de parure. Etude technologique: les percements des incisives de bovinés et des canines de renards. *Méthodologie Appliquée à l'Industrie de l'Os Préhistorique* (H. Camps-Fabrer, ed.), 303–10. CNRS: Paris.

TABORIN, Y. 1982. La parure des morts, in *La Mort dans la Préhistoire, Dossier de l'Arch*. 66, 42–51.

TABORIN, Y. 1985. Les origines des coquillages paléolithiques en France, in *La Signification culturelle des Industries Lithiques* (M. Otte, ed.), 278–301. British Arch. Reports, Int. series No. 239, Oxford.

TABORIN, Y. 1993. *La Parure en Coquillage au Paléolithique*. 29e Suppl. à Gallia Préhistoire. CNRS: Paris.

TARASSOV, L. M. 1971. La double statuette paléolithique de Gagarino. *Quartär* 22, 157–63, 1 pl.

TARDOS, R. 1987. Problèmes posés par les mains mutilées de la grotte de Gargas, in *Archéologie et Médecine* (7e Rencontres Int. d'Arch. et d'Histoire, Antibes 1986), 211–18. Editions APDCA: Juan-les-Pins.

TARDOS, R. 1993. Les mains mutilées: étude critique des hypothèses pathologiques, in *La Main dans la Préhistoire, Les Dossiers d'Arch*. 178, Jan., 55.

TAYLOR, T. 1996. *The Prehistory of Sex. Four Million Years of Human Sexual Culture*. Fourth Estate: London.

THEVENIN, A. 1983. Les galets gravés et peints de l'abri de Rochedane (Doubs) et le problème de l'art azilien. *Gallia Préhistoire* 26, 139–88.

THEVENIN, A. 1989. L'art azilien: Essai de synthèse. *L'Anthropologie* 93, 585–604.

THIEME, H. 1996. Altpaläolithische Wurfspeere aus Schöningen, Niedersachsen. Ein Vorbericht. *Archäologisches Korrespondezblatt* 26, 377–93

TOSELLO, G. 1983. *Inventaire des Sites Paléolithiques d'Art Pariétal et Mobilier Figuratif*. Maîtrise de Préhistoire, 2 vols, Univ. de Paris I, Panthéon-Sorbonne.

TURK, I., DIRJEC, J. and KAVUR, B. 1995. Ali so v

sloveniji nasli najstarejse glasbilo v europi? (The oldest musical instrument in Europe discovered in Slovenia?). *Razprave IV. razreda SAZU (Ljubljana)* 36, 287–93

TWIESSELMANN, F. 1951. *Les Représentations de l'Homme et des Animaux Quaternaires découvertes en Belgique*. Mémoire 113, Institut Royal des Sciences Naturelles de Belgique: Brussels.

TYLDESLEY, J. A. and BAHN, P. G. 1983. Use of plants in the European Palaeolithic: a review of the evidence. *Quaternary Science Reviews* 2, 53–81.

TYMULA, S. 1995. Figures composites de l'art paléolithique européen. *Paléo* 7, 211–48.

UCKO, P. J. 1977. Opening remarks, in *Form in Indigenous Art* (P. J. Ucko, ed.), 7–10. Duckworth: London.

UCKO, P. J. 1987. Débuts illusoires dans l'étude de la tradition artistique. *Bull. Soc. Préhist. Ariège-Pyrénées* 42, 15–81.

UCKO, P. J. and LAYTON, R. 1984. La subjectivité et le recensement de l'art paléolithique. Typescript for Colloque Int. d'Art Pariétal Paléolithique, Périgueux, 39 pp.

UCKO, P. J. and ROSENFELD, A. 1967. *Palaeolithic Cave Art*. World Univ. Library: London.

UCKO, P. J. and ROSENFELD, A. 1972. Anthropomorphic representations in Palaeolithic art, in *Santander Symposium* 149–215.

UTRILLA, P. 1979. Acerca de la posición estratigráfica de los cervidos y otros animales de trazo múltiple en el Paleolítico Superior español. *Caesaraugusta* 49/50, 65–72.

VALLADAS, H. *et al.* 1992. Direct radiocarbon dates for prehistoric paintings at the Altamira, El Castillo and Niaux caves. *Nature* 357, 68–70.

VALLOIS, H. V. 1961. The social life of early man: the evidence of skeletons, in *Social Life of Early Man* (S. L Washburn, ed.), 214–35. Viking Fund Public. No. 31, New York.

VALOCH, K. 1970. Oeuvres d'art et objets en os du Magdalénien de Moravie (Tchécoslovaquie). *Bull. Soc. Préhist. Ariège* 25, 79–93.

VAN NOTEN, F. 1977. Excavations at Matupi Cave. *Antiquity* 51, 35–40.

VANDERWAL, R. and FULLAGAR, R. 1989. Engraved *Diprotodon* tooth from the Spring Creek locality, Victoria. *Archaeology in Oceania* 24, 13–16.

VANDIVER, P. 1983. *Paleolithic pigments and processing*. Master Science Thesis, Dept. of Material Science and Engineering, MIT (USA).

VANDIVER, P. B. *et al.* 1989. The origins of ceramic technology at Dolní Vestonice, Czechoslovakia. *Science* 246, 1002–8.

VAQUERO TURCIOS, J. 1995. *Maestros Subterráneos*. Celeste: Madrid.

VASIL'EV, S. A. 1985. Une statuette d'argile paléolithique de Sibérie du sud. *L'Anthropologie* 89, 193–96.

VAYSON de PRADENNE, A. 1932. *Les Fraudes en Archéologie Préhistorique*. E. Nourry: Paris (re-issued in 1993 by Jérôme Millon: Grenoble).

VAYSONNE DE PRADENNE, A. 1934. Les figurations d'oiseaux dans l'art quaternaire. *I PEK* 3–17.

VERTUT, J. 1980. Contribution des techniques photographiques à l'étude et à la conservation de l'art préhistorique, in *Altamira Symposium*, 661–75. Min. de Cultura: Madrid.

VEZIAN, J. 1956. Les utilisations de contours de la roche dans la grotte du Portel. *Bull. Soc. Préhist. Ariège* 11, 79–87, 4 pl.

VIALOU, D. 1981. L'art préhistorique: questions d'interprétations. *Monuments Historiques* 118, Nov/Dec,

75–83.

VIALOU, D. 1982. Une lecture scientifique de l'art préhistorique. *La Recherche* 13, 1484–7.

VIALOU, D. 1982a. Niaux, une construction symbolique magdalénienne exemplaire. *Ars Praehistorica* 1, 19–45.

VIALOU, D. 1984. Les cervidés de Lascaux, in Bandi *et al.* 1984, 199–216.

VIALOU, D. 1986. *L'Art des Grottes en Ariège Magdalénienne.* XXIIe Suppl. à Gallia Préhistoire. CNRS: Paris.

VIALOU, D. 1987. D'un tectiforme à l'autre, in *Sarlat en Périgord. Actes du 39e Congr. d'Etudes Régionales, Suppl. to Bull. Soc. Hist. et Arch. du Périgord* 114, 307–17.

VIEIRA, A. B. 1996. Mythe et magie dans l'art pariétal: la logique de la caverne. *Trabalhos de Antropologia e Etnologia* 36, 169–91.

VILLAVERDE BONILLA, V. 1994. *Arte Paleolítico de la Cova del Parpalló.* 2 vols. Servei d'Investigació Prehistòrica: Valencia.

WAKANKAR, V. S. 1984. Bhimbetka and dating of Indian rock paintings, in *Rock Art of India* (K. K. Chakravarti, ed.), 44–5. Arnold-Heinemann: New Delhi.

WAKANKAR, V. S. 1985. Bhimbetka: the stone tool industries and rock paintings, in *Recent Advances in Indo-Pacific Prehistory* (V. N. Misra and P. Bellwood, eds), 175–6. Oxford and IBH Publishing Co.: New Delhi.

WALKER, N. J. 1987. The dating of Zimbabwean rock art. *Rock Art Research* 4, 137–49.

WALKER, N. J. 1996. *The Painted Hills. Rock Art of the Matopos.* Mambo Press: Gweru, Zimbabwe.

WALLER, S. J. 1993. Sound and rock art. *Nature* 363, 501.

WALLER, S. J. 1993a. Sound reflection as an explanation for the content and context of rock art. *Rock Art Research* 10, 91–101.

WALSH, G. L. 1979. Mutilated hands or signal stencils. *Australian Archaeology* 9, 33–41.

WATCHMAN, A. 1993. Evidence of a 25,000-year-old pictograph in northern Australia. *Geoarchaeology* 8, 465–73.

WATCHMAN, A. 1995. Recent petroglyphs, Foz Côa, Portugal. *Rock Art Research* 12, 86–103.

WATCHMAN, A. 1996. A review of the theory and assumptions in the AMS dating of the Foz Côa petroglyphs, Portugal. *Rock Art Research* 13, 21–30.

WATCHMAN, A. and HATTE, E. 1996. A nano approach to the study of rock art: 'The Walkunders', Chillagoe, north Queensland, Australia. *Rock Art Research* 13 (2), 85–92.

WEINSTEIN-EVRON, M. and BELFER-COHEN, A. 1993. Natufian figurines from the new excavations of the El-Wad Cave, Mt Carmel, Israel. *Rock Art Research* 10, 102–6.

WEISSEN-SZUMLANSKA, M. 1951. A propos des gravures et peintures rupestres. *Bull. Soc. Préhist. fr.* 48, 457–9.

WELTE, A-C. 1975/6. L'affrontement dans l'art préhistorique. *Travaux de l'Inst. d'Art Préhist. Toulouse* 17, 207–305 and 18, 187–330.

WELTE, A-C. 1989. An approach to the theme of confronted animals in French Palaeolithic art, in *Animals into Art* (H. Morphy, ed.), 215–35. Unwin Hyman: London.

WENDT, W. E. 1974. 'Art mobilier' aus der Apollo 11-Grotte in Südwest-Afrika. *Acta Praehistorica et Archaeologica* 5, 1–42.

WENDT, W. E. 1976. 'Art mobilier' from the Apollo 11 cave, South West Africa: Africa's oldest dated works of art. *South African Archaeological Bulletin* 31, 5–11.

WHITE, R. 1986. *Dark Caves, Bright Visions, Life in Ice Age Europe.* Exhibition catalogue, American Museum of Natural History: New York.

WHITE, R. 1992. Beyond art: toward an understanding of the origins of material representation in Europe. *Annual Review of Anthropology* 21, 537–64.

WHITLEY, D. S. and DORN, R. I. 1987. Rock art chronology in eastern California. *World Arch.* 19 (2), 150–64.

WHITLEY, D. S. and DORN, R. I. 1993. New perspectives on the Clovis vs Pre-Clovis controversy. *American Antiquity* 58, 626–47.

WHITLEY, D. S. *et al.* 1996. Recent advances in petroglyph dating and their implications fo the pre-Clovis occupation of North America. *Proceedings of the Society for California Archaeology* 9, 92–103.

WILDGOOSE, M., HADINGHAM, E. and HOOPER, A. 1982. The prehistoric hand pictures at Gargas: attempts at simulation. *Medical History* 26, 205–7.

WILLCOX, A. R. 1990. Comment on paper by A. Marshack. *Rock Art Research* 7, 60–62.

WORSAAE, J. J. A. 1869. Communication…relative à l'authenticité des os de renne présentant des dessins et trouvés en France dans les cavernes du Périgord. *Congrès Int. d'Anth. et d'Arch. Préhistoriques,* 4e session, Copenhagen, 127–34. (Publ. 1875).

WRESCHNER, E. 1975. Ochre in prehistoric contexts. Remarks on its implications to the understanding of human behavior. *Mi'Tekufat Ha'Even* 13, 5–11.

WRESCHNER, E. 1980. Red ochre and human evolution: a case for discussion. *Current Anth.* 21, 631–44.

WRIGHT, B. 1985. The significance of hand motif variations in the stencilled art of the Australian Aborigines. *Rock Art Research* 2, 3–19.

WRIGHT, R. S. V. (ed.) 1971. *Archaeology of the Gallus Site, Koonalda Cave.* Australian Institute of Aboriginal Studies: Canberra.

YATES, F. A. 1966. *The Art of Memory.* Univ. of Chicago Press.

ZAMPETTI, D. 1987. L'arte zoomorfa del Paleolitico Superiore in Italia. *Scienze dell'Antichità. Storia Archeologia Antropologia* 1, 9–35.

ZERVOS, C. 1959. *L'Art de l'Epoque du Renne en France.* Cahiers d'Art: Paris.

ZEUNER, F. E. 1954. The colour of the wild cattle of Lascaux. *Man* 53, 68–9.

ZHUROV, R. I. 1977/8. On one hypothesis relating to the origin of art. *Soviet Anth. and Arch.* 16, 43–63.

ZHUROV, R. I. 1981/2. On the question of the origin of art (an answer to opponents). *Soviet Anth. and Arch.* 20 (3), 59–82.

ZILHAO, J. 1995. L'art rupestre paléolithique de plein air. *Dossiers d'Archéologie* 209, Dec. 1995/Jan. 1996, 106–17.

ZILHAO, J. 1995a. The age of the Côa Valley (Portugal) rock-art: Validation of archaeological dating to the Palaeolithic and refutation of 'scientific' dating to historic or protohistoric times. *Antiquity* 69, 883–901.

ZILHAO, J. 1995b. The stylistically Palaeolithic petroglyphs of the Côa Valley (Portugal) are of Palaeolithic age. A refutation of their 'direct dating' to recent times. *Trabalhos Antropol. Etnol.* (Porto) 35, 3–49.

ZÜCHNER, C. 1975. Der Bison in der eiszeitlichen Kunst Westeuropas. *Madrider Mitteilungen* 16, 9–24.

Index

Where names are preceded by an article (e.g. Le, La, El etc.), the article is printed but ignored for alphabeticization purposes.